ETERNAL GOD/SAVING TIME

Eternal God/Saving Time

GEORGE PATTISON

OXFORD
UNIVERSITY PRESS

OXFORD
UNIVERSITY PRESS

Great Clarendon Street, Oxford, OX2 6DP,
United Kingdom

Oxford University Press is a department of the University of Oxford.
It furthers the University's objective of excellence in research, scholarship,
and education by publishing worldwide. Oxford is a registered trade mark of
Oxford University Press in the UK and in certain other countries

© George Pattison 2015

The moral rights of the author have been asserted

First Edition published in 2015

Impression: 1

Published in the United States of America by Oxford University Press
198 Madison Avenue, New York, NY 10016, United States of America

British Library Cataloguing in Publication Data
Data available

Library of Congress Control Number: 2014948607

ISBN 978-0-19-872416-2

Printed and bound by
CPI Group (UK) Ltd, Croydon, CR0 4YY

Acknowledgements

The substantive part of this book was written on research leave during the academic year 2012–13. The first six months was leave granted by the University of Oxford and the second six months was funded by an Arts and Humanities Research Council Fellowship and I am deeply grateful to both bodies, as well as mindful of the privilege of such opportunities. I am also grateful to Dartmouth College for the generous award of a Montgomery Fellowship in this period.

Several elements in this book have been tested in various seminar and conference presentations and I am grateful to those institutions that have made this possible. These include the Faculty of Theology at the University of Copenhagen (several times), Dartmouth College, the Institute for Theology, Imagination and the Arts at the University of St Andrews, and the philosophy department at the University of Leeds, as well as my new home base in Theology and Religious Studies at Glasgow. I am also indebted to all who have advised on reading, commented on parts of this book, or otherwise contributed in manifold ways to its genesis and development.

Hilary has lived with this even more intensively than with some of my other work, but the gratitude could not be greater.

George Pattison
University of Glasgow, May 2014

Contents

Abbreviations

References to Kierkegaard of the form (for example) 'SKS7, 306ff/CUP1, 335ff' are to the recently complete scholarly edition of his works, *Søren Kierkegards Skrifter*. The volume and page number of the Danish text are followed by a reference to the English translation in the series *Kierkegaard's Writings*, edited by Howard V. and Edna H. Hong, or (rarely) to the Hongs' edition of the *Journals and Papers*. The relevant abbreviations are as follows:

CA: *The Concept of Anxiety*
CD: *Christian Discourses*
CUP: *Concluding Unscientific Postscript*
EO: *Either/Or*
EUD: *Eighteen Upbuilding Discourses*
FTR: *Fear and Trembling and Repetition*
JP: *Journals and Papers*
LD: *Letters and Documents*
M: *The Moment*
SUD: *The Sickness unto Death*
TDIO: *Three Discourses on Imagined Occasions*
UDVS: *Upbuilding Discourses in Various Spirits*
WA: *Without Authority*
WL: *Works of Love*

References given as 'SKSK' are to the volumes of notes accompanying each volume of Kierkegaard's text when these are not yet translated into English.

Most other works are cited in footnotes, with full details in the work's first citation in a chapter. However, second and subsequent citations of the following works are in parentheses in the main text, using the abbreviations shown:

A Martin Heidegger, *Gesamtausgabe*, LII: *Hölderlins Hymne 'Andenken'*
(Frankfurt am Main: Klostermann, 1992)
AR F. H. Bradley, *Appearance and Reality: A Metaphysical Essay*
(Oxford: Clarendon Press, 1930 [1893])
BE Nicolas Berdyaev, *The Beginning and the End*, tr. R. M. French
(London: Geoffrey Bles, 1952)
ER Karl Barth, *The Epistle to the Romans*, tr. Edwyn C. Hoskyns
(Oxford: Oxford University Press, 1933)
F F. W. J. Schelling, *Philosophische Untersuchungen über das Wesen der menschlichen Freiheit* (Frankfurt am Main: Suhrkamp, 1975)

FE Krzysztof Michalski, *The Flame of Eternity: An Interpretation of Nietzsche's Thought*, tr. Benjamin Paloff (Princeton, NJ: Princeton University Press, 2007)

HD John McTaggart Ellis McTaggart, *Studies in the Hegelian Dialectic* (Cambridge: Cambridge University Press, 1896)

HH Martin Heidegger, *Gesamtausgabe*, xxxix: *Hölderlins Hymnen 'Germanien' und 'Der Rhein'* (Frankfurt am Main: Klostermann, 1989)

II Jean-Louis Chrétien, *L'inoubliable et l'inéspéré* (2nd edn, Paris: Desclée de Brouwer, 2000)

IN V. Jankélévitch, *L'irréversible et la nostalgie* (Paris: Flammarion, 1974)

MH Nicolas Berdyaev, *The Meaning of History*, tr. G. Reavey (London: Geoffrey Bles, 1936)

NTZ Michael Theunissen, *Negative Theologie der Zeit* (Frankfurt am Main: Suhrkamp, 1991)

P Michael Theunissen, *Pindar: Menschenlos und Wende der Zeit* (Münich: C. H. Beck, 2000)

PH Ernst Bloch, *The Principle of Hope*, tr. Neville Plaice, Stephen Plaice, and Paul Knight, 3 vols (Oxford: Blackwell, 1986)

S Franz Rosenzweig, *Der Stern der Erlösung* (Frankfurt am Main: Suhrkamp, 1988), 139; Eng., *The Star of Redemption*, tr. William W. Hallo (Notre Dame, IN: University of Notre Dame Press, 1985)

Introduction

Church congregations across the English-speaking world regularly sing of 'the one, eternal God' and prayers, liturgies, and sermons are not slow to take up the refrain. Although philosophers and theologians may argue as to just what divine eternity means, the testimony of believers has until recently been nearly unanimous in affirming that God is eternal and that God's being is eternal being. James Moffatt's early-twentieth-century translation of the Bible used 'the Eternal' for the divine name revealed to Moses at the burning bush (translated in the King James Version as 'the LORD'), and this translation has also been favoured by some Jewish scholars. This implies that 'The Eternal' is not just one divine name amongst others. God simply *is* the Eternal. That is who God is. Affirming that God is eternal seems, then, to be central to Christian belief, as it does to other major religious traditions. And, in this case, popular culture seems to have continued to use the older language even as it abandons the doctrinal framework in which that language was embedded. Lovers declare eternal loyalty and speak of their love as 'an eternal flame'. Funeral notices announce 'eternal memory'. The billboard for the movie *Gladiator* declared that 'What we do in this life echoes in eternity'. Even in the atheist Soviet Union it was common to speak of eternal human truths and values.

But, if they are not mere hyperbole, what do such invocations of eternity mean and what might be meant by calling God 'eternal'? What could it possibly mean to speak of God or any being as *eternal*, when our human experience and knowledge seem in manifold ways to be bounded and pervaded by time? What are we human beings if not 'mortals', that is, beings that have come into being in time and that will, soon enough, pass out of being in death? Whether or not we follow Heidegger any further in an approach to existence that focuses on thrownness towards death as the hallmark of being human, there seems to be something intuitively right about his assertion that 'time is the horizon of the meaning of being', that is, there is nothing that we human beings can experience and know that is not marked by time. But there's the rub, since to be defined and circumscribed by time is to be handed over to change, decay, and annihilation.

A classical view of what is at issue here is found at many points in the many writings of Augustine of Hippo, perhaps the most influential figure of the Western tradition outside the Bible itself. The following quote from Book 11, Chapter 29 of his *Confessions* may be taken as epitomizing the Augustinian view as a whole—William Watts's seventeenth-century translation conveying something of the visceral force of Augustine's language:

> But now are my years spent in mourning, and thou, my Comfort, O Lord, my Father, art everlasting [*aeternus*]; but I fall into dissolution amid the changing times, whose order I am yet ignorant of: yea, my thoughts are torn asunder with tumultuous vicissitudes, even the inmost bowels of my soul; until I may be run into thee, purified and molten by the fire of thy love.[1]

But (to repeat) what does it mean—what *could* it mean—to speak of God as eternal? And how, if we can in fact conceive of an eternal God, might this God enter into such a relationship with us that, to echo Augustine, we might 'run into' and be 'purified and molten by the fire' of that God's love?

In what follows I shall propose a view in which time might come to be experienced otherwise than as a sequence of 'tumultuous vicissitudes', to the point at which it might even be meaningful to speak less of our needing to be saved from time as of 'saving time'—that is, a kind of time capable of saving us from all that is wretched and empty in the experience of time. Nevertheless, this study is premised on the conviction that the powerful attestation offered by Christian tradition to the importance of the theological category of the eternal cannot simply be abandoned without significant loss. There is something here that 'calls for thinking'. Yet mainstream theology has seemed progressively to have lost interest in this once-central topic, and that despite the continuance of appeals to eternity in popular culture and in liturgical formulae. In this regard, this study seeks not only to vindicate the importance what everyday religious life is saying about the eternal but also to provide a new framework within which to interpret what is being said in popular usage.

The theologians' uncertainty and even embarrassment with regard to the eternal is not accidental, however. I have already noted Heidegger's claim that time (and therefore death) is the all-encompassing horizon of the meaning of being and, in this regard at least, Heidegger seems to have spoken for a broad consensus amongst modern thinkers. This is perhaps most obvious with regard to the human side of the question, namely human beings' claims to an immortal soul and the possibility of eternal life. Of course, there is nothing like an absolute consensus here. Traditional Christian ideas of resurrection and immortality have been supplemented by the teaching of nineteenth-century spiritualism, doctrines of reincarnation derived largely from Hinduism, and

---

[1] Augustine, *Saint Augustine's Confessions*, tr. William Watts (London: Heinemann, 1968 [1912]), 281.

new ideas combining elements of still other older or contemporary sources. Yet sustaining such beliefs in the face of medical and biological science's increasingly all-encompassing account of human functioning seems more and more difficult. Often individuals seem simultaneously to hold to one or other belief about the continuation of human life beyond death whilst also accepting the findings of reductionist science—'When you're dead, you're dead'.

At the same time, throughout the modern period and with ever-gathering pace from the Romantic movement onwards, there has been an increasingly positive valuation of life in time. Whereas older sources continued to share Augustine's largely negative view of time as leading through change and decay to death, newer traditions—for which Nietzsche served as an eminent advocate— have affirmed time as the sole medium of human existence and, as such, to be affirmed and celebrated rather than denied or fled. But to the extent that time is thus embraced as the condition of any meaningful experience or knowledge, the idea that human fulfilment is found in an eternal and timeless life with God becomes emptied of significance. 'Eternal life' becomes a symbolic expression for a fulfilled life in time, to which Nietzsche's fascination with an 'eternal recurrence' seems also to bear witness. 'Eternity' becomes a quality of lived time. This philosophical conviction is further underwritten by a distinctively modern kind of mysticism, in which, in Blake's phrase, we seem to experience 'eternity in an hour' and, for a moment, seem to step into timelessness or to experience the eternal as *now*, in a radically intensive sense. In this way, the decline in belief or, perhaps more accurately, a declining interest in some kind of post-mortem life has as much to do with a positive evaluation of life in time as with the intellectual or imaginative weakening of the idea of eternal life itself.[2]

If uncertainty as to the meaning of eternity with regard to human destiny is more obvious than any wavering in the idea of divine eternity, the latter too has been exposed to a sequence of significant attempts at revision in the modern period. From Hegel onwards, philosophers, poets, and others have begun to think of God's own being in temporal terms. Anglo-American process theology is one example of this, but it is only one example.[3] Whereas classical Christian doctrine followed Augustine in defining God's eternity in terms of immutability and timelessness, much modern speculation sees God as also, somehow, temporal, whether that is understood in terms of a kind of celestial temporality other than the historical temporality of human existence or of God 'descending'

---

[2] On this development see my article 'Death', in Nicholas Adams, George Pattison, and Graham Ward (eds), *The Oxford Handbook of Theology and Modern European Thought* (Oxford: Oxford University Press, 2013), 193–212. On Nietzsche, see Chapter 3.

[3] In Denmark, Niels Grønkjaer's *Den Nye Gud* (Copenhagen: Anis, 2010) would be another example. I say less about process thought in what follows than some might wish. However, I believe that the substantial issues it raises are reflected in other ways at many points in this study and occasional allusions will indicate to those who know the material that it has probably been a more important element in the background of my argument than may appear on the surface.

into and participating in human history—even to the point of suffering and dying in it. Such a temporal God may even become a finite, and limited God, first emerging perhaps from within the flux of historical time.

Such developments may be seen as religious thought's own intuitive and symbolic response to the view that time is the horizon of being and that it is only as temporal that any experience, idea, or even metaphysical viewpoint can be meaningful to us. Yet, unsurprisingly, other theologians see all this as evacuating religious teaching about God and about human destiny of what is most valuable in it. If our hope is for this life only, they respond, then what kind of hope is it? What can the pastor say to the bereaved mother of a murdered child—a question that repeats itself even more devastatingly when asked in relation to the mass killings or natural disasters of modern times? And how can a God who has become temporally relativized provide any point of reference for human beings confused by the relativity of modern moral and political frameworks? If there is nothing but time, aren't all our lives, loves, and achievements rendered finally meaningless and pointless?

As such questions indicate, what is at issue here is not just a matter of theoretical interest but engages who we are, how we feel ourselves to be, and what we imagine ourselves to be living for. The loss of eternity induces feelings of existential distress—'vanity, all is vanity'—and will theology, will Christian teaching, have nothing to say?

Like all good theological questions, the question of time and its relation to eternity is not just a question about what may be the case in some hypothetical metaphysical dimension but is also and in the first instance a question about salvation, about how our lives might be delivered from the negative experiences associated with existence in time. In the technical language of theology this means that it is a *soteriological* question. But how or by whom might we be saved and for what? Can there really be any kind of being that is not subject as we are to time and that—or who—would therefore be capable of saving us?

This study belongs to a line of revisionary approaches to the question of the Eternal. At the same time it is rooted in a certain disquiet about the development and outcome of some versions of this revisionary tendency. It seems to me doubtful that traditional symbols such as eternal life can simply be translated into statements about the quality of human life without remainder. Something gets lost in translation. It is of course difficult to say just what this something is—even if it may turn out to be the most important thing of all. When the eternal God of the burning bush becomes one historical player amongst others, it is easy to start wondering why we still insist on the name 'God' for such a being. A 'god' perhaps—but 'God'?[4]

---

[4] For another, non-theistic, way of approaching what it might mean to speak of god or gods after the 'death of God' see Hubert L. Dreyfus and Sean D. Kelly, *All Things Shining: Reading the Western Classics to Find Meaning in a Secular Age* (New York: The Free Press, 2011).

But where might we begin and how might we proceed?

I have set the question up with reference to the testimony of religious life and the recent history of theological reflection, but although I shall continue to draw on religiously committed sources this is also a philosophical work—if only in the loose sense that it does not presuppose any specific dogmatic beliefs on the part of the reader. That it does, nevertheless, presuppose that it makes sense to talk about God without pausing at every breath to justify such references does not (in my view) put it beyond the bounds of reasonable discourse. Experience of many intellectual encounters persuades me that where there is a genuine spirit of enquiry and intellectual openness, those who are not themselves believers or are even avowedly hostile to religious beliefs can nevertheless go a long way in participating in what used to be called 'God-talk' and are fully capable of seeing such talk as meaningful and internally differentiated in complex and significant ways. Lack of belief in Zeus or Athena does not render the literature of ancient Greece a closed book and between the conversant who silently qualifies each mention of God with a 'were he to exist' and the one who equally silently qualifies that same mention with a 'blessed be he' (or, in each case, their feminist variants) the distance is much less than controversialists like to pretend. Many atheists have shown that they understand that how God has been and is talked about can prove revelatory of basic human orientations and commitments and that such talk therefore merits the effort of interpretation.

In developing my argument I make use of a number of approaches, from analytic philosophy, through historical and textual exposition, to what may loosely be called a phenomenological approach. In this regard my 'method' is intentionally 'ecumenical' and presumes upon a certain suspension of hostilities between what are labelled 'analytic' and 'Continental' approaches to the philosophy of religion. In any case, as we shall see, neither can be neatly correlated with particular positions regarding the substantive issue of God's eternity.

Although the question of time and eternity is (perhaps) nearly as old as the earliest written sources of human civilization this book is limited largely to the modern (post-Reformation) period—although, like most of those who have contributed to recent discussions, I shall also refer back to earlier medieval, classical, and biblical sources as and when appropriate. There is clearly work still to be done in developing a rigorous and thorough account of the emergence and development of the idea of eternity/the Eternal from ancient times and its relevance to contemporary discussions, but that is not my task.[5] This work—whilst hopefully not ignoring the legacy of the past—is rather about

---

[5] However, see Walter Mesch, *Reflektierte Gegenwart: Eine Studie über Zeit und Ewigkeit bei Platon, Aristoteles, Plotin und Augustinus* (Frankfurt am Main: Klostermann, 2003). Despite the

how contemporary believers might think of God under the aspect of 'the Eternal'. At the same time, I do not think it possible to abstract our attempts to think the Eternal from a certain intellectual history and context. Even the analytic debate about divine eternity refers repeatedly and deliberately to key texts from the ancient and medieval sources that did most to shape the Christian view of time and eternity. We can only think our thoughts on the basis of having been inducted into a conversation that long preceded our arrival on the scene, and hermeneutical retrieval is therefore a necessary accompaniment of philosophical analysis and critique. In the present case this retrieval will focus primarily on selected key figures of the modern tradition, from Spinoza through to the present.

I begin in Chapter 1 by examining some of the basic options in conceptualizing eternity by focusing on work in the analytic philosophy of religion from the 1970s onwards. The key issue that emerges is whether, following the dominant interpretation of Boethius' classic definition, divine eternity is to be conceived of in terms of timelessness and, if so, whether such a timeless God can be conceived or spoken of by temporal beings or can enter into significant relations with them.

Chapters 2 and 3 review a sequence of philosophers who have attempted to argue that human beings have the capacity to know themselves and their world *sub specie aeternitatis*, that is, in the light of eternal, unchanging, and (in some cases) timeless acts of ideation. In the course of a development that is tracked from Spinoza and Boehme through German and British Idealists we see the question move from the domain of theory or contemplation to that of will or volition, as in Nietzsche's will to eternal recurrence. A final move beyond Nietzsche is the distinctively modern appeal to ancient and oriental mysticisms as offering a will-less experience or intuition of eternity as sheer presence, 'the eternal now'.

The opposite position is presented in Chapter 4 through a succession of modern European thinkers who have contested the possibility of thinking time and eternity together. This includes both the anti-theistic philosophies of Schopenhauer and Heidegger as well as Kierkegaard's and Barth's attacks on the conflation of time and eternity driven by their respective readings of the New Testament itself—a network of positions that is further complicated by the fact that Kierkegaard becomes a significant source for both the Christian theologian Barth and the 'atheist' philosopher Heidegger.

The situation described by Heidegger seems to be one of entire nihilism in which all that we can expect is oblivion. But does human time-experience really point towards death as the sole measure of life in time? Do we not have resources from within time-experience itself that help us resist time's solvent

---

title this is also a work that sets up a creative dialogue between its classical sources and the kind of approach found in Heidegger and post-Heideggerian philosophy.

power? In Chapter 5, this question is addressed in the first instance with reference to Vladimir Jankélévitch and Jean-Louis Chrétien, leading us through nostalgia and remembrance to what Chrétien calls the immemorial and the unhoped-for. After seeing how Lévinas's emphasis on the other entails a kind of messianic hope, the chapter ends with a reading of an episode from *The Brothers Karamazov* that exemplifies the triumph of memory and hope over the oblivion to which, according to the nihilism portrayed in the novel, all human life is condemned.

It is the further development of the theme of hope that is explored in Chapter 6. Against the background of the nineteenth-century discovery of biblical eschatology, I examine the eschatological Marxism of Ernst Bloch, the theological ontology of Paul Tillich, and the 'tragic' eschatological philosophy of N. A. Berdyaev. This leads to reflection on the themes of promise and messianic time, discussed with particular reference to Jürgen Moltmann and Jacques Derrida. Acknowledging that a hope that relies solely on messianic promises can scarcely count as 'knowledge', I nevertheless argue that it may have the capacity to transform our time experience from endless vanishing to 'saving time', time that reaches out beyond the present to embrace those who have gone before and those who are to come. As such, this new time-experience is the actualization in time of what faith has looked for from the one eternal God.

In the last three chapters, I focus on three thinkers, Michael Theunissen, Søren Kierkegaard, and Franz Rosenzweig, each of whom made the issue of time and eternity central to their intellectual projects. Although these might be taken as reflecting, respectively, a pagan, Christian, and Jewish, view,[6] they also—despite all the significant differences between them—offer a convergent view as to how time-experience itself might yield the kind of transformation of life that religious believers refer to as 'salvation'. At the same time a comparison of their positions invites reflection as to how such transformation might also find social expression, how it relates to the conflicting claims of Christian and Jewish theologies, and how, more precisely, we are to understand the call that emerges as the ultimate ground of hope.

In the early modern period 'the triumph of time' became a popular theme in poetry, drama, painting, and music. But it could be developed in two quite opposite directions. In the one case, exemplified by Handel's Oratorio *The Triumph of Time and Truth*, it meant that time exposed the vanity of all earthly values and taught us to turn our gaze to eternity.[7] In the other, it is time that reveals truth and vindicates those who have been wrongly accused or

---

[6] Theunissen's contribution is most fully developed in his interpretation of Pindar's doctrine of time.

[7] This is also the teaching of Handel's contemporary, Edward Young, especially in his influential poem *Night Thoughts*. Countless hymns and prayers bear witness to the same line of thought. Cf. also Petrarch's 'The Triumph of Time' and 'The Triumph of Eternity'.

punished. This is the approach of Robert Greene's play *Pandosto: The Triumph of Time*, a work that became the basis for the plot of Shakespeare's *A Winter's Tale*. This study can be seen as an attempt to explore the tension between these two approaches and to show how the second might, in fact, respond to rather than merely contradict the cry to which the first gives utterance.

My book *'The Heart Could Never Speak'* and Chapter Six of my *Heidegger on Death* were originally conceived in the context of the project of which this work is the most substantial product, and they provide important supplements to what is said here.[8] Heidegger is discussed in the following pages, but Edwin Muir, the subject of *'The Heart Could Never Speak'*, is not. Nevertheless, his poetic output could, I think, be placed alongside the works of Kierkegaard and Rosenzweig as both illustrating and grounding the view towards which this book fights its way. This present work can also be seen as picking up on and taking further the question concerning the relationship between time and being that was broached in my *God and Being*.[9] Especially, it presupposes the view I arrived at in the earlier work, namely, that whilst 'being' has not been rendered redundant by the Heideggerian critique of onto-theology, it can only be meaningful to us in terms that can be related to our experience of living as fragile and vulnerable temporal and embodied beings—the 'bare, forked animal' that Shakespeare long ago saw as 'the thing itself' (*King Lear*, Act 3, Scene 4). However, this present study is by no means the end of the road, and, as the closing chapters in particular suggest, further clarification as to how our understanding of the eternal is grounded in the call to love will require further work. But that must wait.

---

[8] George Pattison, *'The Heart Could Never Speak': Existentialism and Faith in a Poem of Edwin Muir* (Eugene, OR: Cascade Books, 2013) and *Heidegger on Death: A Critical Theological Essay* (Farnham: Ashgate, 2013).

[9] George Pattison, *God and Being: An Enquiry* (Oxford: Oxford University Press, 2011).

1

Eternity, Timelessly?

THE QUESTION

I shall begin this study by taking up aspects of a debate about the eternity of God that has rumbled on through the last forty or so years in analytic philosophy of religion.[1] Why? In the first instance because I take seriously analytic philosophy's claim to offer (or, at least, to help us to seek) clarity in formulating our basic concepts and in the uses to which we put them.[2] As several of the contributors to this debate acknowledge, the concepts themselves are taken from Christian teaching and, to that extent, are what in the late eighteenth or early nineteenth centuries might have been called 'positive' teachings, namely teachings that are not derived from purely rational premises but are inherited from historical traditions. Such pre-philosophical commitments are likely to affect the positions taken regarding the possibility of interpreting divine eternity as a kind of temporality. Yet despite the diversity of views, the common aim is to achieve a maximum of rational clarity and consistency and through rigorous analysis to better justify our claims to know what we are saying in taking up and making our own the positive teachings of our traditions.

To anticipate the outcome of my discussion, I find that whilst the analytical approach does indeed help us to get clear about some basic options regarding divine eternity, it does less than justice to what we might call the religious 'interest' in naming God 'the Eternal'. This has to do with (1) its failure to provide an adequate account of the experienced need of human beings for a

---

[1] It is probably too formal to call the sequence of books and articles I shall be drawing on 'a' or 'the' debate (such as the so-called University Debate about the existence of God in the 1960s). It is rather more like a network of contributions that do not always reference each other directly but that mostly relate to a similar field of historical sources and, in general, show a high level of mutual awareness, as well as giving space to approaches that branch out in more individual directions.

[2] In his Introduction to a reader in analytic theology, Rea cites the attempt to 'prioritize precision, clarity, and logical coherence' as distinctive of the analytic tradition (Michael C. Rea, 'Introduction', in Oliver D. Crisp and Michael C. Rea (eds), *Analytic Theology: New Essays in the Philosophy of Theology* (Oxford: Oxford University Press, 2009), 6).

transformation of time experience that could appropriately be called 'salvation'; and (2) an under-estimation of the human need for God to be 'other' and the consequent neglect of the necessary apophatic element in all theology.[3] With regard to both these points (the former for fairly obvious reasons, the latter less so), I shall suggest that we are better served by what can loosely be called a more phenomenological approach. This, however, is not said in an adversarial sense, but rather by way of suggesting a kind of complementarity that is, perhaps, increasingly reflected in the practice of younger scholars, less bound by what have sometimes been the sterile and acrimonious divisions of the past and which are sadly perpetuated by those who seem inexplicably compelled to deny the title of 'philosophy' to those who philosophize otherwise than in the manner in which they have themselves been schooled.[4] But philosophy itself may prove more generous in its favours than the schools that dispute its possession. We can but hope. We turn, then, to consider the question of divine eternity as that has been presented amongst a number of leading philosophers of religion, largely based in British and American universities, in the course, mostly, of the last forty or so years—but, as I have suggested, invoking at many points the ancient and medieval positions and arguments that have historically contributed to what we might call the Christian view.

OPENING THE DEBATE

As nearly all contributors to the analytic debate acknowledge, the classical statement regarding divine eternity is found in Boethius' *The Consolation of Philosophy*, a work probably written c.524 AD while the author was in prison and awaiting execution. Boethius' definition of eternity both repeats the central intention of Augustine's view (which in turn incorporates elements from Platonic metaphysics) and is itself taken up in the arguments of medieval thinkers such as Anselm and Aquinas and is a constant point of reference in the modern discussion. I shall quote the key definition in Latin before giving it and its following explanation in translation. '*Aeternitas igitur est interminabilis vitae tota simul et perfecta possessio*'.

---

[3] The rejection of apophaticism is also flagged by Rea as a distinctive feature of analytic theology that he defends. See Rea, 'Introduction', in Crisp and Rea, *Analytic Theology*, 19–20.

[4] Rea, once more, demonstrates something of this 'ecumenical' spirit in encouraging analytic theologians to at least read and take seriously those such as Schleiermacher or Heidegger or Marion who seem most antithetical to the analytic project. Unsurprisingly, he also acknowledges the differences that remain or even show up all the more sharply once such approaches have been set in motion. But if agreement is not likely to be immediately attainable (it doesn't in any case occur *within* either analytic or Continental traditions), increasing our understanding of where disagreement lies must be a gain.

Eternity, then, is the complete, simultaneous and perfect possession of everlasting life; this will be clear from a comparison with creatures that exist in time. Whatever lives in time exists in the present and progresses from the past to the future, and there is nothing set in time which can embrace simultaneously the whole extent of its life: it is in the possession of not yet possessing tomorrow when it has already lost yesterday. In this life of today, you do not live more fully than in the fleeting and transitory moment. Whatever, therefore, suffers the condition of being in time, even though it never had any beginning, never has any ending and its life extends into the infinity of time, as Aristotle thought was the case of the world, it is still not such that it may properly be considered eternal.[5]

A key issue that Boethius' definition raises is precisely whether an eternity that has the character of a 'complete, simultaneous and perfect possession of everlasting life' is at all thinkable. Boethius himself continues in true Augustinian style to remind us of the contrast between creatures such as we are, living 'in the fleeting and transitory moment', and the eternal life of God, forcibly confronting us with the *difference* between an eternal kind of life and our kind. It is a difference that seems, at once, epistemological and ontological: how could we ever come to know about such an eternity, and how could we even envisage our own lives as being in some kind of relation to it? These are perhaps the central questions of the recent analytic debate, with the emphasis mostly on the epistemological aspect. However, the passage is also important to recent discussions by virtue of the way in which it draws a sharp distinction between eternity in the full sense of the opening definition and eternity in the sense used in speaking of 'the eternity of the world' in Aristotelian philosophy. This latter, even when it is conceived as infinite both in the direction of the past and of the future, is never (as Boethius says) total, simultaneous, and complete in its self-possession. Rather than being called 'eternity' it is therefore generally referred to as 'sempiternity', which we might render as 'going on for ever'—a usage perhaps suggested by some doxological formulations regarding the God 'who was in the beginning, is now, and ever shall be'. On a strict interpretation of the Boethian definition of eternity it is precisely the application of the tenses 'was', 'is', and 'shall be' that reveals these as having an idea of God as sempiternal rather than eternal. But this distinction between two kinds of eternity, the true or proper eternity of the timeless God and the sempiternity of unlimited time, serves all the more to underline the radicality of what is required in thinking or attempting to think what it could mean to call God eternal.[6]

---

[5] Boethius, *The Consolation of Philosophy*, tr. Victor E. Watts (Harmondsworth: Penguin, 1969), 163–4.

[6] Aquinas also speaks of aeviternity or 'the measure of immaterial substances' and which 'is neither time nor eternity [but] lies somewhere between the two' (ST. 10.5). We shall touch on the question as to the possibility of intermediate states between time and timeless eternity below.

But is a timeless being—any timeless being, not just God—actually conceivable? As Brian Leftow puts it, 'Only if there can be timeless beings and truths can the claim that God is timeless be possibly true'.[7] So: can there be such a being and can we conceive of timeless truths? Leftow's own answer to these questions is developed in a commendably cautious step-by-step way. First of all he notes that contemporary science allows for the existence of time series other than our own. Such an 'extrinsically timeless' series would be one that at no point intersected with ours and would, from our point of view, be timeless. Yet it is nevertheless a time series. Our own temporal frame of reference, therefore, cannot be determinative for all time series. But the existence of multiple time series that are all extrinsic to each other is something different again from positing an intrinsically timeless being, as Leftow acknowledges. In response he offers several possibilities. If, as astrophysics suggests, space-time begins with the Big Bang, then it would seem that space-time is not necessary and nor, consequently, is time.[8] 'No time exists' and 'no temporal things exist' are not contradictions or in conflict with any necessary truths.[9] Thirdly, time is a physical reality and therefore not necessary.[10] But are there in fact any actual timeless beings and, if so, then what kind of beings might they be. As Leftow points out, mathematical truths are 'the most popular candidates for timeless truths that can co-exist with time'[11] and he goes on to argue for the timeless existence of both mathematical truths and the numbers that are the truthmakers for mathematical truths. Such timeless existence does not make numbers other than they are and they are therefore timeless whether or not time exists; consequently, they are necessarily timeless. Timeless entities therefore can exist, and therefore there is at least a prima facie possibility that God might exist as timeless and as co-existing with time.[12]

    [7] Brian Leftow, *Time and Eternity* (Ithaca, NY: Cornell University Press, 1991), 20.

    [8] Leftow, *Time and Eternity*, 32. I am not clear as to how this argument might encompass the possibility that the Big Bang might itself prove to be only an event occurring within a larger cosmic framework.

    [9] Leftow, *Time and Eternity*, 33. However, the assertion that 'no time exists' seems possible only as an event in time. In the 'Sololiloquies' Augustine argued that if the universe ceased to exist it would be true that the universe had ceased to exist, therefore truth endures through and beyond all possible states of affairs. But without a universe there would be no one for whom such truth would be true. Nor could there be any alternative claims, such as 'the universe does exist' against which the truth of the claim as to its non-existence might be measured. In short, the question just couldn't arise in the absence of any universe—although, from within an existing universe, it might be true to say that at some future point the universe will cease to exist and that there will thereafter be no universe—but 'thereafter' is already a temporally inflected judgement. See Augustine, 'The Soliloquies', in *Augustine: Earlier Writings*, ed. and tr. J. H. S. Burleigh (London: SCM Press, 1963), I.29 (40–1) and II.28 (56).

    [10] Leftow, *Time and Eternity*, 34. Again, this might not be much consolation to 'existing human beings' who are also necessarily physical beings. However, we should note that, at this point, Leftow is considering only the general possibility of there being timeless beings and, more specifically, what he will call negative timeless truths.

    [11] Leftow, *Time and Eternity*, 40.   [12] Leftow, *Time and Eternity*, 40–9.

But why might we want to regard God as timelessly eternal? As I have just indicated, much popular religious language might seem to go no further than affirming some kind of sempiternity. However, there is strong support amongst Christian philosophers for the view that in order for God to be God, God has to be timelessly eternal. We have already seen how, for Augustine, this has a clear soteriological motive: life in time is life subject to 'tumultuous vicissitudes' that we experience as destructive of our happiness and that carry us irresistibly towards annihilation.[13] In the light of such experiences, God's elevation above the world of time and change is the guarantee of his power to save and the gift of participation in the divine eternity is perceived as a necessary feature of salvation. We shall return to questions relating to the soteriological aspect of eternity, but there are also more narrowly theological questions that play an especially prominent part in the analytic discussion, that is, questions having to do with the concept of God itself.

'Eternal' (understood as meaning 'timeless' in the sense of Boethius' definition) is regularly listed amongst the defining attributes of God, such as 'immutable', 'omniscient', 'omnipotent', 'impassible', and 'simple'. As such, and on the basis of the classical theist view that God's essence is, quite simply, to exist, to be the being that he is, being eternal is construed as intrinsic to the way in which God is God. In his seminal 1970 study of *God and Timelessness*, Nelson Pike suggests that although 'God' is often used as if it were a name (such that we can easily assume that 'God does x' functions similarly to 'John does y' or 'Jill does z'), it is better understood as 'a descriptive expression...of a special kind, a "title-phrase"'.[14] Another example of such a title-phrase, Pike says, is 'Caesar': such words not only point out a certain individual (God, John, Jill, or Gaius Julius), but tell us what or who that individual is; they communicate a specific content that is inalienable from the individual concerned being thus named. In this case, then, 'If x is God, x is perfectly good, omnipotent, etc.' or 'it is a logically necessary condition of being God' that God is 'perfectly good, omnipotent, etc.'[15] Is it then the case that 'if x is God, then x is eternal = timeless'?[16] On the classical Christian view, yes it is. Timelessness is an essential property of the individual who is God and 'the possession of timelessness is a

[13] See the discussion in my *God and Being: An Enquiry* (Oxford: Oxford University Press, 2011), 21–32.

[14] Nelson Pike, *God and Timelessness* (London: Routledge and Kegan Paul, 1970), 29. Of the various divine attributes discussed by philosophers, 'eternal' seems to have a special claim to function as a name of God and is more likely to occur in prayerful or hymnodic address than, for example, 'immutable' or 'omniscient' (or their English equivalents, 'unchangeable' and 'all-knowing'). As mentioned in the Introduction, James Moffatt's 1926 translation of the Bible used 'the Eternal' to translate the divine name rendered in the King James Version as 'the Lord'. It (and its parallels in other modern European languages) is also used in several Jewish translations. See also the discussion of Rosenzweig's article 'The Eternal' in the section 'The Eternal' in Chapter 9.

[15] Pike, *God and Timelessness*, 33.   [16] Pike, *God and Timelessness*, 33.

logically necessary condition for bearing the name God', two formulations that Pike combines in the assertion that 'It could not fail that the individual that is God could not fail to lack temporal position and temporal extension'.[17] Again, on the classical theist view (that Pike will later seek to modify), this is not least because of the interrelationship between the divine attributes. Most obviously, timelessness 'entails and is entailed by immutability'[18] and the same goes for incorruptible, ingenerable, and incorporeal, although not, Pike thinks, immortal, since it is not impossible that there may be immortal beings—perhaps we should say 'entities'—that either lack life or that have endless duration.[19]

But if we can speak of timelessness as interdependent with at least some of the other negative attributes of God, the difficulties start to multiply when we come to the issue of divine omniscience. At its simplest, the question is whether a timelessly eternal God could actually know anything temporal? At first glance, this may seem somewhat forced. That God is eternal and all-knowing seems to flow effortlessly from the tongue of praise and prayer. As so often, however, matters are less straightforward than they seem, since any knowledge of what occurs in time would seem to have to be temporal, wouldn't it? For what could events happening in time possibly mean to a being that had no experience or sense of time? If I now get up and go to the kitchen to make a cup of tea, what kind of knowledge could there be of such a mini-history that did not reflect the flow of events by virtue of which we experience it as both a unified act (going-to-the-kitchen-to-make-a-cup-of-tea) and temporal? I cannot get up, go, boil the kettle, prepare the pot, and make the tea all in a trice—and, even if I could, a 'trice' is still an instant of time, no matter how minuscule. And if that is true of such a simple everyday act, surely it is all the more true of a whole human lifetime and yet more so when it comes to complex historical events stretching over several years and involving many human lives.

A particular, and religiously important, sub-set of issues that follow on from this question relates to the whole matter of divine foreknowledge and human free-will, since the claim that God knows all temporal events seems to undermine any possibility of free action on the part of any temporal beings, including human beings. Deferring consideration of these issues, however, let us begin by considering whether any knowledge of temporal events is possible for a timeless being. As we do so, let us also remind ourselves of just what might be at stake here. For if timelessness and omniscience are both supposed

---

[17] Pike, *God and Timelessness*, 33.

[18] Pike, *God and Timelessness*, 44. Aquinas asserts that eternity follows from immutability (ST q. 10.2. reply), as does Paul Helm in *Eternal God: A Study of God Without Time* (Oxford: Clarendon Press, 1988), 19–22. Leftow points out that Aquinas elsewhere argues for God's eternity on the basis of his necessary existence. See Brian Leftow, 'God's Impassibility, Immutability, and Eternality', in Brian Davies and Eleonore Stump (eds), *The Oxford Handbook of Aquinas* (Oxford: Oxford University Press, 2012), 180.

[19] The question of the compatibility of 'life' and timelessness will be discussed further below.

to be defining attributes of the God whose essence is to exist, then amending or dropping one or the other will change the very concept of God—that is, will entail that God is otherwise than classical theism says he is. And if the defender of theism is not to resort to paradox or simple dogmatic assertion (and the working assumption of any version of philosophical theism is presumably that such emergency measures are the end of any meaningful philosophical discourse)[20] this must call either for a major revision of the concept of God or—as those hostile to theism might conclude—that the concept of God is itself incoherent and cannot therefore be a true concept of any actually existing being.

CAN A TIMELESS GOD KNOW TEMPORAL BEINGS?

The stakes are large, then—but the philosophers characteristically coolly open the game with seemingly trivial examples. A favourite is to take the case of someone mowing their lawn. What can God know of that? The question is further refined if we consider the difference between God knowing that I am mowing my lawn now, that I mowed it yesterday, or that I will mow it next spring—or that I mowed it on such-and-such a date, let's say 1 April 2013.

Starting with the simplest of these cases, God knowing that I am mowing my lawn right now—knowing it in the moment that I am actually performing the action called 'mowing the lawn'—would already seem to involve God in time in a manner that compromises the divine timelessness. Why? Because mowing the lawn is a temporal event and knowing that someone is engaged in this temporal event would seem to be possible only by means of a mental act that was itself temporal. But even in relation to this isolated event, the theistic claim is that the God who knows that I am mowing my lawn now is the same God who knows that I mowed it last week and will mow it again next week. And this would seem to involve quiet a lot of time-oriented mental activity. Boethius' answer, followed by many subsequent Christian theists, is that the simultaneity of divine eternity means that God embraces all these time-events in a single all-at-once vision. An analogy put forward to explain the possibility of such a simultaneous knowledge of time-events is that of a person standing on top of a mountain and watching a traveller progress along a winding road far below. From the point of view of the traveller, it is impossible to see the road behind or ahead beyond the last or next corner. From the spectator's vantage point on the mountain, it is possible to see all stages of the journey from the same position. Of course that example can be seen as involving the seeds

---

[20] As a rhetorical move, pleas of ignorance are possible *at any point* in talking about God, but there have to be better motives for making them than the desire to get out of a tough spot.

of its own destruction, since to see the traveller at point A and at point B the observer must wait until the distance between the two has been traversed. The observer might see the entire stretch of road that the traveller walks along at once but does not and cannot see the traveller at the different stages of the road all at once.[21] Like is known by like: only a temporal being can know temporal events.

There are several possible responses to these objections that have received extensive discussion in the literature. One is the postulate of what its proposers, Eleanor Stump and Norman Kretzmann, call 'ET-simultaneity'. Starting out from the Boethian definition, they assert that affirming God's atemporality does not involve denying the reality of time. Time and eternity, they say, are 'two modes of existence' (namely, 'temporality' and 'eternality') and although an eternal being is not internally temporal 'it is appropriate to say that it has present existence in some sense of "present"'.[22] Of course the question then is whether we can give any more definite content to what is otherwise the weasel phrase 'in some sense'. Their answer is the notion of ET-simultaneity, which they introduce with the following example. A moving train is struck by lightning at the front and at the rear. From the point of view of a passenger seated in the train these twin lightning strikes are not experienced as simultaneous. From the point of view of someone standing by the track and able to see the whole train at once, however, they are simultaneous. But this is not just an industrial age reworking of the observer-on-the-mountain argument, since, they say, it illustrates a basic claim of modern relativity theory, namely, that simultaneity is relative to the reference frame of a given observer. Although two events will appear as sequential in one reference frame, they will be seen as simultaneous in another. Nor is this special pleading on behalf of God, since the notion of time as relative to diverse reference frames was not developed to answer a theological question but in the context of physics. Nevertheless, they claim, it is perfectly adaptable for use in relation to questions of time and eternity. And just as the principle may apply to events within time (such as the lightning strikes on the train), so too may it apply to the relation between two distinct eternal events. The outcome is that we are entitled to posit a reference frame in which both an eternal and a temporal event appear as simultaneously present, thereby by-passing the problem of a direct contact between the eternal and the temporal.

This argument, however well-grounded in relativity theory it may be (I am not competent to judge), nevertheless raises as many problems as it answers. On the least charitable view it merely reformulates the basic issues, since, if

---

[21] See Martha Kneale, 'Eternity and Sempiternity', in Marjorie Grene (ed.), *Spinoza: A Collection of Critical Essays* (Garden City, NY: Anchor Books, 1973), 230–1.

[22] Eleonore Stump and Norman Kretzmann, 'Eternity', *Journal of Philosophy* 78/8 (August 1981), 434.

the claim is that we are entitled to posit a reference-frame in which an eternal *x* and a temporal *y* are both present, then we must begin to wonder just what such a reference-frame might be and who might have access to it and under what conditions.[23] Is it a reference-frame that a (temporal) human mind actually has access to? Furthermore, the argument seems to involve a certain oscillation between the notions of 'is' and 'is observed'. If we are able to access a reference-frame in which an eternal entity is observed as present in or to time and a temporal entity observed as present in or to eternity, do we then have a sufficient basis for saying this is how things are—or are we limited, even within this reference-frame, to saying this only that is how they *seem* to be? In a study that, like Stump and Kretzmann's article, wants to defend the notion of divine timelessness against what he calls 'the contemporary onslaught against the predominant received concept of "the eternity of God"',[24] John C. Yates nevertheless points to the question as to whether what we are talking about here is something that can be 'observed' at all. As he puts it, 'That God is timeless or totum simul is surely a deduction of reason and not something that could conceivably be given in experience'.[25] Furthermore, if we are to suppose that such 'observations' do in fact occur in time, Stump and Kretzmann offer no concrete example of what this could mean.[26]

A different objection, also given in the context of a defence of God's timelessness, is that of Brian Leftow. Leftow questions Scotus' assertion that temporal events occur only at the location of their occurrence and suggests that they may also occur as present and actual in eternity ('What we do in this life echoes in eternity'!). It is not a matter of God beholding all the unfolding events of time as occurring at once (they are temporal, and therefore don't all occur at once), but of God seeing at once, in the 'totum simul' of eternity, the

---

[23] Cf. Paul Helm's view that their 'solution' amounts to no more than a re-wording of the problem (Helm, *Eternal God*, 33). Helm also points out—which, indeed they do not hide but explicitly advert to—that Stump and Kretzmann regard the Boethian timelessness that they claim to be defending as a kind of super-time or eternal duration, whereas Helm wants to defend a view of God as simply 'time-free' (Helm, *Eternal God*, 36). Stump and Kretzmann's argument is that Boethius' term 'illimitable' more naturally suggests some kind of 'eternal duration' than it does the kind of Plotinian view that sees 'eternity' as being without extension or interval and therefore as having a point-like character. They also argue that this is the view that Christian tradition has generally taken (Stump and Kretzmann, 'Eternity', 432–3). They regard the charge that the idea of 'eternal duration' is incoherent as arising from confusing such eternal duration with temporal duration ('Eternity', 446). The question is further explored in Leftow, *Time and Eternity*, 112–34. Leftow concludes that 'duration' as used by Stump and Kretzmann is nevertheless only duration 'ad nos' but not 'in se'. I think he is right.

[24] John C. Yates, *The Timelessness of God* (Lanham, MD: University Press of America, 1990), 1.

[25] Yates, *The Timelessness of God*, 128.

[26] Yates, *The Timelessness of God*, 129. However, it might be countered that Christian tradition does offer examples of eternity being experienced in, with, and under the conditions of time, namely, in the kind of mystical experiences enjoyed by Augustine and his mother shortly before her death. See Augustine, *Confessions*, tr. William Watts (London: Heinemann, 1912), Book IX.x, ii. 47–53.

temporally spread out events of time.[27] If there is an eternal being, all temporal events are really simultaneous—in eternity, though not (clearly) in time.
In this sense, then, the 'now' in 'God knows what I am doing now' is univo-
cal as regards both time and eternity in eternity, i.e. for God, although time
and eternity remain distinct as 'modes of existence' (to borrow Stump and
Kretzmann's phrase).[28] However, although this approach certainly has the benefit of elegance in comparison with Stump and Kretzmann's ET-simultaneity,
it still seems to beg the question as to how this might be experienced or (if we
are not minded to regard this as a matter of possible experience) *conceived*
within the continuum of temporal life.[29] Of course, Leftow's aim is not at this
point to gloss how human beings might know divine eternity but, instead,
how a timeless God might know 'at once' all temporal occurrences. Yet the
absence of extended reflection or reference to or discussion of how human
beings might know that God knows this leaves a deficit that, in the view of the
present writer, needs to be made good.[30]

The simplification offered by Leftow also implicitly draws attention to a
formal feature of Stump and Kretzmann's argument that has a certain analogy to some developments in later Platonism. The problem there was how
to reconcile the reality of temporal events with the life of a timeless divine
being. As we have seen, what Stump and Kretzmann do is, in effect, to provide
a *tertium quid*, a theoretical construct (the frame-reference) that allows for
time and eternity to enter into a relationship but which is not itself temporal or eternal although hospitable to both. So too, but differently formulated,
the later Platonists sought ways of thinking worldly time and divine eternity
together. According to Samburksy and Pines, Iamblichus identified two different kinds of time: the indivisible, permanent, and stable time of the intellectual world and the perpetually moving time of the sensible world. The relation
between these is then described as that between a line (the unchanging time of

---

[27] Leftow, *Time and Eternity*, 218. Cf. Paul Helm's view that the fact that an idea is present
to the mind of God does not entail that God is temporally present to anything (Helm, *Eternal
God*, 29).

[28] Leftow, *Time and Eternity*, 228–44. One of the drivers of Leftow's argument is the need to
defend divine simplicity: if God is construed as a direct realist perceiver of events in time then—
precisely on the basis of the like-is-known-by-like argument referred to above—divine simplicity
is compromised. God does not know beings by observing them other than in the unified and
simple light of the eternal act of divine causation.

[29] Of course, whether we can conceive of anything that is entirely lacking in any experiential
base is a controversial question, but not one I want to explore now. Nevertheless, if some things
that fall entirely outside the range of possible experience are conceivable it would seem that they
can only meaningful for human beings if they, in some sense, 'show up' in significant ways in
experience itself and therefore in time.

[30] Leftow notes at the outset that he is simply making the assumption that we can speak affirmatively and literally of God on the basis of arguments provided by Swinburne and Alston. Leftow,
Time and Eternity, 19. However, for reasons that will, I hope, become clear, I regard this as a
rather rash assumption.

the intellectual world) and a series of events that arise and pass away in time, achieving their maximum degree of reality in the 'now' in which they touch and fall away from unchanging intellectual time, each of these 'nows' occurring at a different point on the line of intellectual time. Consequently, there are two kinds of 'now': the single and constant 'now' of the unchanging line and the sequence of 'nows' that occurs as each new event hits against it. Yet even the line of the unchanging intellectual 'now' is only 'a pattern of eternity...in motion with regard to Eternity, but at rest with regard to our time that participates in it'.[31]

This may seem a far-fetched analogy and it is clear that Iamblichus knew nothing of modern relativity theory. Nevertheless, his schematization of the different levels of time, mind, eternity, and the One (since, on this model, eternity is not itself ultimate, being situated between the level of intellect—the intelligible order—and the One) provides a rather sophisticated way of identifying a medium in which what is purely temporal can be related to what is eternal without suggesting a direct encounter such that one or the other would have to succumb to its opposite, the eternal being compromised by time or time vanishing into eternity. Time remains—and always will remain—time; eternity will (by definition!) always remain eternity—but yet the twain shall meet: in the dimension of intellectual or intelligible time. The suggestion is attractive, but both the idea of reference-frames in which time and eternity can be co-present and the postulate of a mediating dimension of intellectual time invite a further regress. For co-presence must presumably mean more than common participation in a third element. Like a social go-between or matchmaker, it is not enough for this mediating element to have a relation to each side—it must be able to bring both together and ensure that genuine contact is made. To put it in visual terms, if you enter the room by one door and I enter by another, we are indeed in the same room. But if, for example, it is the ball-room of Elsinore Castle, we may still be so far away from one another as not to be able to make out our respective features. All the work of effecting a real introduction of the one to the other remains to be done. The problem is restated rather than resolved.[32]

Another line of response to the question as to how a timeless God might know temporal events suggests that although events may themselves be

---

[31] S. Sambursky and S. Pines, *The Concept of Time in Late Neoplatonism* (Jerusalem: Israel Academy of Sciences and Humanities, 1971), 16.

[32] This objection has some analogy with Kierkegaard's view of Hegelian mediation as being a kind of conjuring trick in which differences were progressively minimized to the point of becoming indiscernible; but, he objected, no amount of quantitative adjustment could bring about qualitative change. As he mockingly put it, the proponents of mediation offered only a series of approximations that could never achieve identity, but only continued 'so long...so long.. until'—until, that is, the reader gave up and ceded the field. See SKS7, 306ff/CUP1, 335ff (see 'Abbreviations' for the form of citations of Kierkegaard's works).

temporal, they can be known in propositions that are not themselves temporal. Often accompanied by a nod towards McTaggart's theory of two series of time,[33] an attempt is made to translate propositions that have a temporal and tensed character into non-temporal and tenseless statements. 'It's now 9 o'clock and I'm on my way to the station', 'I went to the station yesterday at 9 a.m.', or 'I will be going to the station tomorrow at 9 a.m.' are all statements that depend on the temporal location of the speaker. If it is now 3 o'clock in the afternoon, the first statement is no longer true since it is not now 9 o'clock. Assuming that I don't go to the station at exactly the same time every day, the second statement becomes untrue with the passing of another 24 hours, whilst the third statement proves to be untrue if, tomorrow, I change my mind and stay at home. However, if we restate what is said along the following lines 'At 9 a.m. on 12 January 2013 the writer of these words was going to the station', it will, if true, always be true. It might be objected that it could only become true at a certain moment in time and is dependent on a temporal sequence of events actually occurring on 12 January 2013. However, let us allow that such statements might be further modified so as to shave away even this trace of temporal contingency and that we could therefore envisage temporal or tensed propositions being restated in non-temporal form without significant loss of meaning, in what Pike calls 'meaning-equivalent formulations of statements utilizing temporal indexical expressions'.[34] Construed in this way, the issue is not whether God is, so to speak, the spectator of events in time by analogy with human sensory intuition, but whether, qua omniscient, God is able to enjoy the knowledge he has of all that is 'all at once'. God does not 'see' temporal events, he *knows* them, and, in his case at least, seeing is not a necessary condition of knowing. God's omniscience is his knowledge of all true propositions—all that is the case—or, as Helm puts it, 'God's eternal existence has no temporal relations whatever to any particular thing which he creates' but he does have a relation of atemporal knowledge: 'the creation is not temporally present to God in his knowing it, nor is it distant. God knows, and that is all'.[35] God doesn't have to be 'in' the 'now' to know what is happening now (to those who experience it as 'now') any more than he has to be me to know what I know.[36] Nor does this result in a 'second-class omniscience', since there is a distinction to be drawn between God

---

[33] The A series is the series of events defined by relations of past, present, and future, as in 'I went there yesterday', 'I'm doing this now', or 'I'll catch you later'. When today becomes tomorrow, it will, of course, no longer be true to say of the events referred to in the first two examples that they occurred 'yesterday' or 'now', and even 'later' will at some time become an event of the past. In the B series, by way of contrast, events are described merely in terms of before/after, which, McTaggart argues, is not sufficient to demonstrate the reality of time. That *x* occurred before *y* will, if true, always be true, whether the reference is to a past, present, or future event. For further discussion see the section 'Hegelian Timelessness' in Chapter 2.

[34] Pike, *The Timelessness of God*, 95.   [35] Helm, *Eternal God*, 36, 37.

[36] Helm, *Eternal God*, 75.

knowing everything (as Christianity claims he does) and God being everything (which Christianity does not claim and which the doctrine of creation very specifically prevents).[37]

But even if we imagine a maximum de-temporalization of tensed statements, is any assertion *actually* timeless? Leftow states that the proposition 'This is a book' can be understood 'all at once'—you 'just see it', he says.[38] Clearly it seems like that to us. But is it true? Apart from the fact that speaking the words or scanning them on a page takes time, no matter how little, my basic capacity for understanding what is said in language is only developed through manifold and lengthy processes of learning. Indeed, the fact that I am a being capable of language is itself the outcome of a process of evolution stretching over many millions of years. And perhaps, in the case of a foreign language I never quite get to the point of understanding '*Das ist ein Buch*' 'all at once' but there is always the briefest of moments in which the words and the connection between them have to settle into place before I can 'see' what is being said. Is that three-dimensional green block on the table really what Germans call a *Buch*? Is that the right word for it? Is that what it is? And perhaps this is even true when I am speaking my 'mother tongue', although, by dint of practice, I have become so adept in speaking it that it seems like second nature, as if it takes no time at all to see what is being said. 'As if'—but perhaps this is simply a perspectival illusion? 'I see what is being said' seems like the simplest and most elementary statement of what is involved in understanding, but it is itself suggestive of a complex and inherently tensed process built up out of the interplay of sense-experience, mental intuition, and the translation (as it were) of these into and back out of language. That I see the object as a book (or, it may be *ein Buch*) never escapes—entirely—the temporal process by which I learned that this is indeed a true statement of what is actually the case. And this, of course, is before we even begin to reflect on the temporal nature of the object itself, an artefact that was written, designed, printed, distributed, purchased, and placed just here so that it is possible for me, now, to see it as the book that it is. Were not those Platonists right who held that timeless truths could only be known timelessly, in pure intuition, without the mediation of words, and therefore beyond even thinking?[39]

---

[37] Helm, *Eternal God*, 78–80.

[38] Leftow, *Time and Eternity*, 286. Cf. Helm's assertion that God knows the temporally ordered creation 'at a glance' 'in rather the way in which a crossword clue may be solved in a flash' (Helm, *Eternal God*, 26). However, this is all the more exposed to the following comments, since—as opposed to just seeing the book on the table—hitting upon the answer to the crossword clue is likely to be the outcome of a complex interaction of conscious and unconscious processes that may go on for some time before the moment of intuition arrives.

[39] See Richard Sorabji's account of Plotinus' 'mystical' account of the self's timeless, non-cognitive, and supra-noetic experience of the One in *Time, Creation and the Continuum: Theories in Antiquity and the Early Middle Ages* (Ithaca, NY: Cornell University Press, 1983), 157–61. However, Sorabji argues that Plotinus also allows for a kind of non-discursive thought that is timeless but nevertheless articulated in propositions and that consists in the self

Leftow, of course, understands the possibility of such objections, but he regards it as accidental that, for human beings, such instantaneous knowledge presupposes long-term biological, social, and other processes. Moreover, the actual act of achieving knowledge need not involve any other processes (even if, for us, it does); 'for perhaps', he muses, 'human or biological organisms are not the only sorts of things that can be alive'.[40] Well, this may be true, but the decisive question is surely what we human beings can know as true or what it is for us to know something to be true. And even if, in a certain metaphysical perspective, the temporality of biological and social life is 'accidental'—and let us admit for the sake of argument that it is accidental that each of us has the particular physical, social, and personal qualities that we have—it is nevertheless not possible to conceive of a human being whose existence and capacity for knowing is not grounded in such temporality. It is not necessary, perhaps, that there are human beings at all, but given human beings, it is necessary for their being human that they have a biological, social, and personal history as integral to their identity. The kinds of knowledge available to non-human entities can, for us, only ever be a matter of speculation and not, in the end, decisive for what we can and cannot know. Conceding such other possibilities may encourage us to take a more humble view of our place in the order of things—as in Christian speculations on the life of angels—but it does little enough to tell us what are our actual possibilities for knowing this and not knowing that. The problem becomes all the more urgent if we once regard the human interest in such questions as not solely theoretical but as driven by the soteriological imperative of seeking a radical transformation of our lives and of our present possibilities. In any case, I have already argued that it is a kind of perspectival illusion to believe that I can understand even the most elementary proposition in a non-temporal fashion, even allowing for my ability to do so as having somehow, magically, appeared fully-formed, like the goddess of wisdom herself.

However, we again note that the issue here is not whether human beings can themselves have timeless knowledge, but whether the concept of timeless knowledge is sufficiently coherent for us to suppose that a timeless God could

---

'contemplating [the definitions of things in terms of genus and differentia] arranged into a unified network' (*Time, Creation and the Continuum*, 153). Perhaps there is some analogy here to the kind of knowledge that more recent philosophers ascribe to God as knowing all true propositions, without needing directly to 'observe' the flux of mutable things.

[40] Leftow, *Time and Eternity*, 287–8. However, in the following chapter on 'Time, Actuality and Omniscience' he does concede that whilst it is perhaps forced to say that God's omniscience is limited because, not being a physical being, he cannot know what it feels like to be a walker, it may be preferable to modify omniscience than to give up on timelessness. Similarly it may be that there may be some propositions that can only be fully understood by one individual, such as 'I am Herman', where only Herman himself really knows just what he means by this. Yet a God who was non-omniscient in this sense could still be a God worthy of worship. See *Time and Eternity*, 317–27.

have knowledge of a temporal universe. In this regard, the kind of approximation to such timeless knowledge that human beings might be capable of—even if they will, in life, always remain bound by the conditions of temporality—is not being asked to do more than provide a remote analogy to how it might be for God. Nevertheless, the more seriously we take the inherent temporality of all language, the less useful the work that the analogy sets out to do—perhaps to the point at which it seems, at best, suggestive and no more. The example of a kind of knowledge that is thoroughly temporal—human knowledge—can scarcely provide the basis for a confident assertion that knowledge as such can be timeless. At the same time, if, as Leftow assumes, we can speak affirmatively and literally of God (see above), the temporality of human knowledge as embedded in true propositions would seem to set a barrier to our knowing just what we are saying when we speak of God as having timelessly eternal knowledge of the temporal world. And even if angels or other entities might be supposed to have the possibility of different kinds of knowledge that, in a way unknown to us, have been purged of their temporal encumbrances this may seem cold comfort, since, evidently, their knowledge is not our knowledge and the possibility of our attaining a true knowledge of God is not helped.

TIMELESS KNOWLEDGE AND HUMAN FREEDOM

Although I have suggested that the analytic approach to the question of the eternal lacks something of the soteriological passion manifest in Augustine's *Confessions*, there is one area of discussion where the question of salvation does make itself at least indirectly felt, namely, in relation to divine foreknowledge. For if God foreknows events that occur in time, this has significant implications for questions of human nature and destiny that are also central to Christian views of creation and salvation. For if God timelessly knows all that is the case, then, many commentators conclude, there is no room for human free-will. Whatever I might now imagine that I will do in the future—whether it is a matter of some trivial everyday action like making yet another cup of tea or a 'world-historical' intervention in the affairs of nations—if God already knows it and if God can only know what is true, then (it seems) I actually have no choice as to whether I will do it or not. I will do it. Nothing else can happen.

Theologically, the question is intensified when it is brought into relation to the classical Christian account of human existence as fallen and exposed to future possibilities of salvation and damnation according as to—well, that's just the question. Is it according as to whether the human being does good or ill, or is it according to the inscrutable counsels of God that have decided from 'before the foundations of the world' whether we are destined for salvation or damnation? In this perspective the question of foreknowledge segues into the question of predestination, a question that became pressing in Augustine's later thought and that would again become central to debates generated by the Protestant Reformation. In this latter context it was Calvinism that went to the furthest extreme of denying freedom to fallen human beings and resting the entire outcome of human existence in the foreknowing and predestining will of God, whilst at the same time (and in common with Catholic theology) asserting that Adam's initial fall was brought about by an act of free-will that, although foreknown by God, was nevertheless free.[41]

As a dogmatic issue this clearly relates closely to such other theological topics as creation, providence, regeneration, and redemption, provoking such questions as: To what extent does the initial act of creation bring into being a world in which creatures might be free to think their own thoughts and do what they will? To what extent and how might God and creatures collaborate in furthering the purposes for which the world was brought into being? Can a creature that has fallen away from its original perfection do anything to restore its lost possibilities? Or must God 'take over'—'all is of grace'?

On all these topics there has been significant variation in Christian tradition, although the broad stream of both Latin and Greek Christianity has generally wanted to affirm both God's sovereignty over the whole world-process as guaranteeing that his purposes will not fail, whilst also allowing for human freedom, including the freedom of even fallen human beings to turn back to God and, assisted by co-operant grace, to contribute to their own sanctification and to bring forth fruits worthy of final reward. As we have just seen, however, there is also a strong minority tradition that holds that human freedom is illusory and that God is not only the sole agent in the creation and preservation of the world but also of human regeneration and action. Paradoxically the same logic seems to underwrite many secular responses to religion, since, it is argued, freedom is such that it cannot be distributed between God and human beings. All must be on one side or the other. As Dostoevsky's fictional nihilist Kirillov puts it ' "If God exists, all is His will and from His will I cannot escape. If not, it's all my will and I am bound to proclaim self-will." '[42] We might think too of Jean-Paul Sartre's aversion to the idea of being the object of a God's all-knowing gaze, a gaze that would, as he believes, circumscribe and annihilate his own freedom.[43] But whether the questions come from within or from beyond the believing community, dogmatic theology may and does primarily appeal to Scripture, to tradition and to the authority of Church teaching

---

[41] On the relationship between foreknowledge and predestination see Calvin, *Institutes*, Book III, 21.5; on the Fall, see Book II, Ch.2.

[42] Fyodor Dostoyevsky, *The Possessed*, tr. Constance Garnett (London: Heinemann, 1946), Part III, Ch. 6.2, 561 (translation adapted). The logic of his position leads Kirillov to the conclusion that he must therefore commit suicide in order to demonstrate that he has total sovereignty over his own life.

[43] See, e.g. Jean-Paul Sartre, *Words*, tr. Irene Clephane (Harmondsworth: Penguin, 1967), 65.

itself, and if it does, for the most part, also use reasoned argument to show the cogency of the claim at issue it is neither ashamed to invoke the limitations of human knowledge nor obliged to put more weight on reason than the exigencies of evangelistic or pastoral work require. What right has the pot to question the potter, who can make whatever he will from the clay! Philosophy, however, even when understood as philosophical theology and therefore as operating within assumptions prescribed by faith, must, at the very least, press the limits of knowledge as far as they can be made to go and put forward arguments and models that are plausible on their own terms.

Boethius himself already inherited a long history of thinking about how God's timeless knowledge can be harmonized with human freedom, a history reaching back through Augustine to the Bible itself (at least to a certain reading of the Bible). His solution (or rather 'Philosophy's'[44]) is to draw a parallel with the way in which the same event can be known as an object of sense-experience, imagination, or reason. We all recognize another human being when we see one coming towards us, but only reason sees what is essential in being human: 'man is a biped rational animal'.[45] But this in no way intrudes upon normal everyday ways of human interaction. So too, God's foreknowledge (infinitely elevated above even reason), does not disturb the integrity of the plane of free human action. Boethius concludes that when at the same time we see the sun rising and a man walking about, we instantaneously recognize the one as happening by necessity and the other as willed. 'In the same way the divine gaze looks down on all things without disturbing their nature; to Him they are present things, but under the condition of time they are future things.'[46] However, few more recent philosophers would find this a sufficient philosophical response—even if it is sufficient for faith.

One much-referenced response is the doctrine known as compatabilism, that is, the view that divine foreknowledge does not compromise human freedom but that the two are mutually compatible and this compatibility can be explained in terms of what is called God's 'middle knowledge'. This theory is associated with the seventeenth-century Jesuit Luis de Molina who set out, partly in response to the debates about predestination set in train by the Reformation, to defend the compatibility of God's foreknowledge and human freedom and to argue that God can have a timeless knowledge of freely initiated events occurring in time.[47] Molina's proposal is to distinguish between different kinds of knowledge in God—natural, free, and middle. God's natural knowledge is the knowledge proper to his nature and is the knowledge

---

[44] The *Consolation* is written as a dialogue between Boethius and Lady Philosophy.

[45] Boethius, *Consolation*, 159.   [46] Boethius, *Consolation*, 166.

[47] See William Hasker, *God, Time, and Knowledge* (Ithaca, NY: Cornell University Press, 1989) Chapter 1 for discussion of the historical context of Molina's views. The following summary draws primarily on this chapter.

whereby he antecedently knows all possibilities. His free knowledge is the knowledge he has subsequent to his willing which of these possibilities will, in fact, obtain and which not. Middle knowledge is the knowledge God has 'in his own essence' of what each free creaturely agent would do if placed in any possible order of things. Such middle knowledge extends beyond omniscience in the conventional sense, since it is not just a matter of God knowing everything that is the case or all true propositions but of God also knowing all possible configurations of how things would have been if, at any point, any free creaturely agent had or were to act otherwise than God's free knowledge foreknows.

At one level such middle knowledge seems fairly plausible. Parents, teachers, partners, politicians, and military strategists often prove highly competent in discerning how those with whom they have to deal might respond in a range of possible scenarios. That we can set up a situation in which another person may have a range of options, each of which we have thought through in advance, does not make the option they choose any less their own. Of course, even within human relationships there is a rapidly rising scale of complexity. If I give my child money to go and buy milk, I perhaps reckon with the possibility that he will actually spend it on sweets or, being of a careless disposition, may lose it, or simply forget to go to the shop and run off to play football with his friends instead. But whichever of these possible scenarios actually occurs is nevertheless a result of what he freely does, even if I have correctly anticipated which of these possibilities is most likely and measured out the money I give him or the timing of my request so as to ensure, as far as I can, the required result. Of course, if we move on to consider a large-scale event such as a political or military campaign, the strategist may (and should) have thought through a much more extensive range of possible sequences of events. Precisely because of this complexity it is always likely that things will not go according to plan, since the opponent may have had options the strategist had not been in a position to predict. As is often said, most battle plans become useless in the moment in which battle is joined. And, of course, it is all even more complicated in the case of God, since God would need to know every possible sequence of outcomes of every possible free act in advance. But if we can allow the analogy to stretch that far, perhaps we can also factor in that, by virtue of his free knowledge, it is also God who decides which of the possible configurations of events actually occurs. I (or the general) can set things up so as to maximize the likelihood of the desired outcome, but only to a certain degree. God, however, can set the universe as a whole up so as to ensure the outcome most pleasing to him and most in accord with his glory.

But—and, differently posed, this was already a subject of debate in the Middle Ages—does this idea of middle knowledge not compromise the nature of God's knowledge? Surely God is not some kind of super supermarket shopper surveying an infinite range of possible scenarios out of which he picks those that seem best to him? For God's knowledge is precisely the knowledge known by the one who has brought all things into existence out of nothing. It is therefore not only incorrigible knowledge of all that is actually the case but knowledge grounded in the ultimate coincidence of knowledge and creative will within the divine simplicity. Therefore, God's knowing x to be x is precisely what makes x be x and it is precisely God's making x be x that gives God the knowledge he has of it.[48] God's knowledge is both necessary and causal, whereas middle knowledge seems to render it contingent and, in a sense, reactive. In the human examples I have given, although the parent or general initiates the sequence of actions, their initiating action is itself responsive to their knowledge of how the child or enemy might act in the scenario brought about by their action. Doesn't the idea of middle knowledge similarly constrain God in deciding which of all the perhaps infinite manifold of possible worlds (or even which human historical events) he will bring about?

One response to this objection is to say that it is allowing itself to be misled by the terms in which the proposal has been set out. Although we have no alternative but to help ourselves out with human analogies and therefore can't help but depict God as if he were some sort of human planner, envisaging future scenarios and selecting from amongst them, this is not how it is for God. If God is timeless, then there can be no question of his adapting his decisions in the light of events in time or in response to actual human acts. His seeing all possible configurations of events and his determining which actually come to pass is, in itself, non-temporal, although it is enacted in time. This, of course, returns us to more general questions as to how we might conceive of the relationship between a timeless God and a temporal world—can God know what is happening now or what will happen tomorrow, etc.?

In the context of discussing the possibility that we might have power in relation to the past (which he believes to be a necessary implication of compatabilism), Hasker draws attention to the view that not every way in which something is 'brought about' necessarily involves a causal relationship. Socrates' death brings it about that Xanthippe becomes a widow, and the fact that at a certain moment in time I do x brings it about that those people who yesterday believed that today I would do x were, in fact, right—but in neither case is the relationship causal.[49] And, as Hasker notes, Molina's compatabilism precisely refrained from holding God's foreknowledge to be causal.[50] That

---

[48] Cf. Gilson's objection that Avicenna conceives of the essences of things as a range of possibilities existing independently of God, with God choosing which are to be actualized. To this Gilson opposes the Thomist view that God's knowledge of essences is not other than his knowledge of them as existing. See Etienne Gilson, *Being and Some Philosophers* (Toronto: Pontifical Institute of Medieval Studies, 1952), 74–96. See also Friedrich Schleiermacher, *The Christian Faith*, ed. and tr. H. R. Mackintosh and J. S. Stewart (Edinburgh: T. & T. Clark, 1928), §55 (on divine omniscience), 219ff.

[49] Hasker, *God, Time and Knowledge*, 107.

[50] Hasker, *God, Time and Knowledge*, 141.

I now mention this is because, as we return to the more general question as to the relationship between a timeless God and a temporal world, we have to ask more precisely just what kind of relationship this is and—if we allow the force of soteriological motivations to play a part in our attempts to think about God—what kind of God-relationship it is we are looking for. Are we going to be satisfied with a God who timelessly knows all true propositions and, in the case of middle knowledge, all possibly true propositions as well? Are we obliged to construe God's relation to the world as causal, bringing about all that comes to pass in time through one single supra-temporal divine act, as much Christian tradition seems to have assumed? Or might we re-envisage it as perhaps a kind of (no less supra-temporal) letting-be, permitting the best of all possible configurations of beings to come into being? And however we answer these and similar questions, how intimately do we expect God to be concerned with, let's say, what's happening in my life right now? Putting it somewhat outrageously, how does God feel about what I'm feeling? Can he feel anything? Do I want him to?

Even philosophers who wish to defend knowledge as a defining element in the divine–human relationship have, nevertheless, acknowledged that—as Helm puts it—we seem to need more than a God who knows propositions.[51] Similarly, those who want a God-relationship at all seem mostly to want a God who acts towards us (and perhaps even reacts to us) other than as the remote or immediate cause of our actions.[52] In other words, the terms of the discussion thus far fail to do justice to the characterization of God in terms of life (an explicit element in the Boethian definition) and personality and, equally, of our relation to God as living and personal. No matter how much the timeless God may be qualified and no matter how nuanced the mediation of the timeless divine presence in relation to temporal human beings, don't the categories of 'life' and 'personality' drive an irreparable hole through the constructions of the philosophers? That, at least, is the opinion of some of the philosophers themselves.

THE LIFE AND PERSONALITY OF THE TIMELESS GOD

That timelessness conflicts with some of the things theists want to say about God becomes a central element in the later chapters of Pike's study of *God and Timelessness*. Having defended the mutual entailment of timelessness and

---

[51] See Helm, *Eternal God*, 48.

[52] On the connection between the problems of an eternally timeless God knowing the world and acting in it see Helm, *Eternal God*, 67.

(a) immutability, (b) ingenerability, and (c) incorruptibility, and having argued that timelessness entails (though is not entailed by) incorporeality, he points out that the predication of immortality does not itself follow directly, since an immortal being must also be a living being and only the supposition that whatever is not corporeal has the character of living soul would justify the further inference that incorporeal beings are also necessarily immortal beings. This is not yet to say that there is a conflict between ideas of timelessness and life, but if the living being in question is also a person (as, it seems, theists want to say God is) this involves, for example, remembering, anticipating, and acting, and therefore 'a timeless person is not possible'.[53] Further considering the limitations of a timeless (and therefore also incorporeal and immutable) being with regard to knowledge of and responsiveness to corporeal and temporal beings (such as we are), Pike ponders suggestively 'How much of a person would a timeless person be?'[54] Similarly he discerns issues around the connection between timelessness and God's superlative greatness, if the latter is understood in terms of evoking worship. Worshippers, he suggests, don't actually behave as if their object of worship were timeless, and although timelessness might naturally elicit admiration there is, as he adds, much scope for taste.[55] It might be better for us if a powerful and benevolent being were timeless, 'but this would not show that the individual in question would be a better individual if he were timeless rather than temporal';[56] that is, timelessness is not in and of itself self-evidently an excellence in the divine nature.

Richard Swinburne, who acknowledges having changed his views from earlier work, goes further, arguing in *The Coherence of Theism* 'that God is completely changeless would seem to be for the theist an unnecessary dogma' since such a claim would seem to compromise the divine freedom: 'a perfectly free person could not be immutable in the strong sense, that is *unable* to change'.[57] These comments both dissolve the link between divine immutability and divine timelessness and politely despatch both if not to the dustbins of history then to its intellectual recycling point. Divine timelessness both conflicts with many of the things theists want to say about God and stretches the sense of simultaneity beyond any possible meaningful use.[58] Even with the help of analogy, this is a step too far.

---

[53] Pike, *God and Timelessness*, 121. Martha Kneale, however, does hold that there is a fundamental contradiction between the ideas of 'life' and 'all at once'. See Kneale, 'Eternity and Sempiternity', 230.

[54] Pike, *God and Timelessness*, 129.   [55] Pike, *God and Timelessness*, 161.

[56] Pike, *God and Timelessness*, 164.

[57] Richard Swinburne, *The Coherence of Theism* (Oxford: Clarendon Press, 1993), 222.

[58] Swinburne, *The Coherence of Theism*, 228. A somewhat different point but one that arrives at the same place regarding God's timelessness is Hasker's comment that faith needs less than some theologians and philosophers have supposed: Hasker, *God, Time and Knowledge*, 199.

Instead of the timeless God of classical theism, then, Swinburne states that everything that happens, 'including the mere existence of a substance with its properties', 'happens over a period of time'[59] and he also affirms that, in accordance with Scripture and the general position of the Church in the first three centuries, God is to be understood as everlasting rather than timeless. This latter view is the consequence of Neo-Platonist influences, as seen in Boethius, that regard anything short of timelessness as making God 'time's prisoner'. However, Swinburne ripostes, not only is the timeless view incoherent (nothing can be that is not temporal), but 'to the extent that the "everlasting" view does have the consequence that God is time's prisoner, this only arises from God's voluntary choice',[60] adding that, since this is so, 'It is God, not time, who calls the shots'.[61] More expansively, he sums up by asserting that 'the unwelcome features of time—the increase of events that cannot be changed, the cosmic clock ticking away as they happen, the possibility of surprise in the future—may indeed invade God's time; but they come by invitation, not by force—and they continue for such periods of time as God chooses that they shall be'.[62]

Keith Ward robustly affirms temporality as a necessary condition of God being the God who is the creator of a temporal world in the well-being of which he continues to take a benevolent and saving interest. This is not just a matter of God getting involved in the temporal life of creation (although Ward asserts that God is 'contemporary with every present'[63]) but of God's own nature. 'May there not be in God an element of creative spontaneity, so that he can freely generate new ideas, just as a human artist creates new tunes or patterns of colour? The archetypal world may not be immutably fixed; it may well be modifiable by the creative intellect' so that 'there is no stock of eternal ideas, but a constantly changing state of imaginatively created ideas'.[64] Most succinctly, Ward states that 'God is unlimitedly potential and therefore ceaselessly changing'[65] and, even more radically than the Molinist position can allow, that 'the future is truly open and undecided'.[66] Ward claims still to

---

[59] Richard Swinburne, *The Christian God* (Oxford: Clarendon Press, 1994), 72.
[60] Swinburne, *The Christian God*, 139.   [61] Swinburne, *The Christian God*, 140.
[62] Swinburne, *The Christian God*, 143.
[63] Keith Ward, *Rational Theology and the Creativity of God* (Oxford: Blackwell, 1982), 163.
[64] Ward, *Rational Theology*, 154.
[65] Ward, *Rational Theology*, 159. Ward states briefly here what I regard as a major revolution in metaphysics. On the Aristotelian view that informs both Aquinas himself and the majority of modern Thomists, actuality will always, so to speak, trump potentiality, since what is actual (a fully-grown oak tree) is what merely potential entities (an acorn) are striving to become and thereby to realize their potential. Famously, this constitutes the basis of the argument for the existence of God based on the phenomenon of motion, namely that nothing can be unless there is one being who is pure act—God. For further discussion see my *God and Being*, Chapter 7 and especially 277–85.
[66] Ward, *Rational Theology*, 154.

'treasure' Boethius' definition of eternity, except for the 'tota simul', which, he says, 'is the source of all the contradictions so tempting to subsequent theologians, so destructive of clarity and coherence in theology'.[67]

Allowing the concept of God to become temporally inflected in such ways promises enormous gains to philosophical theology, allowing notions of divine life, personality, and involvement with human affairs to be more fully and plausibly articulated than when God is regarded as strictly timeless. Nevertheless, and by the same token, it seems also to be marked by a readiness to surrender the kind of precision that might be desirable when we are tasked with the construction of time-free propositions in favour of a suggestive, looser, and more uninhibited resort to metaphor. Swinburne's references to God 'calling the shots' or inviting 'the unwelcome features of time' into his own time are charmingly vague, whilst Ward projects onto God a highly Romantic account of creativity that, attractive as it is, is massively challengeable at the human—let alone the divine—level.[68] Certainly, this seems to bring philosophical theology into closer proximity to the language of living religion (and, for that matter, of normal human discourse about important life-issues), but at the same time such a move exposes it all the more to the projection of culturally and psychologically variable human estimations of greatness and goodness onto the divine being. And here we might say—anticipating a fuller discussion below—that although asserting God to be timeless weighs against some of the things that religious persons want to say about God it does at the same time draw attention to and underline something that they do want to say about God, namely, that (to use Kierkegaard's phrase) God is infinitely, qualitatively different, he is 'other', 'he that has made us and not we ourselves'. Does a temporal concept of God then risk bringing God so close to human temporality as to obscure the proper ontological difference between divine and human?

Yet the language of Christian faith's source documents—the Bible—seems at many points also to speak of God in temporal terms and as doing all those things that Pike says belong to a temporal being, such as remembering, anticipating, and acting, and much more besides. Admitting a temporal dimension into our concept of God, then, may not just be a way of responding to the philosophical problems that are generated by defining God as timeless; it may

---

[67] Ward, *Rational Theology*, 169. Whatever view one takes of the Boethian definition, one might regard the elimination of the 'tota simul' as effectively destroying its entire point.

[68] Similar comments could be made about many features of the God of process theology. Perhaps the most widely quoted theological remark from the process literature is Whitehead's statement that God is 'a fellow sufferer who understands', but the ways in which God can be a 'fellow' to those who are entirely (and not only in their consequent aspect) given over to biological death and exposed to the tragic ruination of their core life-projects needs more specification than Whitehead at least seems to give it, and some account needs to be made also of just what is being surrendered and what retained from more classically theistic views.

also be a way of bringing philosophical theology closer to what, for Christians, is the paradigmatic source of all theological language: Scripture.

BACK TO THE BIBLE

Faithfulness to Scripture is a major element in Nicholas Wolterstorff's call to prefer (with Swinburne) an everlasting God to a God who is timelessly eternal. On the basis of the hermeneutical principle that 'an implication of accepting Scripture as canonical is that one will affirm as literally true Scripture's presentation of God unless one has good reason not to do so' and, since Scripture in fact presents God 'as having a history', 'the burden of proof is on those who hold that God is outside of time—as those who hold that God is timeless, eternal'.[69] And this despite what he calls the 'massiveness of tradition' pointing the other way (i.e. as represented in the Boethian definition).[70]

In his influential 1974 article on 'God Everlasting', Wolterstorff follows through the implications of rethinking the temporality of God in the light of the biblical witness. Since the Bible presents God's redeeming work for human beings as the work of 'an agent...acting within *human* history' it follows that 'God the Redeemer *cannot* be the God eternal'.[71] Just as the timeless tradition regarded God's timelessness as integral to his identity qua God, so Wolterstorff conversely holds God to be fundamentally non-eternal.[72] Not only simultaneity, understood as an attribute of temporal existence, but succession and, precisely, 'changeful succession' can be said of God.[73] A further consequence of this is that God can be the object of human action and can be responsive to the temporal and free actions of human beings.[74] That God is represented in the Bible as unchanging cannot be taken as implying immutability in the sense of classical theism but is understood by the biblical writers in terms of faithfulness.[75] Yet, as for Swinburne, it doesn't follow that God is in any sense

---

[69] Nicholas Wolterstorff, *Inquiring about God: Selected Essays, Volume 1* (Cambridge: Cambridge University Press, 2010), 158–9. This raises enormous questions about the legitimacy of such a use of Scripture in philosophy or even in theological reflection conducted in the context of a public university open to students and scholars of all religious persuasions and none. However, in the context of the present discussion, in which all participants acknowledge some normative role for scripture, Wolterstorff's claims are exceptional chiefly for their directness. Consequently, I will not here pursue the question as to their legitimacy further.

[70] Wolterstorff, *Inquiring about God*, 159.

[71] Wolterstorff, *Inquiring about God*, 133; my emphasis. This and the following quotations are from the chapter 'God Everlasting' which is a reprint of the 1974 article.

[72] Wolterstorff, *Inquiring about God*, 139.

[73] Wolterstorff, *Inquiring about God*, 149.

[74] Wolterstorff, *Inquiring about God*, 150.

[75] Wolterstorff, *Inquiring about God*, 154.

'time's prisoner', although Wolterstorff's way of putting it is appropriately more biblical than Swinburne's talk of God 'calling the shots': 'Though God is within time', he writes, 'yet God is Lord of time'.[76]

But if all this is true, and if Christian tradition has from the beginning claimed to base its teachings on the Bible, how could the divergence between the everlastingly temporal God of the Bible and the timeless God of the Christian philosophical tradition have been overlooked? One reason, according to Wolterstorff, is rooted in a generally insuperable element of human existence itself, namely, 'the feeling…that the flowing of events into an irrecoverable and unchangeable past is a matter for deep regret'.[77] The second is the influence of classical Greek philosophy on the development of Christian thought, even though its fundamental patterns 'are incompatible with the patterns of biblical thought'.[78] Within Christian literature, both of these elements, existential exigency and the influence of Platonic thought-forms, are paradigmatically exemplified in the writings of Augustine. As regards the first of these, we shall be returning to it at many points as we come to explore further the phenomenological evidence for a human orientation to divine eternity and whether or how this might demand construing the eternal God as, specifically, timelessly eternal. As regards the second, the duality of biblical and Greek thought is one of the commonplaces of twentieth-century theology, not least as regards the conception of time. Where Greek thought is said to be limited to an essentially cyclical concept of time, 'prophetic' biblical thought sees linear historical time as the proper realm of God's action; where Greek thought ultimately sees time as illusory or lacking substantial being, biblical thought sees what is said and done in time as decisive for human identity and human destiny; where Greek thought, in its soteriological aspect, sees salvation in terms of deliverance from time, biblical thought looks to the fulfilment of time; and where Greek hope focuses on cultivating the non-temporal life of the immortal soul, biblical hope looks to the flourishing of human community.[79] Although these two streams came gradually to converge prior to the emergence of the

---

[76] Wolterstorff, *Inquiring about God*, 156.

[77] Wolterstorff, *Inquiring about God*, 134.

[78] Wolterstorff, *Inquiring about God*, 134–5. However, in the version of the article reprinted in *Inquiring about God*, Wolterstorff acknowledges that in a subsequent 2001 article, 'Unqualified Divine Temporality' he partially retracts this second view, referring to the reprinted version of the second article, also in *Inquiring about God*, 179.

[79] For example, see Paul Tillich, *The Interpretation of History* (New York: Charles Scribner's Sons, 1936), 243–8 and 'The Struggle between Time and Space' in *Theology of Culture* (London: Oxford University Press, 1959), 30–9; Alan Richardson, *History Sacred and Profane* (London: SCM Press, 1964), Chapter 2 'The Two World Systems'; Mircea Eliade, *The Myth of the Eternal Return: Or, Cosmos and History*, tr. Willard R. Trask (Princeton, NJ: Princeton University Press, 1971), esp. 118–24. In Chapter 7 we shall see that, although it has been a widespread view amongst theologians and some philosophers, this is a much too simplistic account of 'the' Greek view. See also the discussion of Plotinus in the present chapter.

Christian Church and the formation of its supplementary scriptures, the New Testament concept of time is still recognizably Hebrew rather than Greek, and even the extravagant expectations of the author of the Book of Revelation seem to look towards a future that is, in some sense, temporal—everlasting felicity rather than absorption into timeless bliss (and, conversely, everlasting torment for the damned). In short, Greek thought construes time in essentially spatial terms, whilst it is uniquely Hebrew thought that opens the way towards a genuinely temporal and historical understanding of human existence.

Now it is undoubtedly the case that, from the apostolic period onwards, Christians reading what became 'the Old Testament' did so in ways that, both from Jewish perspectives and in the light of modern biblical scholarship, seem to go against the plain sense of the text. Hermeneutical procedures were developed and authorized that allowed the text to be used for purposes that would in many cases have been incomprehensible to the original authors, not least with regard to the way in which readers such as Augustine read Greco-Roman conceptions of timelessness into biblical texts in ways entirely alien to the authors' original intentions. But does it follow that 'Greek' and 'Hebrew' attitudes to relations between time and eternity had nothing in common? After all, the affliction of what has been called 'essential loss' resulting from what Wolterstorff describes as 'the flowing of events into an irrecoverable and unchangeable past' would seem likely to be an experience equally available to Greeks and to Hebrews, as it was and is to humans living in other widely divergent cultural contexts. But what about the concept of eternity itself?

Half a century ago, Ernst Jenni concluded that the term *ʿōlām* and related forms are not found with any distinct theological usage (i.e. as applied directly to God) in the pre-exilic period. More salient are their use in relation to the constant and reliable cycles of nature.[80] Applied to the Israelite kings (as in 1 Kings 1:31: 'May my Lord...live forever'), and if said as more than a mere courtly formula, 'the king's "eternal life" embraces all the guarantees of divine help for the people that are provided through the king'.[81] Yet the content of such hopes is still derived from the cyclical life of nature rather than from speculations about world-history. In the historical books it is used in relation to the succession of generations, suggesting the possibility of a transition from the order of nature to a historical time-consciousness, but this is still at this point implicit.

The post-exilic period, especially as manifest in the writings of Deutero-Isaiah, marks a new stage in theological application as time 'acquires its direction and non-repeatability'.[82] Yet God's eternity is not an abstract metaphysical attribute but remains rooted in the history of Israel and the peoples; as

---

[80] Ernst Jenni, 'Das Wort ʿōlām im Alten Testament', *Zeitschrift für die alttestamentliche Wissenschaft* 65/1 (1953), 1.

[81] Jenni, 'Das Wort ʿōlām', 8.   [82] Jenni, 'Das Wort ʿōlām', 16.

such it means 'God's freedom from becoming and transience, his being Lord over all temporality, his activity in all times' and 'his faithfulness'.[83] Yet, given God's uniqueness, if eternity is his attribute, then ' "eternal" thereby came to approximate ever more closely to "divine" and so the tendency to reserve the term for religious language grew'.[84] Thus follow such usages as the eternity of God's word, eternal salvation, eternal justice, and eternal goodness—all in contrast to the transitory nature of the world. The term is particularly prominent in idealizing the past ('from of old') and in relation to the eternity of God's wrath. In subsequent apocalyptic literature it becomes a constant attribute of the world beyond.

Yet if the conclusion is that there are no grounds for rigorous philosophical ideas of either timeless eternity or everlastingness (sempiternity) in Hebrew biblical literature, it is clear that a simple opposition between a cyclical Greek conception of time and a linear Hebrew conception will not do. If the covenant with Noah inscribes the never-ceasing cyclical pattern of 'seedtime and harvest, cold and heat, summer and winter, day and night' (Gen. 8:22) at the foundational level of the covenantal theology of the final recension of the Old Testament scriptures, the correlation between the feasts prescribed in the Mosaic law and the agricultural cycle further endorsed this connection. In the prophetic vision of a Jeremiah, the testimony of the cycles of sun, moon, and stars underwrites the prophecy of a new covenant (Jer. 31:35). To the exiles in Babylon his advice is not to rest their hopes on any sudden divine intervention, but rather to 'build houses and live in them; plant gardens and eat the produce; marry wives and rear families...' (Jer. 29:4–6), that is, to reconstruct their lives around and on the basis of the natural cycles of human life. Of course, it is axiomatic—certainly for the later literature—that God is 'Lord' of these temporal cycles, although the exact way in which he is Lord remains undefined.

Habits of ecclesiastically mediated reading make it only too easy to read many biblical texts as if they were speaking either of a timeless or an infinitely everlasting God, but in the light of Jenni's caution we should hesitate. Perhaps biblical references to God as 'everlasting' say no more than that mortals whose days are 'as grass' (Ps 103:15) have no measure with which to conceive of God's Lordship over time—as expressed in Psalm 90: 'Before the mountains were brought forth or the earth and the world were born, from age to age you are God...in your sight a thousand years are as the passing of one day or as a watch in the night...so make us know how few are our days, that our minds may learn wisdom' (Ps. 90:1, 3, 12). An Augustine would interpret such a verse in terms of timelessness, but this is once more to read more out of it than it actually requires. The point is more simply that God's time or God's relation to time is beyond all measure, as is especially emphasized in God's challenge

---

[83] Jenni, 'Das Wort 'ōlām', 17.   [84] Jenni, 'Das Wort 'ōlām', 17.

to Job: 'Where were you when I laid the earth's foundations? Tell me, if you know and understand. Who fixed its dimensions? Surely you know! Who stretched a measuring line over it?' (Job 38:4–5).[85]

Human error in relation to time and eternity is therefore to judge and to act on the basis of events of which the causes and outcomes are uncertain, an error that is compounded in the case of false prophecy by assigning short-term political twists and turns to the divine will rather than seeking this in the constancy of the divine promise—a promise that, as we have been seeing, is repeatedly underwritten by appeal to the cosmic order. As the example of false prophecy shows, the motivation driving such confusions is for the most part the self-will and self-interest of those who welcome it. The challenge, then, is to shift from basing judgement and action on what suits the promptings of individual or collective egoism and the measure that such egoism is able to impose on cosmos and history, and to acknowledge instead the immeasurability of the basic structures of life from a human perspective. For although these structures—seed-time and harvest, day and night, etc.—are assuredly reliable, they precede and encompass any human structuring of life and in that sense are 'immeasurable'. We can never get behind them, as it were, and can only trust that they are under the Lordship of the God for whose constancy they are themselves prominent witnesses. Their true measure is known only to God, not to us.

As we shall be seeing in Chapter 7, when we come to consider Michael Theunissen's philosophical interpretation of the time-experience revealed in early Greek poetry, this view of time and (in a non-philosophical sense) divine eternity is certainly not altogether alien to the world of the Greek poets, whatever we make of its relation to the Greek philosophical tradition.[86] But is it in fact entirely alien to the philosophical tradition?

---

[85] Consider also the following examples from Isaiah: Isa. 40:8, 12, 18, 25, 28 (or, indeed, chapter 40 as a whole); 41:4, 26; 42:5; 43:10; 44:6, 24; 46:10; 48:3, 12. I suggest that these, and the whole context of this sequence of chapters, points to a God who is beyond all possible comparison with or measurement in terms of humanly experienced time, a God who already 'is' in the immemorial past and who will outlast the ultimate future, but there is no suggestion of sheer timelessness. And whilst the promise of being with Israel may seem to suggest duration through time, it is also clear that God's way of, as it were, 'experiencing' this covenantal companionship will not itself be measurable in Israel's own human perspective.

[86] James Barr, for example finds parallels for the sense of *'ōlām* as 'the remotest time' or 'in perpetuity' in Hesiod and Plato. See James Barr, *Biblical Words for Time* (London: SCM Press, 1962), 69–70. The scholarship of Jenni, Barr, and others lends little support to Paul Helm's view that 'there are biblical data of a non-technical kind which can reasonably be understood as countenancing timelessness and that a satisfactory explanation can be provided of data which appear to go the other way' (Helm, *Eternal God*, 11). That said, Helm's assertion of the principle 'that biblical writers did not consciously use ideas of timelessness does not mean that they did or would have to reject it' (*Eternal God*, 4) is surely correct. Scripture, as a body of texts arising from and addressed to a civilization very different from our own, has to be interpreted—which means, often, being translated into an idiom not its own. This last comment may be taken as registering further uneasiness with Wolterstorff's apparent view that we can take as 'literally true' scriptural statements about God.

Plotinus postdates the later books of the Hebrew Bible by half a millennium, and that distance in time—and culture—is not to be underestimated. He also writes as one for whom the earlier Greek philosophical tradition and its key figures from the Presocratics through Plato and Aristotle is available as a body of 'classical' literature. In Treatise III.7 of *The Enneads*, 'Eternity and Time', he gives an account of divine timelessness that expands on some rather undeveloped hints in Plato, shaping a line of thinking that ultimately finds expression in Boethius' definition of eternity, and taking issue along the way with the understanding of time put forward by Aristotle.

Eternity, Plotinus starts by saying, belongs to 'everlasting nature' (*aidios physis*), whilst time belongs to the world of becoming, 'this universe' (*to pan*). It is connected to the intelligible being (*ousia*) but whereas knowledge of the intelligible being allows for the differentiation of the whole and its parts, 'eternity includes the whole all at once, not as a part',[87] it is 'a life abiding in the same, always having all present to it, not now this, and then again that, but all things at once', 'as if all together in a point'.[88]

In the continuation of his argument, Plotinus offers a refutation of the Aristotelian approach. Aristotle had explained time as a kind of counting applied to the movement of physical bodies with regard to what is, in the first instance, the 'before' and 'after' of their spatial location; time itself is not the movement from one location to another, but a kind of container embracing temporal events 'as things that have locality are embraced in their place' (*Physics* 221.a.29). As Heidegger points out, in a critically sympathetic reading of Aristotle's text, time, as 'containing' temporal events, is 'before beings, before things moving and at rest' and although more objective than all objects 'exists only if the soul exists', although to call it 'subjective' would be to read back into Aristotle a view that is not really relevant to his thought-world.[89]

---

[87] Plotinus, *Ennead III*, tr. A. H. Armstrong (Cambridge, MA: Harvard University Press, 1966), III.7.2; III.7.7.

[88] Plotinus, *Enneads*, III.7.3. It is perhaps important here to note the 'as if'. Plotinus is not stating a doctrine of the point-like nature of eternity, but in accordance with his comments in the treatise on 'The One' on the necessary imperfection of all language applied to the One is consciously speaking, as it were 'under erasure'.

[89] Martin Heidegger, *Gesamtausgabe, xxiv: Die Grundprobleme der Phänomenologie* (Frankfurt am Main: Klostermann, 1975), 356; Eng. *The Basic Problems of Phenomenology*, tr. A. Hofstadter (Bloomington, IN: Indiana University Press, 1982), 252, 254–5. Still more provocatively Heidegger defends Aristotle against Bergson's view that he subordinates time to spatiality. Motion, he says, is treated as a structure or dimension of away-from/towards that is purely formal and of which the spatial extension is only one possible modification. See *Grundprobleme*, 327–88 and especially 344–9 (Eng. tr. 231–56, 242–7). In this way Heidegger grounds the possibility of an existential account of time in what might otherwise be naively seen as a treatise on purely physical questions of time, more akin to cosmology than to existence in Heidegger's distinctive sense. This is characteristic of his attempt throughout the 1920s to read Aristotle as a kind of original phenomenologist.

Plotinus, however, entirely rejects the association of time with movement and number or measure. The measure of magnitude will only ever give us a rule for identifying a certain length of time; and if we base our measurement on the sequence of 'before' and 'after' then this already presupposes some knowledge of time, since (against Aristotle) 'before' and 'after' are primarily temporal in meaning. Number cannot of itself make a movement temporal and if time itself is unbounded (*apeiros*) it cannot in fact be numbered. Nor does saying that time accompanies movement explain anything. To know time, then, we have to return to eternity so as to see how time arose from 'that quiet life, all in a single whole, still unbounded, altogether without declination, resting in and directed towards eternity'.[90] Conversely, it is precisely by turning away from eternity that the soul, *psychē*, brought the temporal world into existence. As Plotinus puts it in vivid language that, in its own way, resonates with the account of temporality in existentialist philosophy, a 'restlessly active nature which wanted to control itself and be on its own, and chose to seek for more than its present [i.e. eternal] state, this moved, and time moved with it'— yet, in giving itself over to time, the soul—we?—became its slaves.[91]

Despite the more than obvious differences from anything we might find in the Bible, I want to suggest two points of convergence. Firstly, as in biblical teaching, the measure of time is beyond human calculation and depends entirely on the divine eternity from which it first arises and to which it ceaselessly strives to return. Secondly, the 'problem' with time is not that it is temporal but that its temporality is the manifestation of a wilful act of self-assertion on the part of *psychē*, the outcome of which, is in fact, enslavement to the order of time. Although *psychē* here refers to something more like a world-soul than an individual soul, Plotinus' argument suggests that the issue of time's relation to eternity is not just a physical or even a metaphysical problem but arises from a certain orientation of *will*. Entanglement in time is a consequence of a wilful insubordination on the part of psychic life in relation to the blissful life of the eternal. And perhaps, therefore, it is a problem to which the only solution can be a kind of redemption. In this way, Plotinus' metaphysics of time shows itself, like the narratives of biblical history, to lead, in the end, to a soteriology of time.

This is obviously by no means enough to close the gap between the Greek and biblical views of time, but, when connected to the continuing presence of cyclical models of time in even the more historicized prophetic texts of the Hebrew Bible, it suggests that this is not a case of simple opposition. It also suggests that one route by which we might seek a phenomenological and hermeneutical reanimation of the concept of divine timelessness might be less by

---

[90] Plotinus, Ennead II, tr. A. H. Armstrong (Cambridge, MA: Harvard University Press, 1966), II.7.11.

[91] Plotinus, *Enneads*, II.7.11.

way of direct defence of the *concept* of eternity and more by way of investigating the meaning of divine Lordship in relation to human time-experiences and what it might mean to suffer from time or *mutatis mutandis* to be redeemed from time. But that is to run on ahead of where the argument has currently arrived. To conclude this chapter, then, I shall now proceed to some general comments on the positions and counter-positions we have been examining.

CLOSING REFLECTIONS

Within the range of what is often referred to as analytic philosophy of religion, it is clear that there is a wide variety of approaches to the question of divine eternity and considerable divergence in the positions reached. Such diversity and disagreement is perhaps what we might expect from philosophers and is a reminder that, whatever else it may be, philosophy is not natural science, which proceeds collectively and consensually and in such a way that a scientifically established truth becomes the uncontested basis for all future research. Even if we add the observation that there has been a broad shift towards accommodating a more temporal account of God (a shift we may see as going back at least to Hegel and paralleled in process thought[92]), this has by no means been universal and is certainly hard to harmonize with the teaching of Aquinas that remains normative for many contemporary philosophers of religion.[93] This last comment indicates another difference between the practice of philosophy and science, in that we can scarcely imagine scientists investigating a question in physics or biochemistry taking their bearings from medieval or ancient sources, even though this is standard procedure amongst the philosophers we have been considering.[94]

Neither of these comments is intended critically. Philosophy is not science and doesn't have to see itself as science. It rarely reaches and perhaps for the most part doesn't aim at reaching conclusions that will command acceptance across the field; and it is entirely proper for it to be committed to the task of attending to the ancient sources of words and ideas that remain current, for

---

[92] Yates makes the interesting comment that the shift away from a purely timeless God towards a sempiternal God of infinite temporal duration can be associated with the Reformation and is especially marked in English thought of the seventeenth century (Yates, *The Timelessness of God*, 44). Purely anecdotally, it is the author's experience that, on the whole, Anglicans and other Protestants are more comfortable with ideas of divine temporality than their Catholic colleagues.

[93] Naturally, it remains open to question whether this is the sole possible reading of Aquinas, or whether another, more 'negative' Aquinas might not also be possible. The issue also has relevance for those who, in the wake of Heidegger, see Aquinas as a proponent of onto-theology, that is, the identification of God with the most universal and highest principle of being.

[94] However, there is also considerable variation in the extent to which these sources are then treated as normative.

good or ill, in contemporary discourse, technical and non-technical. This is especially unavoidable in the case of the philosophy of religion, in which the matter at issue—the content of religious beliefs—is embedded in the affirmations of believing communities that explicitly identify themselves as continuing traditions founded in the ancient world and continued through normative interpretations of their sources established in the medieval and Reformation periods. If there is a problem, then, it is not the difference between science and philosophy but the possibility of confusing the spheres in such a way that philosophers assume that their formulations are binding on all subsequent research or that there is only one demonstrably reliable procedure for dealing with the kinds of questions we are considering here.

Connected with these remarks is the observation that, as Helm complained of Stump and Kretzmann's postulate of ET-simultaneity, some of the arguments proposed essentially do little more than to reformulate the question. This might, for example, seem to be the case with arguments relating to God's 'middle knowledge', which, whatever their religious or apologetic merits (and these are, arguably, significant), seem in the end to say no more than that God's foreknowledge can be reconciled with human freedom in a way that is ultimately unfathomable. But if such knowledge, like God's foreknowledge, is unfathomable, it doesn't really advance our (human) understanding other than that we are enabled to articulate our ignorance in a more complex way.

It may also be said that for all the clarity on which analytic philosophy prides itself, there is much in what we have been considering that is far from clear. When Swinburne says that a temporal God is not time's prisoner but continues to call the shots, being the one who invites time and its accompanying negative features in (and not vice versa), this is certainly clear in the sense that the language used is not grammatically complex or excessively burdened with neologisms. But it is far from clear what it could possibly mean or why, beyond the assurance that God can still call the shots, the reader should go along with it. How can God 'call the shots' if—to continue in this metaphorical vein—he can't 'call time'? The will to affirm God's Lordship of time is clear—but what, exactly, does this Lordship mean? Is he the supreme surfer of time, constantly riding the crest of its forward-moving wave? Or is he perhaps a champion tennis-player or boxer, constantly responding to what comes at him from whatever direction and with whatever force? Or maybe the creative entrepreneur, who both initiates a new business and can be relied on to steer it through all the peaks and troughs of market turbulence? Or what, exactly? But exactness is precisely not what's on offer.

Problems can, of course, only be dealt with one at a time, and most of the contributors to the discussion we have been staging have written across a range of problems in philosophical theology and philosophy of religion. If we were to examine how they have approached these other problems this would doubtless extend and deepen our understanding of their approach to questions of time and eternity.[95] Nevertheless, it seems not too harsh to say that there is a certain anthropological and soteriological deficit in many of the arguments we have been considering. That is to say, the God being considered seems often remote from the God to whom human beings turn in their hour of need and whom some, at least, experience as closer to them than they are to themselves. As we have seen, this is a point of which some philosophers are very much aware, and it is likely to come to the fore in complaints that a timeless God can scarcely be either living or personal, let alone responsive to prayer—and, if the Bible is to be believed, is not able to redeem (thus Wolterstorff).

This is not to be taken as implying that the idea of divine timelessness is redundant. Augustine remains an eloquent witness to how the postulate of a timeless God can relate to the needs of a soul that experiences its temporality as enslavement to a succession of tumultuous vicissitudes. Whether, in fact, we need to go all the way to a timeless God to respond adequately to such an experienced need or whether some kind of sempiternal yet unchanging God would do as well is another question, but a temporal God, who is constantly adapting his actions in response to the constantly changing whirl of events, would scarcely seem capable of doing what Augustine, at least, would want him to do. This may, of course, be Augustine's problem or a problem arising out of an ancient world-view and may not need to be our problem. Nevertheless, there is something unsettling in the vision of a thoroughly temporalized God continually responsive not only to the multi-billion chaotic strings of events in the physical universe but also to the seething mass of individual and collective human acts and errors that range from the miseries of childhood and adolescence to global crises of manifold kinds. If we stop to think about just what this could mean in concrete terms, the vagueness of what is meant by 'calling the shots' is soon apparent. Even if we imagine such a God as the most proficient multi-tasking political operator, who always seems to get his way in the end, won't he always be at risk of becoming like Jacques Derrida, when he became consumed with worry that while busy feeding his own cat he was neglecting all the other cats in the world?[96] This is not to say that such a God is impossible, merely that it is not at all straightforward to assume that a temporally inflected God will be better at being God, that is, being the Christian God, than will a timeless God, although the ways in which they fall short of being fully God-like differ.

Here—as whenever we attempt to speak truthfully and appropriately about God—we find ourselves stumbling at the cliff-edge of meaningful discourse.

---

[95] We have, for example, noted how, at one point, the relations of time and eternity raise questions about the nature of divine action, whilst Leftow indicates that this argument requires further clarification and justification of the notion of divine simplicity.

[96] See Jacques Derrida, *The Gift of Death*, tr. David Wills (Chicago: University of Chicago Press, 1995), 71.

If a timeless God seems to fall short of all that is required of a living, personal being who answers prayer and is with us in our suffering, a temporal God might come to seem, well, perhaps like a science-fiction super-being but not quite *God*—and perhaps no less exposed to the challenges of theodicy than the timeless God, since what assurance do we have that he really does and can 'call the shots'? Boxers, surfers, and dancers, whose skills involve constant attentiveness and responsiveness to their opponent, environment, or partner and who, at their best, might provide some remote analogy to the interactive God, will all sometimes fumble and slip. What, other than the assertion that he can't, makes us think it's otherwise with God?

Perhaps we just can't have it all, whichever route we choose. We have, for example, noted how Pike sees timelessness as involved in relationships of mutual entailment with one series of divine attributes whilst contradicting attributes of life and personality. Likewise, we have noted that Leftow is prepared to concede some ground on God's omniscience if this is the only way to preserve his timelessness. Lacking from the discussion, however, is what we might call a functional awareness of some of the basic ground-rules of talking about God that are standard in more specifically *theological* discourse. Leftow, as we have seen, relies on Alston and Swinburne as having demonstrated the legitimacy of speaking affirmatively and literally about God, and other participants to the debate seem, in practice, to do the same, with or without acknowledging this to be the case. However, in slightly varying formulations, theology has long understood that whatever it says about God is said under the condition of a threefold manner of speaking, that is, (1) as involving the negation of its standard human meaning (God's goodness is not goodness as we know goodness in human relationships), (2) as attributing the source of its human meaning to God (God is the source of the human capacity for goodness), and (3) as the eminent instance of meaning (God is the perfect instance of goodness).[97] The scope for literal attribution of predicates is therefore limited from the outset.

How might this illuminate the present discussion? For a start, it is striking that it is often unremarked that 'timelessness' is not actually a divine attribute but a negative concept that says nothing at all about God except that God does not have time in any sense that human beings have time. But rather than timelessness in an absolute and non-negotiable sense, this does not preclude a time that is other than time as we know it. More will be said in subsequent chapters of both the theological and anthropological imperatives that drive the need to speak of God as 'other' or as 'infinitely qualitatively different', and for now I merely flag this as needing to be kept in mind. To say that God is

---

[97] For a good exposition of these three ways and their interdependence, see Schleiermacher, *The Christian Faith*, 197. What Schleiermacher calls 'the way of causality' is what, following Aquinas, I am calling the way of attribution.

timeless may therefore have some initial justification in registering the *difference* between time in human life and time in the divine life, but, strictly as a negative attribute, timelessness says nothing at all about how God is and therefore cannot of itself be in either agreement or conflict with other attributes of God. It is a mark of difference, not part of a substantial definition. It is somewhat different when Leftow, Stump and Kretzmann speak of 'eternality' as a divine mode of being. Here, as opposed to mere timelessness it seems that something positive is being said (although it is then confusing when the negative term 'timelessness' continues to be used to say it). In this case we most likely have an example of the way of eminence: that divine eternality is a way of being temporal that is supremely above, beyond, and surpassing all possible human meanings of 'temporal'. But if this is so, then something more is being said than that God is, simply, tim*eless*. In this case, it would seem equally legitimate to say that eternality itself is a mode—the eminent mode—of temporality, although, as Plotinus insists is always the case in speaking of the One, only under the caution of 'as if', that is, God's eternality is 'as if' temporal.

It would scarcely be possible and would certainly be tedious to trawl back through the discussion thus far and comment on how the terms 'timeless' and 'eternal' are being used at each point with regard to the threefold structure of theological language. However, keeping such distinctions in mind would certainly contribute to greater clarity as to just what was being said and what denied at any moment. And even if such caution might considerably slow the discussion down, it might be regarded as a case of 'mind your foot when you enter the house of the Lord' (Eccl. 5:1). It is especially important not treat the requirement of negation as if it were merely a part of the preamble of theology and not a continuing and abiding part of the whole theological endeavour.

These last comments lead on to another suggestion, which was briefly touched on in the discussion of possible analogies between biblical and Platonist views of the relationship between time and eternity. Here, especially clearly in the biblical testimony, we encountered the view that human beings fundamentally lack an appropriate measure with which to measure divine eternity; for Plotinus (and I think this is implicitly true also of many biblical texts), we are unable even to measure the entire span of time that encompasses us: we take time with us wherever we go and in such a way that its horizon is always receding beyond our view. Eternity may be timeless or may be a kind of eminent time, but either way and most importantly, as the ground and precondition of our time-experience, it is not time that we could ever measure or whose depths we could ever plumb.

Swinburne interestingly suggests a possible relativization of the difference between an entirely timeless God and the temporal God that he himself proposes. Having argued that the measurement of time is possible only in the context of a world in which there are laws of nature such that there exist some constant means of measuring intervals of time, he suggests that a God who existed by himself, unattended by laws of nature, could not be measured by temporal intervals. 'There would', he says, 'be no difference between a divine act of self-awareness which lasted a millisecond and one which lasted a million years'.[98] Similarly, there would be no difference between 'a divine conscious act that was God's only conscious act and was qualitatively identical throughout, that was of finite length, and one that was of infinite length' and also genuinely identical throughout.[99] Swinburne goes on to suggest that, in the light of this, a theory of God as temporal could give 'timelessness-theorists' what they were essentially looking for. In a sense I think he is right—only the argument cuts both ways. From a human perspective, a properly timeless act (if such an act is at all conceivable) and an act that could not be measured by the metric supplied by the laws of nature would be indistinguishable, and both would have the character of immeasurability. But that would also mean that insisting on the temporality of God has no intrinsic advantages over claims of timelessness, since we have no measure by which to measure the divine temporality. He might as well be timeless. Or, to put it more positively, the criterion of immeasurability might give both timeless theists and temporal theists what they are looking for.[100]

Immeasurability, I am suggesting, does the essential work that the notion of timelessness sets out to do. That is, it marks the fundamental difference between divine and human being and reminds us that God can be neither bound nor limited by what, for us, are the necessary (and in this case necessarily temporal) conditions of understanding—and that if we do, nevertheless, want to speak of God as, in an analogical or eminent sense, 'temporal', then this will always have to be qualified by significant negation. He is temporal—but not in ways that we know or can measure. Moreover, compared with 'timelessness', immeasurability more directly indicates that what is at issue is a limitation of human capacities rather than a positive feature of God and that, when it comes to God, we have no scale with which to measure his relation to time. But this further leaves open whether or how there may be anything positive to say of this relation or in what way God may be Lord of time, capable not only of calling the shots but of calling time—or perhaps, more importantly, creating and giving time. By saying less, it does not constrain—as timelessness constrains—what more we may want to say of God and time. Yet it also warns

---

[98] Swinburne, *The Christian God*, 140.   [99] Swinburne, *The Christian God*, 140–1.

[100] I see some convergence between what I am saying regarding immeasurability and Alan Padgett's notion of 'relative timelessness'. See Alan G. Padgett, 'The Difference Creation Makes: Relative Timelessness Reconsidered' in Christian Tapp and Edmund Runggaldier (eds), *God, Eternity, and Time* (Farnham: Ashgate, 2011), 117–25. I similarly note Christian Tapp's concerns that affirming the temporality of God 'faces the danger of putting Him on the level of worldly entities' (Christian Tapp, 'Eternity and Infinity', in Tapp and Runggaldier, *God, Eternity, and Time*, 104) and his suggestion that affirmations of divine temporality be qualified by the comparative infinity of the *via eminentiae*.

us against attempting to model, in more detail than we have to, how God may be considered as temporal, since, whatever his temporality, it will always—from our point of view—be immeasurable.

It might be objected that Swinburne's argument is precisely formulated as hypothetical: in other words, if there were no natural laws, then there would be no temporal difference between a timeless and a temporal God. But, as things stand, there are natural laws (indeed, they are ordained by God himself), so that, in his relation to time, God's time does become, in some sense, measurable. Don't we measure God's time whenever, for example, we say that God acted in AD 30 or 'yesterday' or that God will act—or just 'be'—tomorrow (and, many Christians would say, we would be right to do so). However, this is once again to elide the difference between the divine and the human and to forget the negative marker that accompanies all theology—not to mention forgetting what the psalmist already knew, that 'in your sight a thousand years are as the passing of one day or as a watch in the night' (Ps. 90:4). If God has time, and even if (as, according to the way of eminence, must be the case) his time is the very best of times, his time is still not our time. As 'other' than the physical universe that is governed by natural laws and precisely as its Lord, God's time, if we can call it time, is beyond all measure.

To parse ideas of divine timelessness in terms of the immeasurability of divine time is, I suggest, conformable to both biblical and classical sources of Western and, more specifically, Christian thought; and in the modern period we can see something like this Platonic division being restated in an inverted form by Bergson's account of the difference between intelligence and life. Intelligence, as manifest in the mathematical basis of natural science, can only know its objects by making them into 'objects' possessed of constant identities and attributes fitted for incorporation into mechanistic and teleological systems. And, as Bergson went on to illustrate with examples including Spinoza, Leibniz, Fichte, and Spencer, this is also how philosophy itself typically proceeds. However, this is to abstract from the living reality of life for the sake of explaining it, although this reality, ineluctably temporal as it is, does not need explaining and cannot be explained. Time is not a deficiency of being to be made good or a problem to be solved. It is indisputably real and is 'the basis of our being'.[101] *Durée*, the quality of lived events to maintain their identity through time, is not something that needs to be accounted for but is a basic given of life. Yet, as such, it also resists calculation. Life is temporally manifest in manifold tendencies, all in motion according to the basic impulsive character of life itself, the *élan vital* or living impulse that is another of Bergson's key terms. All attempts to understand or explain life therefore reach their limit in what Bergson calls their cinematographic character, that is, they offer a construction of reality as a sequence of still frames that may be run

---

[101] Henri Bergson, *L'évolution créatrice* (Paris: Presses universitaires de France, 1941), 39.

together in such a way as to give the impression of movement but always, in reality, falling short of reproducing that movement itself. In these terms—although this is not immediately Bergson's problem—what I am suggesting is that the question of the eternal be addressed not by way of attempting to model ideas or ways of being that are elevated above the flux of time but by seeing the depths of our time-experience's inner immeasurability as potentially opening out towards a further level or dimension of time that would merit being named 'eternal'— 'God's time', perhaps (which, as Bach reminds us, is the very best of times).

It is in the spirit of this suggestion that the main thread of the following enquiry will look to human beings' time-experience itself as pointing us towards what a religiously significant sense of eternity might mean. However, as this comment implies, a specifically theological enquiry will not be one that comes to the question of time-experience with a blank sheet. Instead, it comes to the question with its own particular agenda, an agenda that, as I have already intimated, concerns the way in which time is experienced in terms of a dialectics of fall and redemption. In other words, the anthropological limitation of the investigation to human beings' time-experience is further qualified soteriologically, that is, with regard to the question as to what it might mean to be saved from whatever is oppressive in time and how such redemption might be construed as by or for eternity. Putting it like this, however, is merely to return to what has always been the practice of Christian theology. For Augustine and Aquinas, at least, it is clear that the question of divine eternity is also a question as to whether or how 'eternity' might be the proper goal of human longings and strivings. So, with regard to any proposal regarding divine eternity, how might such a view be experienceable as saving—as good news—for a heart seeking God?

I have acknowledged the unavoidable division of labour in philosophical as in other kinds of work; and if one is to deal with one set of problems or apply one set of tools, this will inevitably be to the neglect of other problems or other tools. But, granted the need to focus on a particular topic, what might be achievable within that focus could still be explored more fully than is typically the case in the discussions we have been following. And here I have two particular complaints. The first concerns language. For if the question is allowed to be a question that arises from the lived context of human time-experience, then greater sensitivity needs to be shown to the often richly metaphorical and culturally contextualized language such as that in which the great historic currents of Christian thought have found expression. Again, this is in part a matter of neglecting the negative markers that accompany talking about God. But it is also a matter of not reflecting the extent to which language itself arises out of, discloses, communicates, and in some cases perhaps transforms human needs, desires, and longings—our sense of self, of who we are.[102] And this then

---

[102] There are perhaps two distinct points here: (1) our language is itself embedded in a long historical development in such a way that it is at many points difficult entirely to enforce the

feeds into my second complaint, namely, that the philosophers sometimes seem to speak as if we are or could be spectators of the divine being, capable of abstracting from the conditions of our own human experience and arguing only over what exactly it is we are seeing 'out there'.[103] But whilst there may be good reason in most other cases simply to trust the deliverances of reason, this cannot this be the case when it is a question of our human God-relationship, when we must necessarily take seriously the irreducible difference between divine and human perspectives. This is especially the case when the question is raised against the background of a Christian tradition that, even when it allows for the possibility of a natural knowledge of God, is constantly alert to the distortions and corruptions that enter into our dealings with God as a result of the fallen nature of human knowledge. What are we looking for, what are we wanting, what is really going on in our experienced need of God? And how might a relation to an eternal God possibly take shape and become meaningful in relation to that religious need?

If I suggest that what may, in a very loose sense, be called a phenomenological approach can throw such light on these questions, this does not mean that this should simply be preferred to an analytic approach but only that, working with a different set of philosophical tools, it might prove to be a helpful supplement to the as yet undecided debate about divine eternity that we have been examining here. It is a presupposition of such a phenomenological approach that it limits itself to what appears within the open sphere of human consciousness. But can the eternal appear in this way? Intimations that it could are found amongst the ancients in, for example, Plotinus,[104] and in the modern period, the view that the eternal is a constitutive dimension of human consciousness emerges in the thought of Spinoza. In the following chapter I shall therefore explore how this idea is developed in Spinoza before being taken over and further transformed in German Idealism.

---

kind of univocity of meaning that many philosophers seek, being encumbered—or, it may be, enriched—by the history through which it has passed; and (2) language is also the mode of our own contemporary self-understanding and, as such, entwined with what is in fact the open and unresolved question as to what time and eternity are taken as meaning to us, today—or, indeed, what the proper terms of debating such a question might be. Who is actually in a position to adjudicate on what might count as an adequate account of the relations of time and eternity? The scientists? The mathematicians? The analytic philosophers (and if so which ones)? The Continental philosophers (and if so which ones)?—Or even the theologians (and, if so, the Christians, the Jews, the Muslims, the Hindus...)?

[103] An aspect of this is the way in which some philosophers develop arguments that depend on speculation as to possible kinds of non-human knowledge, as it these might help further the question of a specifically human knowledge—which, I am assuming, is only ever going to be the best we can hope for.

[104] See Plotinus, *Enneads* III.7.5–7.

2

Time, Eternally: Theory

In Chapter 1 I complained of the soteriological deficit characteristic of the debate about divine eternity amongst recent Anglo-American philosophers. For Augustine and Aquinas, by way of contrast, the importance of asserting the timeless eternity of God was inseparable from the promise that we human beings might be delivered from the 'tumultuous vicissitudes' of temporal existence and from final extinction in death. Knowing God as eternal was the way in which we ourselves would be transformed into a likeness of the divine eternity. Nor did they play down the dramatic nature of the transformation this would involve.

That philosophers must, of necessity, practice a certain division of labour is unavoidable but in this case, I suggest, the human correlate of divine eternity is integral to whatever sense the idea may have. This, however, generates a problem that is more than the problem that arises from the necessary specialization of all academic work. It is also the problem as to whether such a timeless future is, in fact, what contemporary human beings actually do wish for themselves. Finding this out cannot just be a matter of gathering statistics and, as noted in the Introduction, the language of eternity remains widespread in popular religious culture, within and beyond formal religious communities. But it also seems that as a result of epochal changes in what we regard as valuable in human life, what people wish for when they wish for eternal life may not be quite what Augustine and Aquinas had in mind. This difference suggests a more general problem in attempting to connect a modern view of the self to a pre-modern view of God. Admitting a dimension of temporality into the idea of God may be one way of dealing with this but it is doubtful whether it amounts to more than a temporary fix. Much of what follows in the remainder of this study is an attempt to lay some of the groundwork for a more long-term solution, although I am certainly not claiming to offer any final 'answer'. Such things are the work of centuries and of generations, not of individuals.

There are doubtless multiple reasons for the declining persuasiveness of medieval eschatology, but amongst these is surely the rise of what might be called the historical consciousness. This plays out at many levels, including that of individual identity and the nature of mind. At every level 'history' has tended more and more to extend into and to absorb all other dimensions of life. History is no longer just the drama of human actions performed upon the stage of a stable telluric and cosmic order, but human life, the earth itself, and the entire cosmos have become revealed as thoroughly temporal, that is, as being what and how they are only as a result of long-term temporal processes. And, as we have seen, the process is sometimes extended even to God. But once everything is seen as temporal through and through, what could 'the eternal' possibly mean?

The question is not merely rhetorical. It may be the case that, in the end, a thoroughly temporalized view of the universe and of human life leaves no place for the eternal, but a number of serious philosophical attempts have been made to show how thoroughly temporal beings such as we are may nevertheless attain a knowledge of themselves and of their world that can appropriately be called 'eternal'. This, to anticipate, is a central part of the agenda of German Idealism and in this and the following chapter we shall be looking at how these Idealists attempted to square the circle of time and eternity, with particular reference to Hegel and Schelling. I shall also argue that the same challenge is integral to the thought of Friedrich Nietzsche, a claim that seems to go against the grain of deeply-rooted habits of reading the thinker of eternal recurrence. Of course—and not least with regard to Nietzsche—even when the great thinkers are aiming at systematic unity it will usually be the case that their thought involves manifold competing tendencies, and I am not seeking to impose any simplistic reading on my sources. First, however, I want to turn to a thinker, Spinoza, who might, at first glance, seem to be amongst the least historically oriented thinkers of modern times, at least in the work we shall be considering here, his *Ethics*.[1] Nevertheless, Spinoza's modelling of the relationship between mind and world was to be paradigmatic for the more emphatically temporalized world-view of the German Idealists, not least with regard to how the flux of worldly events may be known *sub quandam specie aeternitatis*.

SPINOZA

Spinoza's decision to present his treatise 'in the manner of geometrical demonstrations' already seems to set his world up as one in which temporal relations are non-essential.[2] This impression is reinforced as we are led by a few

[1] That Spinoza's thought has important implications for history and for political life is, of course, more apparent in other works, such as his *Theological-Political Treatise* (Cambridge: Cambridge University Press, 2007).

[2] Whether we are to understand the geometrical method as simply a way of presenting a unified holistic vision or as a means of advancing knowledge has divided commentators. Leon Roth,

short steps to the view that there is only one substance, which is God (Prop. XIV), and, as such, *causa sui* (Def. I), 'absolutely infinite' and 'consisting of infinite attributes, each of which expresses eternal and infinite essence' (Def. VI).[3] Although this substance is subject to manifold modifications, 'God is the indwelling and not the transient cause of all things' (Prop. XVIII). As such, he is 'not only the effecting cause of the existence of things, but also of their essence' (Prop. XXV). He is the cause of both essences and existences necessarily, i.e., from the necessity of his own nature (Prop. XXXII) and in such a way that 'Things could not have been produced by God in any other manner or order than that in which they were produced' (Prop. XXXIII); as existing, they must 'exist for ever and infinitely' (Prop. XXI). Whereas the metaphysics of the Middle Ages saw substances other than God as lacking the intrinsic capacity to remain unchanging in existence, Spinoza (who, as we have seen, only allows the one substance: God), consequently sees existence and eternity as internally connected. As he states in Definition VIII: 'I understand ETERNITY to be existence itself, in so far as it is conceived to follow necessarily from the definition of an eternal thing.' But is eternity here strictly timeless, or is it to be taken in the sense of sempiternal duration? Spinoza himself immediately glosses the point by saying that existence 'cannot be explained by duration or time, although duration can be conceived as wanting beginning and end' (Def. VIII, *Explanation*). However, this seems to leave some scope for interpretation.

According to Martha Kneale (for whom the notion of timelessness understood as the '*totum simul* of time' is in any case 'self-contradictory'[4]), timelessness and sempiternity are mutually entailing; moreover, although sempiternity does not entail necessity, necessity entails sempiternity, a point which is directly relevant to Spinoza, since, as we have just seen, God is cause of himself and of the world *necessarily*. An order of things grounded in necessity is therefore also necessarily sempiternal, that is, it is characterized by continuing duration in time. In this sense she argues that whilst there is no point in asking 'when' it is true that 2 + 2 = 4, it is not meaningless to say that today 2 + 2 = 4 has been true for a day longer today than it was yesterday. So too, Spinoza's substance does not change, but it does endure. This view is contested by Diane Steinberg, who argues that the combination of *Ethics* I definition 8 and II definition 5

in his *Spinoza, Descartes and Maimonides* (Oxford: Clarendon Press, 1924) argues that whereas for Descartes the mathematical method was conceived as a means of advancing knowledge (14), Spinoza adopted it purely for presentation (40ff.). For discussion of how contemporary readers might regard this presentation see Genevieve Lloyd, *Routledge Philosophy Guidebook to Spinoza and the Ethics* (London: Routledge, 1996), 19–22.

   [3] Spinoza is quoted using standard references from Benedictus de Spinoza, *Ethics; and On the Correction of the Understanding*, tr. Andrew Boyle (London: J. M. Dent [Everyman's Library], 1977).

   [4] Martha Kneale, 'Eternity and Sempiternity', in Marjorie Grene (ed.), *Spinoza: A Collection of Critical Essays* (New York: Anchor Books, 1973), 231.

mean that Spinoza's God 'must be eternal in a sense which excludes all temporal relations, a sense which is incompatible with sempiternity'.[5] Time, she suggests, relates only to existing things whose existence is not immediately identical with their essence (as is the case with God), whereas the identity of essence and existence in God excludes duration: 'eternity', that is, timeless eternity, 'is [therefore] an attribute under which we conceive the infinite existence of God'.[6] The question is further complicated if, as Alan Donagan says, Spinoza 'had no term for "time" in what I take to be its fundamental ordinary sense'.[7] Instead, time is subordinated to existence in such a way that whilst it is plausible to speak of a *continuatio existendi* (indeed, Donagan calls it a fundamental concept), this does not involve reference to the passage of time. Donagan himself argues for divine eternity as involving the 'necessary continuation of existing' or what he calls 'omnitemporal' existing, but, since this is not to be understood in terms of any linear time series it is not exactly duration. Yet, drawing on evidence from Spinoza's letters, Donagan also proposes that whilst the divine substance is characterizable in terms of non-durational but continuous and omnitemporal existing, the modes in which substance is manifest at the finite level may be seen as having duration. The point is God's constancy and self-sameness but there is no 'when', 'before', or 'after' with regard to God's eternal decrees even if these can be predicated of the modes in which those decrees are manifested.[8]

We are slightly running on ahead of ourselves, however. For whilst we have thus far spoken of God, we have not spoken of his relation to the world nor, more particularly, of his relation to human beings as we know them living, moving, acting, and suffering in the world. But when we do turn to consider human life in the world in a Spinozan perspective two questions come immediately to the fore. The first is how we in fact experience and know ourselves in relation to divine eternity; and the second, closely related (indeed, on Spinoza's distinctive premises these are really two aspects of one question), is whether or how we have a share in the divine eternity. It is precisely in this regard that Spinoza (a) shows what we might call the soteriological aspect of his thought and (b) lays down some of the principles that will be central to the German Idealists. With regard to (a), it is worth noting that the work we are considering is, after all, entitled *Ethics*, which is to say that it is not a work of theological metaphysics detached from consideration of the human subject but is very pointedly a work about human life and what constitutes human beings' ultimate happiness. However, before we come to the 'last things' of the *Ethics*,

---

[5] Diane Steinberg, 'Spinoza's Theory of the Eternity of Mind', *Canadian Journal of Philosophy* 11/1 (March 1981), 55.

[6] Steinberg, 'Spinoza's Theory of the Eternity of Mind', 56.

[7] Alan Donagan, 'Spinoza's Proof of Immortality', in Grene, *Spinoza*, 242.

[8] Donagan, 'Spinoza's Proof of Immortality', 244–6.

we need first, however briefly, to consider how Spinoza's one divine eternal substance relates to a world that we experience as variform and changing.

Spinoza allows that the divine substance may have an infinity of possible attributes. However, human beings are only capable of conceiving two of these: thought and extension. These attributes (def. IV) constitute the essence of the divine substance (albeit in collusion with the other attributes that lie beyond our possible knowledge) and consequently each 'expresses eternal and infinite essence', 'necessarily exists' (Prop. XI), and is immutable (Prop. XX. Cor. II). However, these attributes, as they exist, exist in a manner that is infinitely determined. Extension, for example, exists as the totality of extended things, which are the modifications of its one eternal essence. By virtue of its intrinsic infinity and eternality, this attribute must of necessity manifest itself in all possible manifestations proper to itself, that is, in the entire universe of spatially extended things. Although extension itself is indivisible and exists as a continuum, the manifold of its possible expressions relate to each other in complexes of determinate existence (the earth goes round the sun, the moon goes round the earth, etc., although—versus the older Aristotelian cosmology—earth, moon, and sun are no longer conceived as distinct substances, each having its own proper form of motion). All are grounded in the one necessary causative infinity of the divine being, but are manifest as an infinite number of finite entities. Unlike God, these entities are not comprehensible through themselves but only in terms of their relations to others—in the first instance, their relations to other finite entities; ultimately, their relation to God.

Starting at the level of pre-philosophical common-sense empiricism, human beings find themselves existing as bodies in constant interactions with other bodies and are aware of complexity and change even in their own bodies, both day-to-day (and even moment to moment) and over a long-term period. As Spinoza fairly uncontroversially asserts in a sequence of postulates regarding the human body, 'The human body...is composed of many individuals (of different nature), each one of which is also composed of many parts...[these individuals] are some fluid, some soft, and some hard...[They] and consequently the human body itself is affected in many ways by external bodies...[and] can move external bodies in many ways, and dispose them in many ways' (EII. Postulates I, II, III, VI). This moveable, composite, and interactive body is, moreover, the exclusive object of the human mind: this and only this is what we are able to know under the natural conditions of human knowledge (EII. Prop. XIII). This may seem unpropitious with regard to our capacity for knowing the eternally immutable one divine substance that is the ultimate cause of corporeal appearances and their ontological basis. However, the complexity of the human body already provides Spinoza with a way forward. The more complex, healthful, and vigorous our bodily life is, the more we are able to experience a multiplicity of relations to other bodies as constituting a single field (my life), the more memory can help us to realize that immediate multiplicity is not the be all and end all but may be expressive of a deeper and more unified order. In particular we can learn to recognize commonalities amongst bodies and to develop adequate ideas of things—of how their own inner complexity is also constituted as a unity. Thus we ascend from imagination to reason (*ratio*) and our first impressions of the world are transformed into well-grounded knowledge.

In all of this, two underlying assumptions play a vital role. The first is that since God is the sole substance and all phenomenal effects are ultimately modifications of the divine attributes co-constituting that substance, everything that we experience and know is, ultimately, God's own self-experience and self-knowledge manifested in or, more accurately, limited by the conditions of human embodiment (EII. Prop. XX). That we see ourselves and other bodies as finite and separate is a perspectival illusion, comparable, as Spinoza says in another connection, to the way in which we see the sun as smaller at noon than at sundown. The second underlying assumption is that 'The order and connection of ideas is the same as the order and connection of things' (EII. Prop. VII). This is because, as we have seen, thought and extension are both equally attributes of God. As Spinoza puts it, 'a circle existing in nature and the idea of an existing circle which is also in God is one and the same thing, though explained under different attributes' (EII. Prop. VII. Note). Human beings are in the first instance limited to seeing things as they encounter them in the sphere of bodily life and do not have direct access to thought or ideas. Nevertheless, through experience, memory, and reason they may acquire some knowledge of the ideas and essences that determine things to appear as they do. In its highest operation, reason sees things according to their necessity and is thus freed from perceiving them as varying through time. Instead, it sees them 'under a certain species of eternity (*sub quadam aeternitatis specie*)' (EII. Prop. XLIV. Cor. II). But this is not just a matter of contemplating, let's say, the idea of a bluebell or an elephant. For 'since every idea of every body or individual thing actually existing necessarily involves the eternal and infinite essence of God' (EII. Prop XLV), such knowledge is also an implicit and, Spinoza says, 'adequate and perfect' knowledge of God—precisely as long as we keep in mind that 'existing' does 'not mean here duration' but 'the very nature of existence, which is assigned to individual things by reason of the fact that they follow from the eternal necessity of the nature of God... I speak, I say, of individual things in so far as they are in God' (EII. Prop. XLV. Note).

Such knowledge is, to repeat, the affair of reason. Images and words remain dependent on bodily motion, but ideas are free from the contingencies of bodily action and interaction. As the highest form of knowledge, knowledge of ideas is simply intuited (*scientia intuitiva*) (EII. Prop XL. Note II). We simply see things as they are, as the necessary and eternal products of the divine mind. And, as Spinoza says in concluding Book II 'On the Nature and Origin of the Mind', this doctrine 'teaches us to act solely according to the decree of

God and to be partakers of the divine nature' and 'not despise, hate, or ridicule any one: to be angry with or envy no one' (EII. XIXL. Note). Here, then, we can again see why this is an ethics and not simply a metaphysics—although on the basis of these statements alone we might not yet see how it could be said that it involves a certain soteriology. What has been offered thus far might seem to propose the ataraxic calm of a Stoic sage, but not yet the blessedness of Jewish and Christian accounts of salvation.[9]

That Spinoza is setting out something that *might* (and for now I stress only *might*) count as a doctrine of salvation becomes clearer as we move into Part Five 'On the Power of the Intellect or Human Freedom'. In the intervening two parts Spinoza has been considering the emotions and, in accordance with his general view as to the relationship between bodies and ideas, has set forth a programme for how we might move from merely 'living' our emotional life to *understanding* it, that is, coming to see our emotional life in the light of the reasons for its manifold fluctuations. Here, too, then, even the emotions can be referred to God and be known *sub quadam aeternitatis specie*. 'He who understands himself and his emotions loves God, and the more so the more he understands himself and his emotions' (EV. Prop. XV). Spinoza adds that the word 'love' is appropriate here, since, because such knowledge is possible only as an active exercise of the mind and such activity is experienced as intrinsically joyful (EIII. Prop. LIII), its relation to God is likewise one of joy. And this 'love towards God', he tells us, is what 'must occupy the mind chiefly' (EV. Prop. XVI).

It is at this point that Spinoza makes the step that might make it sound as if he was offering a doctrine of salvation. For whilst all human knowledge is limited by the experiential field of bodily life and therefore also time-bound, there is a kind of knowledge of the body that is not thus limited, namely, the kind of knowledge enjoyed by God: 'In God, however, there is necessarily granted the idea which expresses the essence of this or that human body under the species of eternity' (EV. Prop. XXII). The body itself is temporal and transient: but the idea of the body is eternal in the mind of God. Thus, Proposition XXIII states that 'The human mind cannot be absolutely destroyed with the human body, but something of it remains which is eternal'. And although eternity is neither defined by time nor has any relation to time, 'nevertheless we feel and know that we are eternal' (EV. Prop. XXIII. Note), a knowledge given in the manner of the intuitive or third kind of knowledge, which is precisely the knowledge we can have of the essence of things—including, in this case, ourselves and our own life in the body. And since this knowledge of ourselves as eternal is

---

[9] However, parallels have been drawn between Spinoza's idea of intuitive knowledge and Maimonides' account of prophetic states of mind that involve 'immediate, super-inferential, and individual yet non-corporeal intuition, apprehending directly through essence' and, as such, partaking of the a priori knowledge that is God's. See Roth, *Spinoza*, 133–4.

knowable only on the condition of seeing ourselves as God eternally sees us, it is dependent on coming to know God and is therefore the supreme instance of 'the intellectual love towards God', which both 'arises from the third kind of knowledge' and 'is eternal' (EV. Prop. XXXIII). As such, this love 'is the very love of God with which God loves himself...in so far as he can be expressed through the essence of the human mind considered under the species of eternity' (EV. Prop. XXXVI). The love with which God loves human beings is the same as the love with which human beings love God, only seen under a different aspect. And, Spinoza adds in one of his rare references to Scripture, 'this love or blessedness is called in the Scriptures "glory"' (EV. Prop. XXXVI. Note). Moreover, since the scope of life enjoyed by a mind that knows its own eternity in God will far outweigh the relatively limited and local life of the body and to such a degree that 'that part [of the human mind] which we showed to perish with the body may be of no moment to it in respect to what remains' (Prop. XXXVIII. Note), knowledge of the divine glory of the mind will tend to diminish the fear of death. Thus Spinoza confirms his earlier affirmation that 'a free man thinks of nothing less than of death, and his wisdom is a meditation not of death but of life' (EIV. Prop.LXVII). The cultivation of this wisdom that is also blessedness is likewise the proper office of piety and religion.

But does this amount to a doctrine of salvation? Perhaps this, or something like it, was the consolation of Rupert Brooke's poem 'The Soldier', contemplating his own possible death in battle and his own subsequent state as 'one pulse in the eternal mind, no less, no more'. But is this 'no less, no more' enough? Despite Spinoza's resort to directly religious language and suggestions as to its continuity with, especially, earlier Jewish philosophers such as Maimonides, Spinoza's account would seem to fall foul of the kind of criticism of philosophical theories of immortality expressed by thinkers such as Kierkegaard, Shestov, and Rosenzweig, namely, that he only 'solves' the problem of the fear of death by turning away from the living, embodied, fearful human being who raised the question in the first place.[10]

Here we can see that there is also need for further philosophical clarification as ancillary to (if not necessary for) any defence of Spinoza's doctrine as potentially religiously satisfying. For what, exactly, is the part of the mind that can be deemed 'eternal'? It may well be the greater part, but what exactly is it and how does it relate to what, in our average everyday way, we experience as our selves?[11] Who, exactly, is the subject of such immortality?

---

[10] In his *Spinoza's Heresy: Immortality and the Jewish Mind* (Oxford: Oxford University Press, 2001), Steven Nadler argues that the denial of immortality was precisely the heresy that led to the philosopher being expelled from the Jewish community.

[11] There are issues here which will carry over into subsequent sections of this chapter and of later chapters, such as the possibility of temporal series being reduced to non-temporal series. However, I am limiting myself now to the question as it arises in specific connection with Spinoza.

In an article broadly defending Spinoza's argument, Alan Donagan emphasizes that it is, precisely, a human being, and more specifically the human
being who enjoys knowledge of the formal essence of the individual body. But
what does this mean?

We have seen that, for Spinoza, the order of ideas and the order of things
are strictly correlated. This seems to mean that, when the body dies, the mind
actually thinking the idea of that body dies with it. There is no longer any
actual body to which the idea could actually correspond. Yet, while actual, the
mind conceived of the body as more than a succession of physical and affective states: it also knew it, more fundamentally, as the formal essence of that
body, that is, as the formal essence of just that particular body and not, as it
were, some essence of humanity locally incarnated in this particular bundle of
biological matter. However, as Donagan argues,

> the idea of that formal essence belongs to God *sub specie aeternitatis*; it is part
> of the infinite idea of God, which is an eternal mode of God in the attribute of
> thought. Therefore the part of man's mind which consists in the idea of the formal
> essence of his body must be eternal: it must have pre-existed his body, and cannot
> be destroyed with it. Q. E. D.[12]

Of course, this begs the question as to whether human beings are capable
of such participation or whether, as Donagan suggests, the fact that our idea
of the formal essence of our human body is never going to be more than an
inadequate idea means that we cannot really be helped by the eternal existence
of its *adequate* idea, since, by definition, this is not and cannot become *our*
idea.[13] Spinoza's reply, he says, has two parts. Firstly, in so far as our ideas have
some positive element they are true: as we have seen, error is only a diminu-
tion of eternal truth, not its opposite. Secondly, whilst even God's knowledge

---

[12] Donagan, 'Spinoza's Proof of Immortality', 255. Donagan's argument involves the assumption that the identity of the orders of ideas and things allows for *and demands* 'actual ideas of the
formal essences of non-existent things' (255). This seems to relate back to the medieval debate
concerning God's knowledge of possibilities that are never actualized. However, it would seem
that Spinoza ought to deny such knowledge to God, since, as he puts it in Part I, 'Things could
not have been produced by God in any other manner or order than that in which they were
produced' (EI. Prop. XXXIII). As all things exist by virtue of the necessity of the divine being,
to suppose that God actually knows non-existent things would introduce into the idea of God's
causative action precisely that element of arbitrary divine will that Spinoza is generally careful
to avoid.

[13] A good illustration of the issue here is A. E. Taylor's account of the way in which, on the one
hand, Ptolemy's and Tycho Brahe's astronomical ideas and, on the other, Euclid's and Newton's
ideas 'survive'. The former, as he says, live on in subsequent science but only as having suffered
'strange transformations', whereas everyone who thinks Euclid's theorem's or Newton's laws has
to think them in exactly the same way that they were thought by their discoverers (A. E. Taylor,
'The Conception of Immortality in Spinoza's *Ethics*', *Mind* NS 5/18 (April 1896), 164–5). Applying
the analogy to individual human lives, we may say that most of us are then more like Ptolemy
and Tycho Brahe than like Euclid or Newton, and that we 'live on' only as strangely transformed
and not exactly as we were.

is, so to speak, dimmed by its participation in our bodily life, God also knows us according to his adequate knowledge 'of all intrinsically possible things'. But this is not Aristotle's vision of 'a life of pure thought, in which all awareness of self-identity has perished'. Even though such a divine idea has neither memory nor anticipation (time is a perspectival illusion of embodiment), what it knows is individual and it is the individual 'me' that I know by participating in this divine idea. As such it is not the idea of myself that I have at a particular moment in my life but 'the ordered totality of those [momentary] ideas'.[14]

Does this do enough to allay doubts? Or is such knowledge so remote from what we ordinarily understand by human life that it is a case of finding one self (the eternal part of my mind) only at the cost of losing another (finite, mortal me)? And here, at the heart of Spinoza's much mooted 'monism' we seem to discern a distinctly dualistic possibility, such that the lived life of embodied human existence really has no intrinsic value for human identity at all.[15] Or else we might equally well see the opposite tendency—for if all adequate ideas are actual in the mind of God and, as such, give a kind of objective immortality to whatever God thinks, should we not also populate eternity with rocks, trees, and animals as well as with humans?[16] What exactly is possible for humans and what is distinctive of humans' relation to God's eternal knowledge of their existence?

Acknowledging the gap between how we mostly take ourselves as being in this earthly life and what such eternal knowledge reveals us truly to be, Don Garrett states that it is essential to Spinoza's entire strategy that attaining such eternal knowledge will only be gained as the outcome of a huge intellectual struggle that will never be complete within the compass of this life. And, in an eloquent appeal to read Spinoza as offering an authentically religious reinterpretation of the Christian doctrine of eternal life, Clare Carlisle points out that such paradoxes are themselves characteristic of the religious life, which typically involves affirming the paradoxical conjunction of transcendence and immanence.[17] It is not only that the *Ethics* shows human life as dependent on and participative in divine being, but Spinoza also offers a narrative of fall and redemption, albeit his version of the fall is not (as in standard Christian accounts) a fall from eternal life to a state of mortality but 'from *desire for life* to *fear of death*' and salvation is, consequently, precisely salvation from that same

[14] Donagan, 'Spinoza's Proof of Immortality', 256–7.

[15] Even if a lived human life is the necessary condition for such eternal knowledge of the formal essence, it would seem—on this view—to be only as the object and not the subject of this knowledge.

[16] See Don Garrett, 'Spinoza on the Essence of the Human Body and the Part of the Mind that is Eternal' in Olli Koistenen (ed.), *The Cambridge Companion to Spinoza's Ethics* (Cambridge: Cambridge University Press, 2009), 298ff.

[17] See Clare Carlisle, 'Spinoza on Eternal Life', *American Catholic Philosophical Quarterly* (forthcoming).

fear of death into which we are fallen. In particular, his modelling of eternal life can be shown to have strong analogies with the teaching of the First Letter of St John with regard to the interweaving themes of love of God, love for others, and eternal life.[18] Eternal life is, in this perspective, less a speculative issue to be resolved in theoretical terms and more a way of living and, as such, an ethical and religious task.[19]

What Carlisle's approach highlights is that Spinoza's task is precisely a double one, namely, to give an account of God as being beyond all possible human imagining whilst arguing for knowledge of God as the highest human possibility and, at the same time, the sole sure ground of human happiness. From the one side, the world and all there is, inclusive of human life, is a manifestation of the one divine substance; from the other, learning the intellectual love of God is the eminent form of human fulfilment and is at once knowledge of God and knowledge of self.

There are, clearly, many connections back to earlier philosophical and religious traditions in all of this, not least to Maimonides, but there are also new elements. Of these, the most important here is that Spinoza's God is no longer, like the theistic God assumed in most of the positions discussed in the previous chapter, external to the world. God—the eternal God—is the true face of the world and the world, when it is known as it truly is, is the true manifestation of God: the divine glory. The circularity of the world in God and God in the world would be central to the theology and philosophy of German Idealism, whose debts to Spinoza are probably incalculable and evidenced not least by the way in which the Idealists' critics regularly accused them of Spinozism. Yet, as I commented at the start of this section, although Spinoza's philosophy certainly allows for time, its eternity—as we have been seeing—seems to abstract from the concreteness of temporal relations, regardless of whether we regard it as a kind of duration or as essentially timeless. In German Idealism, however, and especially in the texts of Hegel and Schelling to which we shall shortly be turning, temporality is an ineluctable feature of the world and of being itself; theirs is a world in which, it seems, nothing *is* except under the condition that it also becomes. But this, as we shall, see further intensifies the paradox that we have already encountered in Spinoza. If the transcendent is immanent and the immanent transcendent, what does this mean if both are qualified as radically temporal? Is it possible even to think of such a paradoxical conjunction, in life, of time and eternity? 'What we do in this life echoes in eternity'—yes, but what if 'what we do in this life' is itself the echo of eternity? However, before

---

[18] As Carlisle points out, Spinoza used 1 John 4:13 as the motto for his *Theological-Political Treatise*: 'By this we know that we remain in God, and God remains in us, because he has given us of his spirit'—and, as John's letter makes abundantly clear, this gift is revealed precisely in the increase of love we have for one another whilst, at the same time, being a pledge or even realization of 'eternal life'.

[19] Carlisle, 'Spinoza on Eternal Life'.

we come directly to the Idealists themselves, we turn first to a thinker who, like Spinoza, was a major historical inspiration for their original project, namely, the visionary cobbler of Goerlitz, Jakob Boehme.

BOEHME

Although Hegel describes Boehme as a 'complete barbarian' with regard to his lack of formal philosophical discipline,[20] he nevertheless gives him a crucial place in the history of philosophy in the aftermath of the Protestant Reformation. As Hegel sees it, the key philosophical implication of this religious revolution is that modern philosophy now takes as given the unification of this world and the beyond, of immanence and transcendence; or, as he also puts it, it is 'the stand-point of actual self-consciousness'.[21] Descartes may have marked the true beginning of modern philosophy, but the issues are anticipated, and in a fundamental way, in the dissimilar twins, Jacob Boehme and Francis Bacon. Boehme is said to be 'the first German philosopher', whose thinking, as Hegel puts it, is '*echt deutsch*'.[22] According to Hegel, 'His basic idea is the attempt to hold everything together in an absolute unity, the absolute divine unity and the unification of all oppositions in God', nor does he draw back from the implication that 'the universe is one divine life and the revelation of God in all things'.[23]

From these few words we can already see that there is a certain congruence between Spinoza and Boehme. Both seem to invite the charge of pantheism by blurring the distinction between immanence and transcendence that seems definitive of classical theism. However, there are no less significant differences between them, starting with their vastly different styles, for where Spinoza proceeds in the manner of a geometer, Boehme writes with the fire of a visionary seer, creating an extraordinary pyrotechnic display of theogonic and cosmogonic speculations in a language that is simultaneously biblical, mystical, and alchemical.

With regard to Spinoza, Hegel rejects the standard charge of 'atheism'—only to accuse him of the opposite error: his system contains 'too much God'[24] so that the world itself in its complex and concrete determination vanishes from sight.[25] Spinoza repeats Eleatic monism, and his grasp of determination as

---

[20] G. W. F. Hegel, *Werke, xx: Vorlesungen über die Geschichte der Philosophie III* (Frankfurt am Main: Suhrkamp, 1971), 92.

[21] Hegel, *Geschichte der Philosophie III*, 63.

[22] Hegel, *Geschichte der Philosophie III*, 94.

[23] Hegel, *Geschichte der Philosophie III*, 98, 99.

[24] Hegel, *Geschichte der Philosophie III*, 163.

[25] For full discussion of Hegel's criticism of Spinoza for 'acosmism', see Yitzhak Y. Melamed, 'Acosmism or Weak Individuals? Hegel, Spinoza, and the Reality of the Finite', *Journal of the*

negation does not go so far as to encompass what Hegel would seize upon as decisive, namely, the negation of the negation and the concomitant possibility of the individual subject transcending itself towards the absolute through its thought, suffering, and action.[26] By way of contrast, Boehme's system is genuinely dynamic and his dialectic is 'fully alive', as shown by the way in which he develops the production of light, the Son of God, and, ultimately, the fully diversified and concrete world out of the relations of the qualities and his preparedness to embrace—as Spinoza cannot—a world-founding self-diremption on the part of and, we might say, *in* the very being of God.[27]

These last comments are probably fairly unintelligible without further explication of Boehme's system. However, a full account of this would take us very far afield from the main line of this study, not least because of the 'barbarism' of his style and the constant difficulty of sustaining the tension between what we might call its mystical or visionary fire and its intellectual structure. The true Boehme requires both, but I shall limit myself to only the merest sketch of some of the basic outlines of his vision, with particular emphasis on what is most relevant to our theme of the eternal.

Boehme's system is fundamentally dynamic. If Spinoza's universe could plausibly if unjustly be portrayed as a kind of frozen waterfall, Boehme's world and even Boehme's God is in perpetual motion. Nothing in Boehme's universe 'is' without being caught up in a complex of relational and temporal processes—including God. The question then is: how did God *become* 'God'? As Alexandre Koyré put it, Boehme starts from 'indeterminate deity', the divine nothing, and seeks to attain God'—God in the fullness of his being, life, and relation to the world, not just all-knowing, but all-smelling, all-tasting, etc.[28] This may already sound excessively mystical, but Koyré sees the task Boehme has set himself as basic to any metaphysical doctrine that sets out to explain the origin of beings. That is to say, the point of departure can assume or presuppose nothing at all, yet 'it is necessary to get something out of nothing', even if only (in the first instance) a hypostatized 'Nothing'—which is then taken as a basis for explaining all subsequent metaphysical and cosmic events. Yet if we can only think such a Nothing in the mode of negation (apophatically), how

---

History of Philosophy 48/1 (January 2010), 77–92. Melamed argues that, following Salomon Maimon, Hegel sees Spinoza as unable to account for the actual diversity of the world and for not allowing independent existence to the modes. Spinoza's individuals are not concrete self-centred selves having substantial identity, but manifestations of events that are not necessarily coterminous with what we regard as selves. This is, of course, a criticism that also relates to the previously discussed question as to Spinoza's account of immortality. Ironically perhaps, this is essentially the same complaint that Kierkegaard will later make of Hegel himself. See n. 41 and the section 'God is in Heaven We are on Earth' in Chapter 4.

[26] Hegel, *Geschichte der Philosophie III*, 164.

[27] Hegel, *Geschichte der Philosophie III*, 118.

[28] Alexandre Koyré, *La Philosophie de Jacob Boehme* (New York: Burt Franklin, 1968 [1929]), 317.

can it then help us to explain anything at all unless it is also knowable in some positive way (cataphatically)?[29]

Boehme's distinctive account of the transition from apophatic to cataphatic knowing involves characterizing the primordial Nothing as the *Ungrund*, that is, not a 'ground of being' in God but an utter, abyssal absence of ground, an ultimate groundlessness. Although, in a sense we shall shortly consider, this *Ungrund* precedes God (in the full, cataphatic sense of 'God'), it is not a 'cause' of God since, by definition, it cannot be a 'ground' for the being of God or of anything whatsoever: it is precisely a *non*-ground (*Ungrund*). In what might be described as something like a proto-phenomenology Boehme proceeds via a series of carefully graded steps to show how God, as it were, 'becomes' and how, from this initial, undifferentiated *Ungrund*, first the Absolute, then the One, then the full Trinitarian God, until, finally, the entire unity of God-and-world comes into being. This is not a necessary logical development or natural process, however, but an expression of the divine will that exists initially only as undifferentiated and object-less and first becomes concrete as will in God's Trinitarian life.[30] Even evil is included in this development, since the creation and fall of Lucifer is integral to the process whereby God becomes fully God as well as being a pre-condition of the creation of the world. As a result, the duality of good and evil is inscribed in every possible manifestation of life.[31] Even though God himself is entirely good, '*all the powers* which you can search out in nature, and which are in *all* things, proceed from him', inclusive of evil.[32]

Even from this briefest of outlines, it is clear that Boehme faces a fundamental challenge with regard to the divine eternity. As Koyré puts it in discussing the stage of emergence of the eternal One, 'the eternal One *is not yet*'.[33] And yet Boehme also affirms that the Father 'is thus from eternity to eternity unchangeable: He never changed himself in His *being* (Wesen), neither will He change Himself in all eternity'.[34] As in Spinoza, there is a level at which God simply *is* and abides eternally self-same, above and beyond change. There is, however, the important difference that, for Boehme, the difference between eternal self-identity on the one hand and change, emergence, and time on the other are now located *within* the life of God himself. As Hegel put it, Boehme effects a diremption in God's own being. If God is 'from eternity to eternity

[29] Koyré, *Jacob Boehme*, 103–5.

[30] For a brief exposition of Boehme's theogony, see Koyré, *Jacob Boehme*, 320–51. Another older study that is still helpful is John Joseph Stoudt, *Sunrise to Eternity: A Study in Jacob Boehme's Life and Thought* (Philadelphia, PA: University of Pennsylvania Press, 1957). For recent discussions of Boehme's thought and its influence see Ariel Hessayon and Sarah Apetrei (eds), *An Introduction to Jacob Boehme: Four Centuries of Thought and Reception* (New York: Routledge, 2014).

[31] See, e.g. Jakob Böhme, *The Aurora*, tr. John Sparrow, ed. C. J. Barker and D. S. Hehner (London: James Clarke, 1960), 40.

[32] Böhme, *Aurora*, 62.   [33] Koyré, *Jacob Boehme*, 325.   [34] Böhme, *Aurora*, 70.

unchangeable' Boehme also speaks boldly of the Holy Trinity as 'a triumph-
ing, springing, moveable being', of the generation of the Son as 'the heart and
lustre shining forth from the powers of his heavenly Father'[35] and, even more
vividly, he depicts the Spirit as 'the moving, flowing, boiling power of God',[36]
all of which suggests some kind of temporality in God.[37] The life of God thus
reveals an inner 'history' and is consequently knowable only in the form of
temporal becoming. If, for classical theism, the problem is how the eternally
timeless God relates to a temporal world, Boehme's problem is therefore how
the eternal God relates to His own temporality. Koyré writes that Boehme's
God 'is precisely the God who "wird und entwird" ["comes into and passes
out of being"] eternally, in a single *nunc aeternitatis* (*ewiges Nu* [eternal now])'
who is not 'outside' time but passes through all the stages of his evolution 'in a
moment'.[38] But is it possible even to imagine, let alone think such a paradoxical
temporal/eternal process?

Boehme himself was not ignorant of the problem, as can be seen from his
emphatic denials of any essential change in God and his affirmation of God's
eternal self-sameness in response to anticipated objections. However, he does
not address these philosophically, but in his own visionary mode. The philosophers do address them—but can they, in the end, do more than restate the
problem? That they aspired to do more is beyond doubt, since they wanted to
use Boehme's imaginative visionary ideas in such a way as to yield sure and
certain *knowledge*. And, importantly, this was not just a question about the
divine being but about human self-understanding—Spirit.

HEGEL

In the opening paragraphs of Hegel's lectures on the philosophy of religion the
philosopher reminds his auditors of the object of their reflections. 'We know',
he says, 'that in religion we withdraw ourselves from temporality and that it is
that region of consciousness in which all the riddles of the world are solved,
all the contradictions that arise for those who think more deeply are laid bare,
and all the sorrows of our sensibility grow dumb; it is the region of eternal
truth, of eternal rest, of eternal peace'.[39]

The object that defines religion itself is, more specifically, God. Listing the
things that make us human—thought, Spirit, science, art, political life, and
everything deriving from human freedom—Hegel says that all of these 'find

---

[35] Böhme, *Aurora*, 76.   [36] Böhme, *Aurora*, 78.   [37] Böhme, *Aurora*, 61.
[38] Koyré, *Jacob Boehme*, 317.
[39] G. W. F. Hegel, *Werke*, XVI: *Vorlesungen über die Philosophie der Religion I* (Frankfurt am
Main: Suhrkamp, 1969), 11.

their final mid-point in religion: in the thought, the consciousness, the feeling of God. God is thus the beginning and end of all things and as everything proceeds from this point, so everything returns again to it. Likewise he is the middle that enlivens, that gives spirit, and ensouls all those existing forms that he sustains'.[40]

Reading these words, we might imagine that Hegel was about to repeat the system of Spinoza—that God was the true being of all beings and that to know God was to know the world as it is *sub specie aeternitatis*.[41] But we already know from Hegel's own criticisms of Spinoza that this would be to indulge in 'too much God' and that simply to reduce the world to how it might be seen *sub specie aeternitatis* is to fail to do justice to the concrete individuality of the manifold forms of life and, especially, of human spirit itself. In other words, the aim of philosophizing is not to turn aside from the manifold of manifestation and to know, simply, God. The aim is rather to know the manifold and temporal world in the light of the eternal and, conversely, to interpret temporal life so as to disclose the animating spirit of the eternal. Philosophy, quite simply, is dialectical. But is such a dialectic thinkable?

If we turn back to Hegel's path-breaking entrance into philosophy, to the *Phenomenology of Spirit* and, in particular, to its programmatic preface, we find a view of philosophy that is, at the very least, in quite a different register from the introduction to *The Philosophy of Religion*. There, polemicizing against the view that the task of philosophy culminates in an instantaneous intuition of the absolute, Hegel proposed that this would be mere absorption in a night in which all cows were black. Philosophy requires the acknowledgement and understanding of the different concrete forms in which the absolute is manifest and there is no other way to achieve this than by a theoretical recapitulation of the processes through which the actual world in which we all live and move and have our being has become what and as it is and, by virtue of thus thinking it, also to know that this actual world is the manifestation of a single and coherent reason. To illustrate his point Hegel compares life and the kind of philosophizing that seeks to comprehend life to the processes of botanical life:

> The bud disappears in the bursting-forth of the blossom, and one might say that the former is refuted by the latter; similarly when the fruit appears, the blossom is shown up in its turn as a false manifestation of the plant, and the fruit now

---

[40] Hegel, *Philosophie der Religion I*, 10–11.

[41] In the next section we shall see how, for some British Idealists, this is precisely the outcome of the Hegelian system. Hegel's own words about the 'region' of religion might even suggest a reversion to some kind of Platonist dualism, such that the world of eternal verities was somehow separated out from the world of time and change. We might also note that it would be a central element in Kierkegaard's criticism of Hegelian speculation that it presumed upon a knowledge of the human condition *sub specie aeternitatis* that, in Kierkegaard's view, was unavailable to existing human beings. See SKS7, 81/CUP1, 81.

emerges as the truth of it instead. The forms are not just distinguished from one another, they also supplant one another as mutually incompatible. Yet at the same time their fluid nature makes them moments of an organic unity in which they not only do not conflict, but in which each is as necessary as the other; and this mutual necessity alone constitutes the life of the whole.[42]

Most succinctly, 'the truth is the whole', and the whole can only be known by those prepared to track its entire development and thus to see that absolute being is inseparable from the process by which it becomes what it is: being is becoming or, as Hegel also puts it, substance is subject,[43] that is, the manifestation of the one divine substance 'is' or occurs as the life of concrete individual existences.[44] These individual existences negate the primordial unity of substance: they exist as different from it and to know them requires knowing that difference.[45] Alienation and otherness are integral to the life and knowledge of the whole, and only through such alienation and otherness can the one substance develop from a state of being in-itself to being for-itself. At its most extreme development this may seem to involve the entire evacuation of being in-itself. As Hegel writes:

> The circle that remains self-enclosed and, like substance, holds its moments together, is an immediate relationship, one therefore which has nothing astonishing about it...But the life of Spirit is not the life that shrinks from death and keeps itself untouched by devastation, but rather the life that endures it and maintains itself in it. It wins its truth only when, in utter dismemberment, it finds itself...Spirit is this power only by looking the negative in the face and tarrying with it.[46]

As Hegel suggests elsewhere, this moment of utter emptying-out may be understood as occurring in the event of the crucifixion, radically interpreted as the death of God.[47] In any case, it indicates that the absolute only ever 'is' as temporal. The evocation of the ethereal regions of eternal peace cited at the start of this section should therefore be understood less as a call to turn away from the temporal development of life and more as a rhetorically inflected call

---

[42] G. W. F. Hegel, *Werke, iii: Phänomenologie des Geistes* (Frankfurt am Main: Suhrkamp, 1970), 12; Eng. *Phenomenology of Spirit*, tr. A. V. Miller (Oxford: Oxford University Press, 1977), 2.

[43] Hegel, *Phänomenologie des Geistes*, 24 (Eng. tr., 11).

[44] Hegel, *Phänomenologie des Geistes*, 23 (Eng. tr., 10).

[45] We have seen that Hegel approves Spinoza's realization that the determination of specific modes is essentially a negation or limitation of the divine substance. But, as was also noted, Spinoza did not (in Hegel's view) take the modes sufficiently seriously to allow for the possibility of a further negation in which they (the modes qua concrete individuals) negated their own particular standpoint and returned to a knowledge of the whole.

[46] Hegel, *Phänomenologie des Geistes*, 36 (Eng. tr. 18–19).

[47] See Hegel, *Phänomenologie des Geistes*, 571f. (Eng. tr. 476). See also the early article 'Faith and Knowledge': G. W. F. Hegel, 'Glauben und Wissen', in *Werke, ii: Jenaer Schriften 1801–1807* (Frankfurt am Main: Suhrkamp, 1970), 432.

to grasp the processes of life in their eternal meaning. The concept of God and our capacity for grasping that concept are inseparable from the history of religion and from the human self-consciousness that this history both manifests and makes possible.

The radical implications of this approach are already laid bare in the closing parts of the *Phenomenology*. Having expounded the history of religion as culminating in the death of God and in the birth of Spirit as the self-understanding of a community (the Church) that takes this death as a salvific revelation of human beings' capacity for becoming freely responsible for their own spiritual existence, Hegel recapitulates what this means for knowledge. Religion knows this process in the manner of representation (*Vorstellung*), which Miller translates as 'picture-thinking'. But philosophy seeks to grasp it as concept (*Begriff*). In other words, religion sets forth the truth in myth, image, and narrative, whereas philosophy grasps the same religious content in the light of its essential truth. It does not think something different from religion, but it thinks it as distinct from the manifold of historically transmitted and historically conditioned stories and symbols that are the distinctive forms of religious life. Consequently, the philosopher can also at this point dispense with religion's characteristic invocation of authority and tradition. Knowing what is true, the philosopher doesn't need to be told it.

Nevertheless, all of this might seem only to reinforce the view that Hegel is making philosophy and the knowledge that philosophy is to lay hold of essentially temporal. How, in the light of all he has said, could philosophy claim any 'eternal truth, eternal rest, eternal peace'? Isn't this precisely to speak in the picture-language of religion, not philosophy?

Hegel is not unaware of the problem, and in some especially contorted paragraphs of the *Phenomenology*'s closing meditation on absolute knowing he addresses the relationship between time and the concept. What religion represents as a narrative of the saving acts of God to which human beings are called to give their consent, philosophy understands as a humanly-achieved act of self-understanding, that is, Spirit's conceptual grasp of itself as, precisely, Spirit. Such knowledge is 'science', in the specifically Hegelian sense. But an act of knowing of this kind must itself be historical, presupposing both the events that religion narrates and that narration itself. Therefore, 'as regards the *Dasein* of this concept, science does not appear in time and actuality before Spirit has arrived at this consciousness of itself'.[48] Yet the original 'knowing substance' that thus comes to know itself as Spirit and as subject is also 'earlier than this form or conceptual Gestalt'.[49] In other words, Spirit knows nothing that is essentially alien to substance and what it knows as having manifested

---

[48] Hegel, *Phänomenologie des Geistes*, 583 (Eng. tr. 486).
[49] Hegel, *Phänomenologie des Geistes*, 584 (Eng. tr., 486).

itself concretely in history is already implicit in abstract form in substance in-itself. We are therefore confronted with a kind of double perspective:

> In the concept, that knows itself as concept, the moments hereby appear as earlier than the complete whole whose becoming is constituted by the movement of those very moments. In consciousness, on the other hand, the whole, though not yet grasped as such, is earlier than the moments. Time is the concept itself that is there ('*da ist*') and that presents itself to consciousness as empty intuition.[50] Consequently Spirit necessarily appears in time and continues to appear in time so long as it does not apprehend its own pure concept, that is, so long as it does not uproot time. It [time] is the pure self intuited as external to the self and not apprehended as such by it, it is the merely intuited concept. In that this [then] apprehends itself, it sublates its temporal form, grasps the intuiting act in such a way that that act is both conceptually grasped and itself grasps [its subject] conceptually. Time thus manifests as the destiny and necessity of Spirit that is not perfected in itself...[51]

As in Spinoza, there is ultimately just one substance. Nothing essentially new ever comes to be. But in and through its temporal transformations this one divine substance can become capable of grasping itself as a living whole—and it does so precisely through acts of human self-consciousness. What is grasped is the permanently abiding whole—being in-itself—yet conceived as manifested in time and as the content of the act of thinking itself.

The aim at least, is clear: to think the experienced flux of temporal existence as it appears in the light of eternity. Such thinking is what Hegel calls *Erinnerung*: the recollection or 'internalization' in thought of what has been manifested on the plane of extension, in time. For Hegel, this is neither the empty circle of a pure intuition that lacks all determinate content nor the vicious circle of mere self-contradiction, but a circle in which Spirit knows its own infinite life—as the *Phenomenology* concludes, adapting Schiller: 'From the chalice of this realm of spirits | foams forth for Him his own infinity'.[52] Again as in Spinoza and Boehme, the split between time and eternity is not between God and the world but *within* an essentially unified structure of a single divine being. But if the aim is clear, what is the outcome and does Hegel really do more than restate the problem?[53] To answer this question, however, we first need to answer another, namely, whether the eternal aspect of this

---

[50] This sentence seems to anticipate the core of Heidegger's argument in *Being and Time*, namely, that if we start our analysis of Being with *Da-sein*, the being that is there, then Being will be revealed as thoroughly temporal.

[51] Hegel, *Phänomenologie des Geistes*, 584–5 (Eng. tr. 487). This, however, points to a kind of non-temporal knowing that Heidegger could never accept.

[52] Hegel, *Phänomenologie des Geistes*, 591 (Eng. tr. 493).

[53] Although this doesn't put him in a worse position than those classical theists of whom Helm complained that their solutions were merely rhetorical rather than substantive.

twofold act of knowing is conceived by Hegel as a fully timeless eternity or more as a kind of sempiternity?

As far as Hegel's own position is concerned, some help in understanding this question (if not perhaps in *answering* it) can be found in his *Philosophy of Nature*, which constitutes the second part of his *Encyclopaedia of Philosophical Sciences*. Here, Hegel sets out a view of nature as occupying a kind of middle point between the initial state of being in-itself and the final infinite self-knowledge of Spirit. The world of nature is not identical with the self-same identity of being in-itself. It is, after all, manifest and must to that extent be regarded as integral to the movement by which Spirit completes the circle of its self-knowing. Boldly, Hegel is even prepared to see it as 'the Son of God, but not as the Son, but as persistence in being other: it is the divine idea held fast for a moment outside [the completed movement] of love'. But it is only thus 'held fast' for a moment, since 'God does not remain fossilized or dead, but the stones cry out and raise themselves to Spirit'.[54] But if nature thus constitutes a middle point in the movement from being in-itself to Spirit, this immediately raises the question as to its metaphysical status and, especially, whether in any sense it is 'eternal'. However, Hegel suggests, 'Eternity is not before or after time, not before the creation of the world, not even when it comes to an end; but eternity is absolute presentness, the now without before or after. The world is created, is now being created, and has been created eternally…Creating is the activity of the absolute idea; the idea of nature, like the Idea as such, is eternal'.[55] The relation between the eternity of being and nature (and, importantly, the philosophy of nature) is therefore not that the eternity of nature constitutes a kind of past that philosophy recollects but that this recollection is more a matter of penetrating the inner, actual, and present Idea that manifests itself in the structured whole of the natural world right now. Philosophy, as Hegel adds, 'is a way of grasping things that is timeless (*zeitlos*), including time'.[56] As he will further argue in the following section on space and time, 'the concept of eternity must not be construed as abstracting from time as if it existed externally to it, still less in the sense that eternity comes after time, which would be to make eternity into the future, into a moment of time'.[57] If human history is necessarily structured according to a sequence of before and after such that 'recollection' can bear (and is required to bear) the double meaning of remembrance and internalization,[58] the laws of nature are constant and unchanging,

---

[54] G. W. F. Hegel, *Werke*, ix: *Enzyklopädie der Philosophischen Wissenschaften II* (Frankfurt am Main: Suhrkamp, 1970), 25. The second quote alludes to Jesus' words recorded at Luke 19:40, when, having been urged by the Pharisees to stop his disciples from proclaiming him Messiah, he responds, 'I tell you, if these were silent, the stones would shout out'.

[55] Hegel, *Enzyklopädie II*, 26.   [56] Hegel, *Enzyklopädie II*, 26.

[57] Hegel, *Enzyklopädie II*, 50.

[58] As in the comment from the Introduction to *The Philosophy of Right* that it is only at dusk that the Owl of Minerva spreads its wings and philosophy is able to gain a retrospective understanding of an era that has reached its final fulfilment. See G. W. F. Hegel, *Werke*, vii: *Grundlinien*

even as regards the changes that clearly do occur within the natural world (such as the movement of planets, geological cataclysms, weather, the life and death of animals—and, we can now add, the evolution of species).

To understand nature—inclusive of the time of nature and of time as a phenomenon of nature—is therefore to grasp the totality of the laws that make nature as it is now. Hegel insists that this is not a question of duration (*Dauer*), which would be a merely relative kind of permanence, but of an eternity that 'does not come to be, that was not, but *is*'.[59] Eternity is 'duration reflected in itself', which, as Hegel's following remarks seem to suggest, means precisely that it is not merely long- or even ever-lasting but constitutes the inalterability of 'the Idea'.[60]

Is Hegel's balancing act credible? Does Spirit—can Spirit—really think time as temporal or will it always, as seems to be the case here, see time itself as manifesting a timeless law? This is Heidegger's conclusion, at the end of his lectures on Hegel's *Phenomenology of Spirit*. Hegel, he says, always—even when speaking of history and Spirit—thinks of time in terms derived from the philosophy of nature. In this regard, Heidegger then contrasts his own position 'that the essence of being is time' with the Hegelian view that 'Being is the essence of time—being, that is, as infinity'.[61] The comment is certainly supported by what we have been reading in the philosophy of nature. But is it the whole Hegel? Or should we give more weight to his claim to differentiate his own philosophy from all the philosophies that have gone before precisely by the way in which he reconfigures truth as a holistic process? But doesn't the holistic element—cf. the remarks at the end of *The Phenomenology* about the differences regarding 'earlier' and 'later' between concept and consciousness—undermine the process element? Can one really have both? Can one have a genuinely dialectical view of the matter? Or is the difference at issue so great—so infinitely and eternally great—that thinking both together is always going to be either impossible or paradoxical?[62]

HEGELIAN TIMELESSNESS

On the Hegelian view we have been examining, eternity becomes manifest as the truth of time. Time is how eternity becomes true—the becoming of the

der Philosophie des Rechts (Frankfurt am Main: Suhrkamp, 1970), 8; Eng. *The Philosophy of Right*, tr. T. M. Knox (Oxford: Oxford University Press, 1967), 13.

[59] Hegel, *Enzyklopädie II*, 50.　　　[60] Hegel, *Enzyklopädie II*, 50–1.

[61] Martin Heidegger, *Gesamtausgabe, xxxii: Hegels Phänomenologie des Geistes* (Frankfurt am Main: Klostermann, 1980), 209; Eng. *Hegel's Phenomenology of Spirit*, tr. Parvis Emad and Kenneth Maly (Bloomington, IN: Indiana University Press, 1988), 144–5.

[62] And, again, these questions should not be taken as suggesting that Hegel is necessarily in a worse case than other philosophies. Maybe the task is itself impossible—for thought.

whole: branch, bud, blossom, and fruit. There is entire reciprocity: the end is the beginning, and the beginning is the end. Thus, for philosophy, knowledge becomes essentially and not accidentally historical: the philosophy of history and the history of philosophy are internal to philosophy itself, and such spheres as aesthetics, ethics, and religion are also understood as essentially historical. Even logic has 'becoming' inscribed in the basic relationship between affirmation and negation.

But is there nevertheless a sense in which, in the end, Hegel's God is a timeless God and logic is, as Hegel famously put it, the thought of God before the foundation of the world?[63] That this is indeed the case was essentially the view of some of the British Hegelians. However, for what are rapidly becoming familiar objections, it seems dubious whether these thinkers succeeded any more than Hegel or the theists in offering a finally satisfying solution to the problem of thinking time and eternity together.[64]

This comment should immediately be qualified by the note that although both Bradley and McTaggart (who will be the main foci of my discussion) acknowledged a significant inspiration from Hegel, neither of them would probably have accepted the designation 'Hegelian' except in a rather broad sense and, as we shall see, McTaggart certainly doesn't refrain from pointing out where he thinks Hegel went wrong. Nevertheless, their approaches offer a way of further clarifying the problem of time and eternity as that is presented within the Hegelian paradigm.[65]

F. H. Bradley's *Appearance and Reality* starts out in the first Book, 'Appearance', with what Bradley takes to be the widespread 'scientific' view that the primary qualities of reality are forms of extension and that all other appearances and experiences (secondary qualities) are reducible to these. However, the primary qualities are abstractions that are never actually given in experience or in reality. 'Extension cannot be presented or thought of, except as one with quality that is secondary'.[66] 'Gravity' is not a natural object of knowledge and we 'know' it only thanks to the existence of falling apples. Consequently, a universe of bare qualities (which Bradley identifies with Aristotelian substances) would scarcely constitute an experienceable world. However, if we redescribed qualities or substances in terms of relations, they

---

[63] G. W. F. Hegel, *Werke, v: Wissenschaft der Logik I* (Frankfurt am Main: Suhrkamp, 1970), 44.

[64] We should note that in McTaggart's case the aim is not to vindicate the eternity of the Absolute/God but the (non-temporal) immortality of the individual.

[65] On the distinctiveness of the British Idealists' interpretation of Hegel, see W. J. Mander, *British Idealism: A History* (Oxford: Oxford University Press, 2011). Mander comments that they generally read Hegel with primary regard to his logic and metaphysics, but without the focus on historical dialectics that would be distinctive of many 'Continental' readings: 'They saw [Hegel's thought] as religious, metaphysically laden, and driven most strongly by logic' (40).

[66] F. H. Bradley, *Appearance and Reality: A Metaphysical Essay* (Oxford: Clarendon Press, 1930 [1893]), 14. (Further references are given in the text as 'AR'.)

would dissolve into these relations and lead to a kind of total empiricism— and without primary qualities or substances relations must collapse in on themselves since they are without relata to relate. Neither quality nor relations are able to provide a firm basis for reality, separately or together. All they yield is 'a makeshift, a device, a mere practical compromise'—*appearance* only and not truth.

The remainder of Book I goes through a sequence of basic metaphysical categories—space and time, motion and change, causation, activity, things, and the self—and finds that none of these are sufficient to ground knowledge of reality as such. All they can collectively achieve is one or other version of phenomenalism, an account of appearances and not of reality.

Regarding time, Bradley comments that 'It is usual to consider time under a spatial form. It is taken as a stream, and past and future are regarded as parts of it, which presumably do not co-exist, but are often talked of as if they did' (AR, 33). However, this is to see time both as a relation (e.g. of past and future) and not a relation (time is the continuum of the stream of which past and future are parts). But is it possible to arrive at a better view by considering time in more properly temporal terms? Bradley thinks not. We cannot experience the present 'now' without implicitly connecting it to a 'before' and 'after' or, at the very least, to a 'before'; but 'any process admitted destroys the "now" from within' (AR, 35). Nor is the situation improved if we suppose that we have a perception of duration, that is, of time arriving in extended units. For if these units can exist independently, without a before or after, then they are essentially outside time and time is just the relation between them. But such a view gives no account of how the composite of such relations could ever be a unity, i.e. 'time' as such, and so time—like the qualities of the earlier chapter—is dissolved rather than explained.

Issues of time are also implicit in much of the discussion of the self, especially when Bradley considers memory as a candidate for grounding the unity of the self (he thinks it proves incapable of doing so). However, I turn now to Book II, where, having demonstrated that the basic metaphysical categories can do no more than organize a world of appearances, Bradley turns to 'Reality'. This distinction should not be taken to imply that the world of appearances is unreal or a veil of illusion. On the contrary 'appearances exist... [a]nd whatever exists must belong to reality' and there is therefore no reality apart from appearances (AR, 114). The Bradleian Absolute, whatever else it is, comprises and is, very precisely, the reality of the totality of appearances. And this totality is constituted as a unity, as one, and the criterion by which we can proceed to characterize it is this: that '[u]ltimate reality is such that it does not contradict itself' (AR, 120). But although this may seem like a merely negative criterion, it also suggests the positive ascription to the Absolute of an essential self-consistency, namely, oneness.

Like Hegel himself, Bradley has a clear soteriological interest. Assurance as to the ultimate unity of reality will also provide assurance as to the perfection of the universe, including its practical perfection (that is, that it is a universe that is not just perfect in itself but is one in which beings such as we are may achieve our own full perfection), and the preponderance of good over evil. Discursive thought cannot arrive at the vision of such a whole, although it might be revealed to some kind of 'higher intuition' (AR, 152).[67] However, the world as we seem to know it now does not present itself as this perfect whole.

Bradley finds himself confronted by the contradiction that time and space are mere appearance but, nevertheless, qua appearance, exist and, as such, are real; both, therefore, 'must somehow in some way belong to our Absolute' (AR, 181). Yet, equally, 'If time is not unreal, I admit that our Absolute is a delusion' (AR, 182). His response to this paradox is to suggest that time itself 'points to something in which it is included and transcended' (AR, 183). It does so, firstly, by relating to what is permanent; secondly, by constituting itself as the unity of past, present, and future (so we experience it as more than simple transience); and, thirdly, by the way in which 'the whole movement of our mind implies disregard for time' in that once we know something to be true we hold to its continuing to be true despite the passage of time (AR, 184).

Bradley concedes that he doesn't know how time comes to appear nor exactly how we might transcend temporal appearance and that these three points amount to little more than suggestions. Yet he does think there are grounds to doubt the apparent 'solidity' of time. Thus, 'we have no reason...to regard time as one succession and to take all phenomena as standing in one temporal connexion' (AR, 186). Where Leftow points to contemporary physics as supportive of the possibility of multiple co-existent time-series, Bradley invokes the different times of dreams and fictions and the incommensurability of, for example, the time of the story of Imogen and the time of the story of Sinbad the Sailor. The ultimate unity of phenomena need not itself be temporal and may incorporate multiple time-series that are not themselves directly mutually connected *in* time. 'There may be many times, all of which are at one in the Eternal—the possessor of temporal events and yet timeless' (AR, 189). But whereas the classical theist would see this as raising the problem of the relationship between a timelessly eternal God and a temporal world, Bradley has made it a problem internal to the timeless Absolute itself—although, to repeat what was said in the case of Hegel, the central problem—the relationship between this identity and this difference—remains arguably unchanged by such relocation.

---

[67] Later (AR, 447), Bradley will quote (without sources) both Western and Indian mystical sayings to this effect.

Similar arguments follow regarding the directionality of time which, again, Bradley sees as dependent on our experience, and in such a way that we cannot exclude the possibility of diverse time-directionalities, each of which is meaningful only in relation to itself and incomprehensible to beings occupying other time-directionalities. The same is true of the connection between time and causality, and the sequence of the causal chain may itself be only appearance (AR, 193).

'*May* be only appearance'—we note Bradley's various hesitations, but a more vigorous assertion of the claim that time is illusory and all true knowledge is therefore eternal knowledge is made by another British Hegelian, John McTaggart. Unlike Bradley, McTaggart approaches questions of time and timelessness with explicit reference to Hegel.[68] His argument is essentially that the succession of categories developed in Hegel's logic is, precisely, a logical and not a temporal succession and that, ultimately, temporal succession is reducible to logical succession. Yet he acknowledges that time needs to be explained. Hegel erred by construing the successive categories as successive events; but 'such a view is incompatible with the system... [and] there seems no doubt that we must reject the development of the process in time'.[69] The Absolute Idea is 'the presupposition and the logical *prius*' of the lower categories and therefore its appearance cannot depend on their appearance (HD, 160). In this respect, at least, Hegel has to be rescued from Hegel. Since Hegel himself maintains that 'the real is the rational', a consistent Hegelian must reject temporal process. 'Can we regard as fully rational a universe in which a process of time is fundamentally real?' McTaggart asks rhetorically (HD, 160). His answer is a clear 'No', since (1) time is infinite only in the sense of the '"false infinite" of endless aggregation' (HD, 161); (2) 'no event in time could be seen as an ultimate beginning' (HD, 162); whilst, equally, (3) a temporal process cannot be produced from a timeless state, so that time can be produced neither by time nor by a timeless being. Thus, although we believe that we experience finite time, what we actually experience is a time series 'extending indefinitely both before and after our immediate contact with it, out of which we cut finite portions' (HD, 165–6). Time, in other words (not McTaggart's) is a perspectival illusion such that it is 'unfit, in Metaphysics, for the ultimate explanation of the universe, however suited it may be to the finite thought of everyday life' (HD, 166).

As intimated in the comment on the priority of the Absolute Idea, the highest term, in which the dialectic ends, is a presupposition of the lower terms and each moment is 'real' only in so far as it is a 'moment' of the Absolute Idea,

---

[68] Mander notes that he is 'the most dedicated Hegel scholar' amongst the British idealists. Mander, *British Idealism*, 47.

[69] John McTaggart Ellis McTaggart, *Studies in the Hegelian Dialectic* (Cambridge: Cambridge University Press, 1896), 159. (Further references are given in the text as 'HD'.)

which must therefore be given prior to its 'development' in time. Otherwise, the contradictions that the system synthesizes would be real contradictions. Hegel *tried* to see the temporal world as if it was perfect, timeless, and changeless, which is not possible—but this is also evidence for the fact that what he was really after was the view that the eternal, timeless and changeless, constituted the real nature of the dialectic (HD, 170). 'We have thus arrived at the conclusion that the dialectic is not for Hegel a process in time, but that the Absolute Idea must be looked on as eternally realised' (HD, 172).

Yet, if the universe is completely rational, it doesn't look like it to us, now. And even if our view is actually untrue, failure to perceive perfection is itself an imperfection and therefore, in a sense, self-fulfilling. McTaggart nevertheless affirms that 'The universe is eternally the same, and eternally perfect. The movement is only in our minds' (HD, 173). But such an illusory movement is not the same as the irrationality of a false thought. It is simply a limitation, not a perversion, of knowledge. However, since Hegel insists that 'reality is only completely rational in so far as it is conscious of its own rationality', Hegel once more undercuts his own essential logic (HD, 175). The point, McTaggart says, is precisely the distortion introduced by attempting to view the whole from within a restrictive time-series. The imperfection of our minds is that we don't see everything '*at once*' but as succession, not *sub specie aeternitatis* but *sub specie temporis*. But where Hegel sees time as a *consequence* of our finitude, McTaggart sees it as *causing* epistemological imperfection. Of course, we cannot see other than in terms of our time-experience, and so we cannot see the ultimate perfection of the universe or the failure of evil, since these are not apparent *now*. But we are capable of seeing that it is not incoherent to understand the universe as a whole as really being timelessly perfect—and, in fact, the view that the universe is imperfect is just as much a matter of judgement and just as little a matter of immediate perception as the view that it is perfect. We cannot yet demonstrate ultimate perfection, but we can argue for its conceivability: 'We cannot expect to see how all process should only be an element in a timeless reality, so long as we can only think of the timeless reality by means of a process. But *sub specie aeternitatis*, it might be that the difficulty would vanish' (HD, 195).

In explicit association with Hegel (for whom, he says, 'all existence is really timeless' is a 'fundamental doctrine'[70]), McTaggart will go on to offer a more technical argument for the unreality of time. This is his theory of the three series, A, B, and C, of time.[71]

---

[70] John McTaggart Ellis McTaggart, *Philosophical Studies*, ed. S. V. Keeling (London: Edward Arnold, 1934), 139–40. This is from a chapter, 'The Relation of Time and Eternity' that is a reprint of an article in *Mind* 18/ 71 (July 1909), 343–62.

[71] This is again reprinted in *Philosophical Studies* from an article in on 'The Unreality of Time' in *Mind* 17/68 (October 1908), 457–74.

The A and B series are, respectively, the series constituted by past, present, and future (A series) and by the distinction earlier and later (B series). Now this latter series, though also temporal, is differently and, we may say, more weakly temporal than the A series. For if I say, 'I saw him after I left you' it will always be subsequently true that I saw him after I left you and I left you before I saw him (whereas, if I say 'I'm going to see him tomorrow, after our meeting today' the event of our meeting today will, in a week's time, be 'last week', and the 'tomorrow' in which I was to meet him will similarly be consigned to the past). Yet that there is an A series is essential to time as we know it. If we could have the B series without the A series, then it would be possible to conceive of change that did not involve time. Furthermore, McTaggart envisages a possible C series in which there is order but not change and, specifically, order of a kind that allows variability in its direction (against the fixed directionality of time). Thus, we can envisage a certain ordered relation between the numbers 17, 21, and 26 such that the sequence 21-17-26 is impermissible but 17-21-26 and 26-21-17 are both equally permissible and equally manifest the fundamental order of the relationship.[72]

Standard ideas of time involve both A and C series. The challenge, then, is to prove that the A series doesn't exist. Applying an implicit criterion of consistency akin to that of Bradley, the key to McTaggart's case is that the arguments in favour of the reality of the A series are mutually contradictory and amount to a vicious circle. The A series 'assumes the existence of time in order to account for the way in which moments are past, present, and future'. In other words, we can only understand what we mean by past, present, and future if we understand what it means to say the present is present, the past was present, and the future will be present. Yet, conversely, the A series is itself required in order to account for time—there is no time unless there is past, present, and future, since a B series of mere before/after is not enough to account for time as we know it.[73] The only alternative would be a 'vicious infinite series' in which we posited a second A series to account for our conventional A series, and so on. But if the A series is contradictory in itself, it cannot, on the principle of non-contradiction, be true of reality.[74] The reasons why we hold to the A series are therefore grounded in a perceptual illusion that arises from the confusion between 'perceptions and anticipations or memories of perceptions'.[75]

In connection with his demonstration of the unreality of time, McTaggart proceeds to argue for the possibility of timeless existence and for the possibility of a timeless being. The 'eternal present' that is much referenced in religious literature is no more than a metaphor, albeit an appealing one that emphasizes the constancy and reality of such a being, as well as its power of causation

---

[72] The letters of the alphabet provide McTaggart with another example.
[73] McTaggart, *Philosophical Studies*, 124.   [74] McTaggart, *Philosophical Studies*, 126.
[75] McTaggart, *Philosophical Studies*, 128. The argument is developed over pp. 128–30.

(since causation occurs only ever in the present).[76] However, the identification of the eternal with the future is more than a metaphor. Events in time happen in a certain order and this order 'is determined by the adequacy with which the states represent the eternal reality, so that those states come next-together which only vary infinitesimally in the degree of their adequacy, and that the whole of the time-series shows a steady process of change of adequacy'.[77] Thus, by analogy with the mathematical postulate of parallel lines that meet in infinity, we can regard the Eternal 'as the final stage in the time-process...as being in the future, and as being the end of the future. Time runs up to Eternity, and ceases in Eternity'. 'Eternity,' he says, 'is as future as anything can be'.[78] We may wonder how 'Hegelian' this still is, although, as we have seen, McTaggart is not averse to correcting Hegel for the sake of what he sees as the essential thought of the system. And here, it is clear, the Absolute Idea is not simply a regulative concept introduced to explain temporal becoming but the timelessness of the Absolute Idea qua eternal is itself what essential. Furthermore, the light of this essential Idea enables us to see that there is a preponderance of good in the overall composition of the temporal universe, despite the illusion of evil's reality now. Seeing the Eternal as future is practically relevant in this regard since, McTaggart suggests, past evil 'does not sadden us like future evil' and if we envisage the future as containing a diminishing quantum of evil, then we will be all the more empowered to strive against it. Thus time becomes 'the process by which we reach the Eternal and its perfection'.[79]

McTaggart believes that he has shown the unreality of time à la A series and the conceivability, at least, of a timeless existence. Going beyond Bradley, McTaggart thus thinks one of the possibilities of the Hegelian approach to the relationship between time and the eternal through to its logical conclusion, namely, the possibility that theoretical and timeless contemplation of unchanging truth will ultimately supervene upon the processes of time and change.

Nevertheless, McTaggart himself acknowledges that his arguments fall short of entire conclusiveness and, like Bradley, appears to admit that as long as we live in time we will not be able to dispense with the illusion of time. But (in a move that will also be reflected in the further development of this present study), both also indicate that, against Hegel's own emphasis on recollection, the aspect of time that is decisive for the possibility of coming to know time *sub specie aeternitatis* is the future. But that also means that we do not

---

[76] McTaggart, *Philosophical Studies*, 136.

[77] McTaggart, *Philosophical Studies*, 144–5.

[78] McTaggart, *Philosophical Studies*, 146. Cf. Kierkegaard: 'The future is the incognito in which the eternal, as incommensurable with time, nevertheless wills to preserve its association with time' (SKS4, 392/CA, 89). McTaggart also notes that the same logic underwrites popular Christian talk of heaven as both timeless and future (*Philosophical Studies*, 149–50).

[79] McTaggart, *Philosophical Studies*, 154–5.

and cannot know time 'eternally'—*yet*. A theoretic contemplation of timeless eternity may be the British Idealists' philosophical goal, but the conditions that would make it possible are lacking to temporal beings such as we are. In the next chapter, I shall examine an alternative route towards a similar goal proposed by a number of other nineteenth-century philosophers, some idealist, some anti-idealist. This alternative sees the will as able to make good the lack we have seen in the Hegelians. First, however, we shall briefly consider a literary approach to the question that offers further illumination of the key religious issues raised by the Hegelian approach.

The Remembrance of Things Past

In the period in which Hegelianism was a major intellectual force, its influence extended far beyond academic philosophy. In broad terms this is connected with the way in which it seemed to assuage nineteenth-century anxieties about the contradictions between the historical relativization of all aspects of human life and religious values focused on claims to timeless, eternal truths. It may therefore be helpful to supplement the arguments of the philosophers with the testimony of a novelist, since both in its subject-matter *and arguably also in its essential form* the modern novel developed as a way of negotiating this same tension, and doing so with an attention to the nuance and detail in which those heights and depths are lived in the concreteness of day to day and flesh and blood that goes beyond what any philosophy is able (or wants) to offer. Philosophy may speak of the limits of knowledge and experience, but the novel might more plausibly hope to show how those limits are lived in terms of despair and hope and, in its way, perform the ever-renewed struggle against loss of memory and abandonment of hope.

Marcel Proust's *In Search of Lost Time* is probably the best-known work of modern literature to take the struggle with time and with the loss that time incessantly brings about as its defining theme. As is well-known, even by those who have not read far into Proust's cycle of novels, it all seems to begin with something as trivial as dipping a biscuit into a cup of tea, from which we are led into a narrative that not only embraces the narrator's own life and loves but also opens out onto a panorama of social events that include the Dreyfus affair and the First World War. That it is also an exploration of what can loosely be called the metaphysics of time-experience becomes explicit when, in the last volume, the narrator finally resolves his crisis of literary vocation. The coincidence of two experiences of taste separated by many years suddenly and unexpectedly releases the narrator into a state that he describes as both transcendence of time and liberation from death. It is the revelation—or 'rebirth'—of a self that lives from the very 'essence of things'. Under the conditions of average everyday life, this self

languishes in observing the present, in which the senses are unable to communicate [this essence] to him, or in consideration of a past that has been desiccated by the intellect, or in awaiting a future that has been constructed by the
will out of fragments of past and present from which it has removed their reality
and retained only what fits with the narrowly human utilitarian goal that it has
assigned to them. But that a noise, a scent, already once heard or smelt may be
heard and smelt anew, being at once in the present and in the past, real without
being actual, idea without being abstract, at one and the same time the permanent essence of things yet hidden by them—this finds itself liberated and our true
'me', which sometimes long seemed dead but was not entirely so, awakens and
is brought to life by receiving the celestial nourishment brought it. One minute
outside the order of time has recreated in us (and in such a way that we might feel
it) the person who is outside the order of time. And, with that, one understands
how such a one would be assured in his joy, even if the simple taste of a madeleine
doesn't seem logically to contain sufficient reason for such joy; and one understands how the word 'death' has no meaning for such a one: being situated beyond
time, how might he fear the future?[80]

A few pages later, Proust will refer to 'resurrections of memory',[81] making explicit the implied symbolism of the madeleine biscuit, not accidentally
named from Mary Magdalene, first witness of Christ's Resurrection. He then
goes on to draw out the ultimate ambition of his literary work, namely, to
decipher the 'inner book' that contains 'the Last Judgement' on human life.
This, as he emphasizes, makes art 'the most austere school of life', since it is not
a matter of literary fantasy but of painful attention to reality itself. 'This book,
the most painful of all to decipher, is also the only one that reality itself has
dictated to us, the only one of which the "imprint" has been made in us by reality itself'.[82] The task of the writer, we may say, approaches that of the phenomenologist. But there is more. The materials of literature, the narrator realizes,
are nothing but the events of his own past life, deposited in the storehouse of
memory for purposes that he only now retrospectively discovers. And now, in
this time of discovery that is also a liberation from death, he can make his own
the words of John 12:24, that 'unless a grain of wheat falls into the ground and
dies, it remains but a single grain; but if it dies it bears much fruit'. Bringing
forth the fruit of the novel is the resurrection to which his life has been secretly
and, for a long time, unknowingly devoted. This, Proust says, is what makes
his life 'a vocation'.[83]

We might think of Hegel's Owl of Minerva beginning her flight at dusk.[84]
For Hegel's eternity, like Proust's, seems to be arrived at through a process

---

[80] Marcel Proust, *Le temps retrouvé* (Paris: Gallimard, 1990), 179.

[81] Proust, *Le temps retrouvé*, 184.   [82] Proust, *Le temps retrouvé*, 186.

[83] Proust, *Le temps retrouvé*, 206.

[84] Several commentators have glimpsed analogies between Proust and Hegel, though not necessarily the same analogies as are suggested by the present focus on the question of eternity. For
example, see Henry Sussman, *The Hegelian Aftermath; Readings in Hegel, Kierkegaard, Freud,*

of recollection. Yet if both great works of retrospection end by elevating the writer above time to a view of life *sub specie aeternitatis*, Proust's vision seems to offer more. Where Hegel aimed but—in the view of many commentators— failed finally to integrate the whole concreteness of lived life into this vision, it is precisely *life* and not life 'desiccated by the intellect' that Proust offers. As such it is much more than the philosopher's 'grey on grey'. Rather, it is a res- urrected world of richly variegated colour, light and shade—a world that has been resurrected into life, and not just into thought.

Proust's work stands on an eminence of greatness that puts it above any mean-spirited criticism, but this does not put it beyond philosophical ques- tioning. Allowing that Proust's literary phenomenology presses more deeply into the folds and cross-currents of lives lived in time than Hegel's panoramic 'world-historical' vistas, we may still question whether his re-presentation of a lived life merits the claim that it is a resurrection or a liberation from both time and death? And if it is, in some sense, a 'resurrection' is it not perhaps a Lazarus-like resurrection, that is, a resurrection back into a life that the relentless forward drive of time will never cease to carry towards death? Is it—can it be—more than a reprieve, a refuge (though what a refuge!) in the storm of time? Is this temporal immortality the best we can do or can we glimpse beyond it the possibility of what we might call an eternal immortal- ity? In a sense, these questions are questions not only to Proust but to Hegel and, more generally, to the Idealist tradition as such. Can the timelessness at which it aims ever be found when we have once conceded some measure of reality to the world of time? To pursue these questions further and to track the further evolution of Idealist thought in relation to an ever-growing awareness of the claims of time, we now turn to Hegel's sometime friend and collaborator, F. J. W. Schelling, who, like Hegel, draws significantly on Jacob Boehme, but who turns away from theoretic contemplation to will as the primary mode in which time and eternity are to be united and, in doing so, anticipates the more radical contestation of Idealism that we shall find in Nietzsche.

Proust and James (Baltimore, MD: Johns Hopkins University Press, 1982); Julia Peters, 'Proust's *Recherche* and Hegelian Teleology' in *Inquiry* 53/2 (April 2010), 141–61. Cf. also Michael Theunissen, *Negative Theologie der Zeit* (Frankfurt am Main: Suhrkamp, 1991), 62–3.

3

Time, Eternally: Will

Despite significant differences in the approaches we considered in the last chapter, the thinkers and positions we examined there were seemingly united in understanding the relationship between time and eternity as essentially knowable. In other words, they continued a long tradition of philosophy that envisages theory or contemplation of eternal truth as the ultimate object of philosophical enquiry. Those we shall now consider also seek to think together the unity of time and eternity. Here, however, the primary focus is no longer on knowledge but on *will*. Hegelian 'Spirit' admittedly incorporated a significant volitional dimension, yet this was seen to be ultimately subordinated to theoretic contemplation. Schelling, however, significantly tips the balance in favour of will, whilst in Nietzsche theoretic activity is itself reconceived as an instrument amongst others in the service of will to power.

UR-TIME AND ETERNITY

Schelling's philosophy has many points of contact with each of Spinoza, Boehme, and Hegel and his philosophical career passed through many stages. Much of his later work is generally regarded—even by those favourable to other versions of German Idealism—as exceptionally obscure and rambling. Kierkegaard, who attended his inaugural series of lectures in Berlin in 1841–2 wrote home that 'Schelling talks endless nonsense both in an extensive and an intensive sense'.[1] Many subsequent commentators have reached a similar view—yet, at the same time (and Kierkegaard himself could be taken as implicitly testifying to this[2]), there is something in these later writings that has stimulated a series of creative

[1] SKS28, Letter 86/LD, Letter 69.
[2] On Schelling's influence Kierkegaard see Lore Hühn and Philipp Schwab, 'Kierkegaard and German Idealism', in John Lippitt and George Pattison (eds), *The Oxford Handbook of Kierkegaard* (Oxford: Oxford University Press, 2013), 64–92.

reactions in both philosophy and theology.[3] Especially influential is the extended essay *Philosophische Untersuchungen über das Wesen der menschlichen Freiheit* (Philosophical investigations on the essence of human freedom), and if his subsequent 'philosophy of mythology' developed further perspectives that go beyond this work it is still widely taken as the most succinct statement of what his later philosophy is about. In relation to the present discussion it is of particular interest both as containing an explicit response to Spinoza and as revealing a significant impulse from Boehme.

From the beginning of his 'investigations' Schelling is in no doubt as to the importance of the topic, emphasizing its centrality to the idea of a philosophical system, which, he says, must be able to explain the contradiction of freedom and necessity. In the aftermath of debates that saw Spinozism branded as 'pantheism' (or simply 'atheism') Spinoza becomes a natural point of reference in relation to the claim that the only possible system of reason is one in which God and world are conflated and human freedom is subordinated to fatalism. Indeed, if we do once allow absolute causality to God, then it is hard to see how there could be any other free being, unless its freedom is somehow 'in' God and its free activity is essentially an aspect or expression of God's own life. But then 'it'—let us say the human being—is no longer an ontologically distinct entity. However, where pantheism is said to confuse God with 'things', Schelling says that, in fact, no one distinguished between God and things more than Spinoza: 'God is that which is in itself and can only be conceived in relation to itself; what is finite, however, is necessarily in another and can therefore only be conceived by reference to this'. Or, which is saying the same thing, God is eternal by nature, things exist only derivatively.[4] But this would seem to confirm the view that 'things' which can only be understood by reference to God have no independent being. Yet, Schelling argues, that A is manifest in A/a does not entail that A/a = A, i.e. is entirely reducible to A. In a relation of antecedent and consequent such as this, the predicate extends and makes explicit what is implicitly given in the subject (but *only* implicitly). Referring to a saying by Leibniz, Schelling notes that the son of a man is no less a man for being 'derived'. But although this shows that Schelling is keen in principle to affirm the distinct (if not independent) reality of things,[5] it falls short of

---

[3] Examples include V. S. Soloviev and subsequent thinkers in the tradition of Russian religious philosophy; Paul Tillich, who wrote both his doctorate and his *Habilitation* on the later Schelling; and Heidegger, who too returned to the later Schelling on a number of occasions from the mid-1930s onwards in the context of his debate with German Idealism.

[4] F. W. J. Schelling, *Philosophische Untersuchungen über das Wesen der menschlichen Freiheit* (Frankfurt am Main: Suhrkamp, 1975), 37. (Further references are given in the text as 'F'.)

[5] Schelling's point here seems analogous to that made against Spinoza by Hegel, namely, that Spinoza does not give grounds for believing in the ontological independence of concrete individual beings. However, it should be noted that Schelling's text (1809) pre-dates the lectures on the history of philosophy referred to above in this connection.

giving an adequate account of what things are or could be for themselves. But, crucially, the key question (in Schelling's view) is not about just any kind of 'things': it is about entities that might be essentially characterized in terms of freedom, that is, human beings. How then does his argument develop with regard to human freedom?

By virtue of its essence, the eternal (Schelling says) is the ground of the temporal and the latter is therefore dependent on it; but this does not exclude self-affirmation or freedom. God's creativity is precisely the power to give being to what is external to him. Indeed, since 'God can only become revealed in that which is like him, in free beings that act by themselves' the relation-ship between creative and created freedom is so far from being a zero-sum game that divine freedom is what makes free creaturely being possible, whilst free creaturely being is precisely the revelation of God as essentially free (F, 42). Freedom is so integral to God that eternity itself has to be conceived as 'self-supporting will [or] freedom' (F, 42–3). Once this is grasped, then it becomes clear that the problem with Spinoza is not that he places 'things' in God but that he thinks of them as *things*. In fact, even the will itself is treated as '*eine Sache*', a 'thing' (F, 44). The issue is not pantheism but determinism, and Spinoza's system is a Pygmalion's statue, needing to have life breathed into it (F, 45). If this could be done, his system would then be transformed into a genuine system of reason 'by means of which the entirety of nature is transfigured in sensibility, intellect, and finally will'. Schelling adds: 'In the last and highest instance there is no other being than willing. Willing is Ur-being and to this alone can all predicates of the same be ascribed: not needing a ground, eternity, independence of time, self-affirmation. All of philosophy strives only to find this highest expression' (F, 46).

But if Fichte had already grasped freedom as the principle of a system of knowledge and explained the construction of the known world as a work of the human mind, Schelling wants to insist on a freedom still more basic than that which is manifest only in the order of knowing. Here the issue is not simply that of self and world and of how the self comes to represent and know itself and its relation to the world, since that fails to account for the duality most distinctive of freedom itself: the duality of good and evil. As Schelling tells the tale, good and evil are not merely two possible applications of a neutral will, but each expresses a definite kind of willing, a good will and an evil will. And, if all he has said about the interdependence of divine and human freedom is true, it follows that the ground of the evil will must be in God no less than the ground of the good will.

This leads Schelling to an exploration of the idea of there being a 'ground' in God himself, echoing aspects of Boehme's theogony. In God, he says, there is a primordial nature or 'eternally dark ground' (cf. Boehme's *Ungrund*[6]) such

---

[6] Though Schelling also uses the expression *Urgrund* ('primordial ground' rather than 'un-ground') and *Ungrund* seemingly synonymously (see F, 98), the metaphorical and even literal

that 'God has in himself an inner ground for his existence, which, in relation to his [actual] existence precedes him; but, equally, God is the prius of this ground in that the ground, as such, could not be if God did not actually exist' (F, 52). The circularity is not dissimilar from that which we have seen in relation to the eternity of substance and the subjectivity of existence in Hegel. If there is a difference, however, it is to do with Schelling's insistent emphasis on this 'nature' or 'ground' in God as 'dark' and, as such, lacking any kind of archetypal rational structure. Instead it is explained as a kind of blind urge: 'it is the longing felt by the eternal One to give birth to itself. It is not itself the One but is nevertheless co-eternal with it' (F, 54). It is not a conscious will, although it is a will that has a 'premonition' of understanding. If order and form are not primary, nevertheless this chaotic power that precedes them is susceptible of becoming endowed with order and form.[7] Remembering that we are still talking about the inner life of God and not yet of the creation of the world, the first birth produced out of this as-yet blind will is a reflexive likeness of God himself qua luminous actual existence—there is no other object for God to know than God—and this is the birth of the divine Word (F, 55).[8] In this Word the darkness is illuminated in what Schelling calls a '*Lebensblick*', a glance that reveals the totality of life; it is the dawn of divine imagination. 'Nature' will subsequently always have a twofold identity: as that which is separate from or other than God and, as such, governed by blind, unconscious urges, and nature as a sphere of universal laws. The synthesis of the two, although given primordially in the divine life, becomes knowable only in human life: 'In man is the deepest abyss and the highest heaven, or both centres' (F, 58). The emergence of self-conscious human life is thus itself a repetition and revelation (and perhaps—it is not really clear—the event itself) of the theogonic process, the

force of the terms is somewhat different. As the following citations suggest, what is primary for Schelling is that it is a ground beyond or behind which we cannot go, the point—metaphysically conceived—at which all explanations stop. In this sense the meaning of *Urgrund* seems to be primary; however, since this primordial ground has no further ground it might also be regarded as 'darkness' rather than as ultimately illuminating and as an *Un-grund*, a ground that is really no ground and for which no grounds can be given.

[7] In this regard, Schelling's position may suggest but is essentially distinct from that of Schopenhauer, whose blind will is ultimately neither truly expressed nor alleviated in the manifold of representation according to the structures of reason. All such representation is, in the end, illusion. For Schelling, however, despite the primacy of the dark chaotic ground, even in God himself, an ordered system of reason is a truthful expression and manifestation of all that is in God. 'Premonition' (*Ahnung*) was a key concept in early romanticism, used, for example, to speak of the hints or intimations of a divine or higher life to be found in the life of unconscious nature.

[8] The Johannine allusion is reinforced by Schelling's quoting from John 12:24, that 'unless a grain of wheat falls into the earth and dies, it remains just a single grain; but if it dies it bears much fruit'—the point being that to bear fruit as the divine Word even the divine life itself must first degrade into the unillumined darkness of earth to be reborn as it truly is. Cf. comments on the use of this same text by Proust (Chapter 2 of this volume) and Dostoevsky (Chapter 5 of this volume).

coming into existence of God. 'Human will is the germ of the divine life that is as yet present only as ground, hidden in eternal longing... Only in [human life] has God loved the world and it is precisely this image of God that the longing has grasped as its central focus when it emerges in its contradiction from the light' (F, 58).

The human being who thus emerges is a particular self whose freedom transcends the life of mere nature (whether this is considered under the rubric of blind will or of universal law) as well as making it independent of God. But although this gives the human being the possibility of being, more or less literally, the centre of the universe, the microcosm in whom the whole is reflected and unified, it also makes it possible for the self to constitute itself as its own centre. This self-constitution is then the primary instance of evil, which is therefore neither a privation of good (as for Leibniz) nor identical with sensuousness (as in Platonism) but an active possibility intrinsic to freedom—a statement which, for Schelling, is equally true of God and of human beings as created in the likeness of God. Existence is consequently and as such a sphere of conflict between evil and love, between the will to self-assertion and the will to harmonize and reconcile. Neither divine nor human life can ever exist other than in relation to their dark ground, so the lure to evil can never be eliminated. Intriguingly, Schelling interprets this lure in terms of an anxiety ('*Angst*') that he, like Kierkegaard, likens to vertigo, commenting that 'the anxiety of life drives human beings away from the centre in which they were created' that fuels the consuming fire that burns away egoism (F, 74).[9]

Whatever we make of this as philosophy, it is fairly easy to read it as a powerful narrative of each human being's emergence from infancy and childhood into the possibilities and conflicts of free adult life. But, in a remarkable development, Schelling suggests that whatever the course of an individual life, the ultimate choice between good and evil that is played out in time is already determined by a kind of transcendent choice that each human being makes for him- or herself. In the primordial creation, Schelling says, the human being has not yet decided for good or evil, that is, we are not created as having-chosen but as having to choose. However, 'this choice cannot occur in time' and 'the act by which [human] life in time is defined does not itself belong to time but to eternity; [however] it precedes neither life nor time but runs through time ([although] not grasped by it) as an action that is, by nature, eternal' (F, 78). Each life involves an original choice that is both distinctive and determinative

---

[9] On similarities and differences between Schelling's and Kierkegaard's views of anxiety and its relation to the Fall, see, e.g. Michelle Kosch, *Freedom and Reason in Kant, Schelling, and Kierkegaard* (Oxford: Oxford University Press, 2006). See also the discussion in my *Kierkegaard and the Theology of the Nineteenth Century* (Cambridge: Cambridge University Press, 2013), 133–6. This is connected with the further question—that arises also for the Boehmian theosophy—as to whether what comes into existence is necessarily always fallen, so that creation and fall coincide.

for that life as a whole, a view that Schelling sees as reflected in the feeling that everyone has that they 'have been from all eternity and have not at all become what they are only in time' (F, 78). This, he goes on to suggest, is also why some children manifest an inclination to evil before they arrive at the age of reason. This reveals the existence of a choice that is 'before this life', albeit not 'before' in a straightforwardly linear and temporal way but in the 'prior' life of the intelligible world. A man is born as he eternally is, Schelling says (F, 79). Whilst rejecting the kind of Augustinian or Calvinist view of predestination that sees each human life as determined from before the foundation of the world by God's inscrutable counsels, Schelling can therefore comment that 'We too assert predestination, but in a quite other sense, namely, that human beings act here as they have acted from eternity and already in the beginning of creation' (F, 80).[10] This also lays the groundwork for the view that true freedom is not incompatible with and may be regarded as a 'holy necessity' in which 'spirit and the heart freely affirm what is necessary, bound only by their own law', which is also a binding of the dark principle to the light, a binding that Schelling sees as the core meaning of *re-ligio* (F, 84). Evil is therefore a necessary concomitant of there being free, personal life at all, human or divine. Good could not simply prevail without struggle from the beginning, since free, personal life is formed precisely by the struggle of good with evil. So God too, like all life, 'has a destiny and is subject to suffering and becoming' and it is only by so doing that he, who eternally is, becomes actual (F, 95). God is God by virtue of or in the act in which he eternally overcomes both the darkness of his own eternal nature and the lure to become an isolated individual.

Just as, in his philosophical beginnings, Schelling sought to show how nature manifested the freedom of the divine idea (appealing then to the instantaneousness of intuitive vision), so now he evinces a particular interest in showing the pre-history, as it were, of human freedom, that is, the stages intervening between a state of primordial nature and the emergence, in philosophy and biblical religion, of fully self-conscious and self-reflecting humanity. Thus, the conclusion of these short 'investigations' becomes the starting-point for the voluminous studies of revelation and mythology that will provide the focus for much of the remainder of his philosophical career. It will not be to our

---

[10] Schelling does not discuss the consequences of this view for, e.g. the administration of justice or the therapeutic rehabilitation of offenders. However, he will add that no human being is ever without some relation to the good principle, so that repentance is possible (F, 81). So a person we meet at, say, age 18 may seem to be given entirely over to evil: but were we able, sixty years later, to see their life as a whole we might see that this commitment to evil was merely a temporary phase, ultimately overwhelmed by a commitment to the good that only appeared in the subsequent passage of time. By definition, the totality of the manifestation of the eternal choice is not identical with its initial manifestation. We are therefore justified in maintaining an attitude of optimism towards our fellow human beings (and ourselves) even when it is an optimism despite appearances.

purposes here to go far into these later developments, but it is worth looking briefly at Book 1, 'The Past' of *The Ages of the World*, where, if he does not exactly clarify the relationship between time and eternity deployed in the treatise on human freedom, he does give more focused attention to it.

Here, even more profusely than in the earlier treatise, Schelling makes abundant use of the prefix *Ur-*, signalling at once that the enquiry is into what is most primordial. The prefix has a twofold sense, since it concerns both what is earliest or most ancient and also what has ontological priority. As we have already seen, and as is also the case for Hegel, this involves Schelling in asserting *both* that the working out of the divine life as process and as manifesting in nature and history is temporal *and* that God, as eternally supra-temporal, is always already what he becomes through history. However (as we have been seeing), there is a significant difference between Hegel and Schelling with regard to the respective weight each gives to reason and freedom or, to be more precise, to theoretical, contemplative reason and to reason as shaped by an original, *pre-rational* (but not, it is claimed, *irrational*) will to selfhood and manifestation.

So, in *The Ages of the World*, Schelling distinguishes the past as what is known (*gewußt*), the present what is cognized (*erkannt*), and the future what is divined (*geahndet*). These distinctions being made, the task he assigns to Book 1 is to trace consciousness back to its earliest beginnings. This will not itself yield what is above time, but the ability to investigate 'the beginning of times' is possible only because we have a relation to what is 'prior to the beginning of time'.[11] We cannot access this supra-temporal principle directly, however, since it is precisely bound to 'another' that is 'by nature unknowing and dark'. However, this unknowing, primordial darkness is not entire emptiness, since, Schelling says, it itself contains an *Ur-bild*, a primordial image, of all things—although, in this initial state, this image too is as if dark and forgotten.[12] Yet, at the same time, this 'dark' image provides the material in which the eternal blessedness of the divine light can become visible and, in this way, understandable to itself. It is this 'secret traffic' with the hidden and forgotten *Ur-bild* that is 'philosophy's secret' and in relation to this dialectic is only a copy.[13] Dialectic is necessary at a certain level of science, but dialectics needs to be underwritten by the 'recollection of the primordial beginning [*Urbeginn*] of things', a beginning that can only be narrated (i.e. in myth) and not deduced or constructed.[14] In fact, Schelling attempts to push the story back still further, to 'the history of the *Urwesen* [the primordial essence]' in 'the time before the world'. 'No tale

---

[11] F. W. J. Schelling, 'Die Weltalter', in *Ausgewählte Schriften*, IV: 1807–1834 (Frankfurt am Main: Suhrkamp, 1985), 216.

[12] In the case of Boehme and several of the Russian religious philosophers this is the jumping-off point for developing the idea of the divine Sophia, figuratively represented as the female companion of the deity.

[13] Schelling, 'Die Weltalter', 216.   [14] Schelling, 'Die Weltalter', 220.

sounds forth from that time, since it is the time of silence and stillness. Only in the discourses of divine revelation is there the light from some far flashes that tears open these primordially ancient [*uralt*] darknesses'.[15]

Investigating these flashes may seem an extraordinarily challenging business, but Schelling sees it as essential to any philosophical system that is going to take time seriously and constitute itself as a 'system of time', albeit only 'in its broadest contours'.[16] A mechanical system that explains everything in terms of causal laws has neither past nor future in the full sense, nor is time thought properly when it is thought as merely repeating what is already eternally complete (a view that could be ascribed to Hegel or, at least, to Bradley's and McTaggart's variations on Hegel). So what is it that puts us in a position to see time for what and as it is (to the very limited extent that we are able to do so)? It is, Schelling says, our own participation in a principle that is above or beyond time, and this is, precisely, freedom. Freedom is above even being (*Seyn*). But whilst most people are inclined to think that nothing is above or more fundamental than being, 'Above Seyn dwells... the true, eternal freedom. Freedom is the affirmative concept of eternity or that which is above all time'.[17] As such it is, in a sense, '*nichts Seyendes*', nothing in being. Up to this point, knowledge is indeed recollection, but recollection can recover only what belongs to the past, and this is an eternity that can never become past.

I shall not now follow the details of Schelling's narration of the origin of time any further, other than to note that his account hinges on the repeated reiteration of a relationship between an eternity that is entire and perfect in itself, that 'is eternally eternity',[18] and a will that is both related to eternity yet distinct from it and that 'first posits the possibility of a time'.[19] Schelling can even speak of two eternal wills, the one that is expressed in infinite self-affirmation and the one that, in positing time, negates this first will. Eternity does not then become the past, but in the eternal will that posits itself as distinct and individual, the eternal 'is there' ('*ist... da*'), i.e. comes into existence. *Seyn* is the manifestation of this second will, so that whereas we are used to thinking of *Seyn* as will-less and as the simple, timeless accompaniment of essence,[20] *Seyn* is in fact and always already a limitation (or as Schelling says 'contraction') in relation to the infinite openness of eternally eternal love. Again we might think back to Spinoza and to the existence of particular things as negative determinations of the one divine substance. However, the difference now is that Schelling rewrites this in terms of dynamic will. In its deepest (most *ur-*) manifestations, *Seyn* is the product

---

[15] Schelling, 'Die Weltalter', 222.   [16] Schelling, 'Die Weltalter', 226.
[17] Schelling, 'Die Weltalter', 226.   [18] Schelling, 'Die Weltalter', 230.
[19] Schelling, 'Die Weltalter', 230.
[20] Schelling doesn't, but might have, referred to the medieval conception of God as *Ipsum Esse*, the being whose being is, simply, to be 'without change or shadow of turning'.

of will and, as such, of an as yet unilluminated urge to come into existence.[21] Schelling likens the life of God prior to the manifestation of existence to that of a child immersed in 'unrestricted delight, entirely full of itself and thinking of nothing, quiet inwardness, rejoicing at its lack of *Seyn* . . . its essence is nothing but grace ('*Huld*'), love, and simplicity. It is what is truly human in humanity, the God in the Godhead'.[22] By way of contrast, its first manifestations in the realm of *Seyn* can be discerned in the archaic, pre-philosophical mythologies that speak of 'Darkness and concealment [as] characteristic of this primordial time (*Urzeit*). It is in the night that all life first comes to be and takes shape. That is why the ancients called this the fearful mother of things and, next to chaos, the oldest of beings (*Wesen*)'.[23]

The next stage is when the play of these two wills generates a sequence of *Ur*-images that reveal the life of this *Ur-Wesen*. These are not, as yet, abstractions, but appear as a glance (*Blick*) illuminating the darkness and as the revelation of a face (*Gesicht*). They are 'ideas' in the *Ur*-meaning of the word, which, Schelling says, is 'the same as the German word "face"', meaning sight (*Ge-sicht*). In this primordial life of nature nothing comes into existence other than as physical and material, which is the necessary presupposition of all subsequent spiritualization and which is also, at the same time, the 'primordially original (*uranfängliche*) eternal nature of God'.[24] But it is all of this, in God, not as actual, but only as possible, qua potency.[25] It is 'nature' in God as the germ or seed of all the nature that God might, in freedom, create and although it is prior to the actual creation it therefore grounds nature in God and, to Schelling's mind, this incorporates the Spinozan God-or-nature into the dynamic God of biblical revelation.

This may be far from yielding clarity, and those who like theosophical speculation might prefer to stay with the shoemaker from Görlitz and not have esoteric visions forced to do philosophical work. On the other hand, as I have already suggested, it may be read as expressive of considerable psychological truth, particularly in relation to the experience of the emergence of self-conscious life in childhood. And, as Schelling himself sets out to show, the project—if not the achievement—seems in some respects compelling, namely, to see whether the forms of philosophical thought can be traced back

---

[21] Subsequent to his own lectures on Schelling, Heidegger adopted the archaic spelling of *Seyn* for *Sein*. This brief summary of Schelling's own usage might suggest a reason for this, namely, that Heidegger too is seeking to reconceptualize being in such a way as to make time fundamental to it.

[22] Schelling, 'Die Weltalter', 247.   [23] Schelling, 'Die Weltalter', 236.

[24] Schelling, 'Die Weltalter', 248.

[25] Schelling, 'Die Weltalter', 255–6. Perhaps confusingly, Schelling uses both *Möglichkeit* (possibility) and *Kraft*, force or might, both of which can be used to translate the Aristotelian *dynamis* (cf. Latin *potentia*), as well as his own *Potenz*, potency—another possible translation for the same term.

to pre-philosophical myths and narratives of human life. But what real work does it do in metaphysics?

Yet to complain that Schelling does not explain or justify the relation of time to eternity is to invite the retort that that is precisely the point: that it cannot be explained or justified, only narrated, as he says. Certainly, the claim that mythology provides the most appropriate means with which to develop such a narrative is interesting and has since, with or without reference to Schelling, been used with not dissimilar aims by a number of disciplines, including psychoanalysis and anthropology. Some recent theologies have also argued for an irreducibly narrative element in religious truth, even if their narratives are less luxuriant than Schelling's. And, analytically, it seems that Schelling is not unaware of the problem that all of the positions we have considered thus far run up against, namely, that if we suppose a dimension of (divine) being that is properly said to be above or beyond time and also allow the reality of the temporal world, it seems impossible to relate the one to the other, except by the introduction of some *tertium quid*, but this (as we have seen) merely pushes the problem back a step.[26] As regards the Kierkegaardian point that any claim to view human life in time *sub specie aeternitatis* inevitably abstracts from the immediate, concrete and pressing concerns of actually existing human beings,[27] Schelling's failure is perhaps not worse than that of other philosophers, even if, critics might say, he takes longer to fail and is not himself clear as to what he does or doesn't eventually achieve. But even if we do, in the end, regard Schelling's project as a failure, he has introduced a new and arguably important emphasis into the line of thinking about time and eternity running from Spinoza to German Idealism. For if Hegel essentially repeated Spinoza's attempt to achieve a timeless theoretical contemplation of human life *sub specie aeternitatis*, albeit giving greater weight to the actual experience of temporal becoming and concrete individual existence, the thrust of Schelling's approach is that neither in the divine life nor in human beings' apprehension of that life is theoretical contemplation ultimate. That the world rises to manifestation and reveals itself as *Gesicht* is possible only on the basis of a more primal urge that is, in a strong sense, simply groundless and, as such, is not available as the object of theoretical contemplation. The will to existence and to revelation precedes knowledge of existence and of what is revealed.[28] Even in relation to eternity, then, and prior (in a now familiarly ambiguous sense) to the possibility of any representation or idea of eternity, knowledge presupposes some kind

[26] Cf. the discussion of Stump and Kretzmann's ET-reference frames or Iamblichus' theory of two kinds of interacting 'Nows' in the section 'Can a Timeless God Know Temporal Beings' in Chapter 1 of this volume.

[27] See the section 'God is in Heaven, We are on Earth' in Chapter 4 of this volume for discussion of this point.

[28] This relates to what Tillich called Schelling's 'a priori empiricism'. See Paul Tillich, *Theology of Culture*, ed. Robert Kimball (New York: Oxford University Press, 1959), 86ff.

of experiential acquaintance that is not (as in empiricist models of knowledge by acquaintance) simply a matter of passively encountering a given object but arises from a basic orientation of will and personality. To know eternity we must want it—and want it in the sense of the kind of longing or desire that religious traditions have long ascribed to the God-seeking soul.

Many commentators have seen Schelling's later philosophy as the culmination and transformation of the basic impulse of German Idealism. From the point at which we have now arrived, we can see why this might be so. Schelling's philosophical ambitions are rooted in the idealist search for a unified system of knowledge that would be adequate both to our knowledge of nature and our experience of freedom. Yet he also shows how the exigencies of freedom will, in the end, undermine the aim of systematic unity. The temporal experience of freedom provides the frame of reference in which and in which alone eternity becomes humanly meaningful. But if human freedom can be spoken of as an act of divine eternity manifest in, with, and under the conditions of existence, this must have a retroactive effect on the concept of divine eternity itself, which, in some sense, now comes to depend on human freedom. With this inversion we seem to be leaving the specific philosophical world of German Idealism behind and finding ourselves on a path that leads to a very different kind of philosophy, namely, the philosophy of Nietzsche— although, as we shall have occasion to see, there are also reasons to question whether, in the end, it is so very different. Certainly, like the late philosophy of Schelling,[29] Nietzsche's is a philosophy that finds its most expressive statement in mythology—only this time, not the reinterpreted mythology of the ancient world but the new mythology of Nietzsche's own fictional prophet and teacher, Zarathustra.[30]

ETERNAL RECURRENCE

That Nietzsche taught a doctrine of the eternal recurrence of all things is probably amongst the half-dozen things that all first-year students know about him.

---

[29] The link between Nietzsche and Schelling was noted by, amongst others, Karl Löwith. See his *Nietzsches Philosophie des Ewigen Wiederkehr des Gleichen* (Stuttgart: Kohlhammer, 1956), 191, where he suggests—rightly, I think—that whereas Schelling thought in an essentially 'speculative-theogonic' mode, Nietzsche's thought is historically positioned by coming after the advent of scientific positivism. Heidegger too seems to have seen a similar connection in notes for his 1941 lectures on German Idealism. See M. Heidegger, *Gesamtausgabe*, xlix: *Die Metaphysik des deutschen Idealismus* (Frankfurt am Main: Klostermann, 2006).

[30] Paradoxically, it is a mythology that seems precisely to invert the mythology (if we can call it that) of the prophetic, eschatological religion inaugurated by the historical Zarathustra. See Laurence Lampert, *Nietzsche's Teaching: An Interpretation of Thus Spoke Zarathustra* (New Haven, CT: Yale University Press, 1986), 257–8.

However, as we would expect in relation to 'knowledge' of this kind, it is neither clear that Nietzsche 'taught' it nor what he might have meant by it. But we should pause to note just how remarkable it is that he gave such extraordinary weight to the idea of *eternal* recurrence. For amongst the other things that all first-year students know about Nietzsche is that he attacked the whole idea of metaphysics and of any eternal order governing human affairs. Furthermore— and rapidly using up our half-dozen things—he surely also taught the utter and ineluctable temporality of all human experience, knowledge, and values. Yet, in the self-accounting of his authorship entitled *Ecce Homo*, he spoke of 'the thought of eternal recurrence' as constituting 'the basic conception' of *Thus Spoke Zarathustra*, saying that it is 'the highest formula for affirmation that could ever be attained', suggesting also that it is the fruit of a momentary inspiration in August 1881: '"6000 feet beyond human beings and time". I was walking that day by the lake from Silvaplana through the woods and stopped by a mighty block of stone not far from Surlei. This thought then came to me'.[31] But in what way is this momentary inspiration 'the basic conception' of *Thus Spake Zarathustra*, and thereby to be counted amongst the defining ideas of Nietzsche's mature thought?

The question has puzzled many of his commentators, and some seem inclined to write it off as, in Walter Kaufmann's words, a 'dubious doctrine, which was to have no influence to speak of'.[32] Few have given much time to arguing either that Nietzsche took it seriously as a contribution to scientific cosmology or, if he did, whether it works,[33] whilst Michael Tanner, having considered alternative explanations of the theory dismissively remarks that 'it does nothing for me either way'.[34] Some, however, do attempt to take the doctrine seriously and try to see why or how it could loom so large in Nietzsche's own view of his work, although, given what are generally acknowledged to be the allusive, often highly literary, and sometimes contradictory statements about eternal recurrence in Nietzsche's works, there is neither consensus nor prospect of consensus. Shestov, for example, saw it as important—but, at the same time, as indicative of a kind of loss of nerve on Nietzsche's part. Having conjured up the spectacle of a universe without goal, unity, or value, eternal recurrence became a means to restore stability to what would otherwise be an anarchic chaos. He comments that 'the important thing...is not the word

---

[31] Friedrich Nietzsche, *Ecce Homo*, in *Werke in drei Bänden III*, ed. K. Schlechta (Frankfurt am Main: Ullstein, 1969), 574. That the 'doctrine' of eternal recurrence translates a personal experience rather than being the outcome of a proper philosophical argument is important to some interpretations.

[32] Walter Kaufmann, *Nietzsche: Philosopher, Psychologist, Antichrist* (Princeton, NJ: Princeton University Press, 1968), 323.

[33] One who does is Arthur C. Danto in his *Nietzsche as Philosopher* (New York: Macmillan, 1965).

[34] Michael Tanner, *Nietzsche*, Past Masters (Oxford: Oxford University Press, 1994), 55.

being defined, but the word doing the defining, i.e., not recurrence, but eternity'.[35] In other words, it is an attempt on Nietzsche's part to draw back from the extreme implications of his view of the all-embracing scope of will to power and to inscribe it (will to power) in a determinate, 'eternal' structure. Others, however, see it as a means of underlining the essential nihilism of Nietzsche's outlook and of existence as 'the epitome of "a tale told by an idiot, full of sound and fury, signifying nothing"', as Kaufmann comments.[36]

But this strange 'doctrine' can also be seen as doing actual constructive work in Nietzsche's overall thought. Karl Löwith saw it as central to Nietzsche's strategy of inverting the Idealist narrative of a progression from Greek to Christian to modern philosophy, returning philosophy to its first beginnings in the Presocratics. As Löwith puts it, 'He repeats antiquity at the apex of anti-Christian modernity'.[37] More specifically, it is central to Nietzsche's constructive attempt to portray human existence as simultaneously culminating in an act of free self-choice and as located entirely within a naturalistically-conceived world: we must both will eternal recurrence but also acknowledge eternal recurrence as the truth of a world that is without aim or purpose.[38] Karl Jaspers, on the other hand, dismissed its intended or unintended cosmological claims, interpreting it as expressing a 'basic human experience' in which the place of the self in its relation to Being as such and as a whole is disclosed.[39] Heidegger used it to counter the view that Nietzsche passed beyond the limits of Western metaphysics, since, he claimed, it basically repeats in an exceptionally clear and powerful form the basic idea that has governed this same metaphysics since the eclipse of the early Greek experience of *alētheia* or the openness of truth. This is because, in conjunction with will to power 'as the permanentizing... of Becoming', it is 'the most constant permanentizing of the becoming of what is constant'.[40] In other words—and perhaps not entirely without some analogy to Shestov's criticism—will to power and eternal recurrence add up to the refusal of an unlimited process in which human life would be caught up and transformed in ways beyond our control. Instead, these twinned ideas reflect

---

[35] Lev Shestov, *Dostoevsky, Tolstoy and Nietzsche* (Athens, OH: Ohio University Press, 1969), 292.

[36] Kaufmann, *Nietzsche*, 327.   [37] Löwith, *Nietzsches Philosophie*, 124.

[38] Yet, Löwith will also argue, it is precisely this attempt to straddle 'Christian' anthropology and 'Greek' cosmology that dooms Nietzsche's thought to incoherence. As Löwith puts it, 'No Greek philosopher thought so exclusively within the horizon of the future or could have regarded himself as a historical fate' (*Nietzsches Philosophie*, 125). Löwith himself chooses the Rousseauesque pole of the dilemma. Human beings, he suggests, do not create the ground of their own existence, which must either be found in God or in nature. If, as he accepts, 'God is dead', then this means that even the free overcoming of nihilism that Nietzsche proclaims is ultimately a work of nature.

[39] See Karl Jaspers, *Nietzsche: Einführung in das Verständnis seines Philosophierens* (Berlin: de Gruyter, 1936), 310–22.

[40] Martin Heidegger, *Gesamtausgabe*, VI.2: *Nietzsche II (1939–1946)* (Frankfurt am Main: Klostermann, 1997), 199; Eng. tr. from Martin Heidegger, *Nietzsche*, III: *The Will to Power as Knowledge and as Metaphysics*, ed. David Farrell Krell (San Francisco: Harper, 1987), 167.

the all-too-human impulse to stamp what Heidegger calls 'anthropomorphic' meanings onto what is otherwise sheer, empty flux. At the same time, and in doing so, both the world and humanity are consigned to meaninglessness, since nothing in the world of becoming is conceived of as having any intrinsic meaning or value except the meaning or value that is more or less arbitrarily stamped on it by will to power. In this regard, will to power and eternal recurrence constitute the pure ideology of global technology and the danger from which Heidegger's later thought seeks to rescue an imperilled humanity.[41]

Like Heidegger and Löwith (whose interpretation he acknowledges as a significant influence), Gianni Vattimo links will to power and eternal recurrence. Eternal recurrence is not just a cosmological reality that we 'will' in the sense of 'accept': it is something that we ourselves will into existence. In other words, Nietzsche is not just offering a theory about eternal recurrence but, in a manner Vattimo calls 'allegorical-prophetic', asking us to will eternal recurrence into being. And, he suggests, this is also why Nietzsche can present his thought as liberating humanity from the history of violence that he connects with the rule of metaphysical and Christian world-views. Why? Because such views presuppose a view of time modelled on the central motif of the Oedipus myth: each passing moment annihilates the one before only to be annihilated in its turn by the next in a perfect image of sons perpetually rising up to kill the fathers that begat them. To freely will the past as my own work is to reject the possibility that my identity is the product of any paternal super-ego, a revolution that Vattimo sees as concretized in Nietzsche's further call for the unmasking of the reifications built in to conventional structures of language.[42]

Alexander Nehamas too seeks to do justice to the literary form and power of Nietzsche's presentation, only he dismisses the cosmological element of the doctrine, insisting that 'the eternal recurrence is not a theory of the world but a view of the self'.[43] On his view, it expresses the consequences of the hypothetical view that 'If my life were to recur, then it would recur only in identical fashion,'[44] inclusive of all the actions, significant or otherwise, that we might wish had been otherwise. But this is not just a matter of recognizing and accepting the empirical facts of my biography and context since, as Nehamas argues,

---

[41] Lampert, however, contends the opposite, namely, that Nietzsche's teaching pre-empts Heidegger's own critique of technology and does so in a more adequate way. Calling Heidegger's a 'lame reading', Lampert writes that 'making the song of eternal return a hymn to machinery runs counter to everything that Nietzsche himself said about it. Eternal return sings not of bodies machined eternal, but of mortal bodies, the things of a day. It does not express a secret desire to achieve permanence but celebrates the impermanent. It is not a song of dominance but of sheltering and letting be'. Lampert, *Nietzsche's Teaching*, 262.

[42] See Gianni Vattimo, *Il Soggetto e la Maschera: Nietzsche e il problema della liberazione* (Milan: Bompiani, 1974), especially Part III.

[43] Alexander Nehamas, *Nietzsche: Life as Literature* (Cambridge, MA: Harvard University Press, 1985), 150.

[44] Nehamas, *Nietzsche*, 153.

Nietzsche does not regard either past or present as fixed quantities independent of the choices I am making vis-à-vis the future. Acknowledging the significant risks of self-deception, how I choose to see myself and my life are therefore integral to being able to will it as eternally recurring. In this regard, eternal recurrence is so far from being an exercise in cosmology as to be the essential expression for a life lived artistically, in which I will myself to be the story I tell about myself—even if 'life', *my* life included, thereby becomes essentially fictional. With Proust as one possible exemplar, Nehamas concludes that 'the model for the eternal recurrence is not to be found in Nietzsche's superficial reflections on thermodynamics but on his deep immersion in writing'.[45]

But if eternal recurrence might—and given the variety of readings it is hard to venture more than a 'might'—offer some kind of aesthetic redemption of what is otherwise the unendurable emptiness of life in time, could we go further and see it as, in some sense, also a *religious* doctrine? One early commentator, Charles Andler, thought so, writing that 'Nietzsche's doctrine of eternal recurrence is a great mystical intuition... the synthesis of all the most profound religions and of all the most noble philosophies... Never has such an audacious attempt been made to gather up the anthropocentric interpretation of the ancient religions with the help of modern science'.[46] A more recent theological interpreter, Alistair Kee, not only describes eternal recurrence as 'a fundamental religious concept, arising from a numinous experience', but goes on to speak of 'Nietzsche's faith'—although he is cognizant that just because a viewpoint is 'religious' by no means makes it compatible with Christianity or any other specific religion.[47] As is well-known, the bitterest of disputes occur within the families of the religious! In this connection, Edwin Muir gave voice to a common Christian response to Nietzsche's doctrine, portraying eternal recurrence as revealing a world of infinite and meaningless time that, if the doctrine were true, would reduce Christ's suffering and death on the cross to a 'counterfeit mortality'.[48] If I now proceed to follow one interpretation that gives full—and positive—weight to the 'eternal' dimension of Nietzsche's teaching, this is therefore not in order to bring him into the orbit of Christian

---

[45] Nehamas, *Nietzsche*, 167. The significance of seeing 'the eternal' as an essentially *aesthetic* redemption of time that otherwise vanishes towards annihilation will be picked up at several subsequent points in this enquiry. For now, I merely suggest that it is worth pondering the difference between Nietzsche's eternal recurrence and Kierkegaard's repetition in something like the following terms: Nietzsche invites us to unconditionally affirm all that we have been, are, and will be, whereas Kierkegaard holds out the rather different hope of being able to be, once more, the best that we have ever been or to live out the best of the possibilities life has bestowed upon us.

[46] Charles Andler, *La dernière philosophie de Nietzsche: Le renouvellement de toutes les valeurs* (Paris: Bossard, 1931), 60–1.

[47] Alistair Kee, *Nietzsche against the Crucified* (London: SCM Press, 1999), 123. 'Nietzsche's Faith' is the title of Chapter 9 of Kee's study.

[48] In 'The Recurrence'. For discussion see my '*The Heart Could Never Speak*': *Existentialism and Faith in a Poem of Edwin Muir* (Eugene, OR: Cascade Books, 2013).

theological ideas—although it is, perhaps, to bring us to a point at which the relation of Nietzschean recurrence to Christian faith might be moved beyond the impasse of mere contradiction.

In a thought-provoking study, Krzysztof Michalski has argued that it is precisely the concept of eternity that unifies Nietzsche's thought as a whole and, indeed, that 'we find the notion of eternity not only where we hear of eternal return; it is present in nearly everything Nietzsche wrote'.[49] Michalski's study also goes a long way towards making striking connections between Nietzsche's thought and aspects of Christian experience and doctrine. I shall therefore begin by surveying Michalski's argument before proceeding to a reading of some of the decisive pages of *Thus Spake Zarathustra*, bearing in mind the context of the whole current of thought running from Spinoza, through German Idealism and on—to Nietzsche himself? We shall see.

Michalski develops the rationale for his interpretation by reference to the centrality of time in Nietzsche's thought. Because it deals with time, it therefore necessarily deals also with eternity, since 'without reference to eternity one cannot understand how [time] flows, what we mean when we say that it destroys or creates'. Eternity, Michalski says, is 'a dimension of time, the core of time, its essence, its engine. Not its infinite continuation, not its refutation... The concept of eternity answers the question of why "today" transforms into "yesterday"; eternity comes to the fore precisely in the flow of time' (FE, vii). But if this is so, eternity cannot be thought of other than in relation to the body, since the body is the mode in which we experience time. But if our bodily experience of time also involves eternity, then bodily life itself is not exhaustively subject to decay and disintegration. There are moments of discontinuity in our bodily experience of the world and in such moments 'all meaning becomes suspended... In this interval, briefer than any moment can measure, in this crack, this fissure, this tear—in the blink of an eye—everything is left to question, and a chance for a new beginning arises. This is "eternity"' (FE, viii).[50] Such moments of eternity mark a breaking-free from the construction of continuity in life and, as such, perhaps allow a place for God—and certainly for a return to innocence and a new beginning, although

---

[49] Krzysztof Michalski, *The Flame of Eternity: An Interpretation of Nietzsche's Thought*, tr. Benjamin Paloff (Princeton, NJ: Princeton University Press, 2007), vii. (Further references are given in the text as 'FE'.)

[50] Here and at subsequent points I am unclear as to how close (or otherwise) Michalski is to Joan Stambaugh's view of eternity as 'the "occurrence" of time itself'. See Joan Stambaugh, *Nietzsche's Thought of Eternal Return* (Baltimore, MD: Johns Hopkins University Press, 1972). We might also compare this idea of a 'tear' in the fabric of time with the moment of discontinuity in the experience of an epileptic seizure, as Dostoevsky described it in his novel *The Idiot*, where he also ascribed to it the possibility of a vision into otherwise concealed heavenly realities and, specifically, of another kind of time (this is discussed further in the following section).

this is not a matter of concepts but of becoming open to experiences of rage, pain, and joy (FE, x).[51]

Relevant here is Nietzsche's use of the child as an image for a life redeemed from the illusions of metaphysics and dogma.[52] Like animals, children seem to live absorbed in the moment with no sense of time: but this state of absorption has to be disturbed if we are to become fully human and enter a world with past, present, and future.[53] Yet the time of children's play continues to exist as a constant and authentic possibility of adult life. The child at play is 'a metaphor for the inalienable dimension of my own life' in which 'reality is temporally undifferentiated' (FE, 20). But in adult life this unity is only rarely experienced. I am as I normally am by virtue of multiple encounters with chance objects and persons that cannot be worked into any coherent or rational pattern, and my constant experiencing of such manifold differences means that I feel myself (and not just my world) 'scattered among all things and accidents' (FE, 46). In this intrinsically painful experience, my sense of personal integrity is constantly being undermined (FE, 51–2). Science responds to this situation by attempting to turn the world into a garden that is 'orderly, stable, and secure, whose fruits and shade one might enjoy at least for a little while' (FE, 70). Nietzsche, however (like Christian apocalyptic), urges a more thorough-going re-envisioning of time so as to open up the possibility of a radically new future. Zarathustra repeatedly urges on us the sense that 'It is time' and that 'shortly' the decisive moment will come to pass. But if our orientation in and towards time is thus decisive for who we ourselves are, what is time and what is the character of the change that time makes possible?

Nietzsche, Michalski says, does not answer such questions with definitions but with images: the river, and the wind that becomes a fire. The image of time as a river comes to us from the earliest days of philosophy, but whilst Nietzsche affirms the implication of time's unceasing flow it is also an image that needs supplementation. For the river of time 'has no borders' (FE, 96), it has no source and no mouth, and ' "Today" does not differ from "tomorrow" the same way as "the Mississippi at St Louis" differs from "the Mississippi at Memphis" ' (FE, 100–1). Time, then, is also something more or other than a river. It is also a wind, a warm wind that melts the ice of concepts produced by inertia and, as such, it is therefore spirit, freedom, and becoming—it 'tears open the gates of the castle of death' (FE, 102). But this wind is not external to me and does not affect me from without: it is I myself; Zarathustra speaks of himself as 'a strong wind' and there is therefore no 'teaching of Zarathustra'

---

[51] See also the discussion of the motif of 'the eternal now' in the following section.

[52] As, for example, in Zarathustra's parable of 'The Three Transformations': Friedrich Nietzsche, *Also Sprach Zarathustra*, in *Werke in drei Bänden II*, ed. K. Schlechta (Frankfurt am Main: Ullstein, 1972), 567f.

[53] Cf. Schelling's remarks on the childlike divine bliss that precedes the assertion of *Seyn*.

apart from his own continuing story of migrations and returns. Wind is 'the unrest that arises from the fact that I am never completely myself...never completely at home, at rest, by the hearth, in my own identity' (FE, 104). Again with reference to the Book of Revelation, Michalski suggests that this 'wind' is a call to 'Come!', 'a call to abandon all houses... [it is] news about the end of the world as it is' (FE, 105).[54] At this point, wind transmutes into the fire that 'is now the name just for what truly is, for life itself' (FE, 111). As a living flame in which the past is continuously consumed 'my life is a chance of a new beginning, a chance to begin all over again, from zero' (FE, 118). And this also means that life, 'my life', 'I', and 'we' are never completely definable. Michalski comments: 'Life as flame, the moment consumed by fire, disconnected from what was or will be. In this sense: eternity...eternity as a dimension of time, without which time is unthinkable, this is the only eternity' (FE, 120). But, citing Cioran's comment that 'saints live in flames', Michalski also asks whether we, who are probably not saints, can live like that, constantly suffering all we know and all we say to be burned and reduced to ashes: 'knowledge is ashes, words are ashes' (FE, 123).

Michalski's answer—which seems to stray from the exposition of Nietzsche— is to see this as integral to our experience of human love, specifically the sexual love of man and woman. Love, carefully regarded, 'is not an attempt to extend the moment...into eternity' but 'is a leap in the dark, into the unknown', that is, into time and what only time can give: 'The touch of love commands me to move beyond myself, to cast aside the person I was and lay it down as a sac- rifice...The wound allows us to see heaven, to get there, but too fleetingly to make room for anything but you and me. To reach eternity' (FE, 136). But if love is thus a kind of 'eternity-sickness' in our lives, the way in which it binds me to time by binding me to the person I love in time means that it also binds me to death. With an implicit critical aside to Heidegger (and perhaps also to Nietzsche), Michalski says that 'being "ready for love" is exactly the same as being "ready for death"...There, under the apple tree, when I am in love, death finds a place in my life' FE, 139). Consequently, 'love, even the happiest love, is tragic' FE, 141).

But what has happened to eternal recurrence in all this? And how does it impact on Michalski's attempt to bring Nietzsche closer to Christian per- spectives on time? After all, as he notes, Christianity, following Hebrew pro- phetic religion, turned away from the cyclical views of time prevalent in the ancient world. In the light of the Incarnation and Crucifixion, Christianity insists on the unrepeatability of time.[55] Importantly identifying the 'over- man' ('*Übermensch*') with the human condition as such, continuously mov- ing on through the all-encompassing unknown (an ignorance that 'the last

---

[54] On the theme of call see also, e.g. the section 'A Certain Messianicity' in Chapter 6.
[55] See the comment on Edwin Muir earlier in this section.

man' tries to conceal), Michalski nevertheless sees a certain parallel between Zarathustra and Jesus. Drawing on Nietzsche's own remarks on 'the evangel' in *The Antichrist*, Jesus, like Zarathustra, is offering or opening a life lived transformatively and 'open to the unexpected' (FE, 175). And if Nietzsche (or, at least, Nietzscheans) criticizes the Christian concept of the Kingdom of God for being an idealistic construction that is alien to life, eternal recurrence is no less prone to being transformed into a doctrine or theory, as it is by the dwarf in Zarathustra's vision or, later on, by his animals.[56] The dwarf and the animals see the 'abysmal thought' of eternal recurrence from the outside, but, as Michalski interprets Nietzsche, the point is that 'one must fight, suffer, and enjoy for oneself... only for the participant... does the *moment* become comprehensible'. The world is constantly coming to be and passing away, it is the whole of that coming to be and passing away, but the moment is the moment in which time is real and the past is differentiated from the future: living the moment, I live the eternity that makes past, present, and future always possible. Living it, really living it, I realize (in the full sense of making real or actual) that 'Time is me; eternity is me, my life' (FE, 187). As Michalski quotes Zarathustra: '"O my soul, how could you not crave eternity"' (FE, 188). The error of the dwarf as representing the 'spirit of gravity' is to attempt to turn this lived moment into knowledge, whereas Zarathustra, the dancing god, is able to experience its lightness (FE, 198). And, having already described Jesus and Paul as teachers of 'lightness' (FE, 175), Michalski finally re-affirms the similarity between this teaching and that of the New Testament, for Paul too teaches eternity as being 'the change of time inscribed in every instant of my life' and, as such, 'the touch of God' (FE, 204). It is 'the hidden dimension of the time of our lives, thanks to which we are not merely children but also children of God' (FE, 205).

There are many details of this original and thought-provoking interpretation of Nietzsche that could call for extensive discussion. Limiting ourselves to how it relates to the preceding discussion of German Idealism, however, we can see that Michalski's interpretation supports the suggestion of significant continuity between Nietzsche and Schelling noted by Löwith and Heidegger. For what he shows as decisive is not the attainment of knowledge *sub specie aeternitatis* but living, willing, and choosing eternity by the way in which I choose my life in time. Both Schelling and Nietzsche also see this choice in terms of the 'timeless' and blissful self-absorption of childhood being brought into creative interaction with the temporal circumstances of worldly or adult life. But no less striking differences also come into view. Where Schelling turns back to ancient myths in order to retrace the emergence of consciousness from the 'dark' life of unconscious nature and to narrate the history of *Seyn* as will,

---

[56] We shall return in greater detail to the figures of the dwarf and the animals shortly.

Nietzsche offers us a new myth, the myth of Zarathustra and the *Übermensch* that we, each of us, may become.[57] Schelling, in other words, remains a speculative philosopher, a teacher of recollection, whilst Nietzsche seeks to move his readers to change their lives and become alert to the possibility of the eternal in the time they are now living and to do so as a way of creatively orienting themselves towards the future. If Schelling tells us that this is how it is, Nietzsche urges us to 'Do it!' But this requires a further, interesting and radical, step that Nietzsche seems to grasp intuitively and that differentiates his work, especially *Zarathustra*, from anything we find in classical German philosophy.

When we are first introduced to the thought of eternal recurrence in *Zarathustra* we are shown Zarathustra ascending a rocky mountain path, bearing on his shoulder a dwarf who is said to embody 'the spirit of gravity, my devil and arch-enemy'. Arriving at a gateway, Zarathustra confronts this burdensome spirit: '"I! Or you!"', asserting that the dwarf cannot endure Zarathustra's 'most abyssal thought'. Indicating the gate, he explains that the path leading to it is an eternity long and the path leading from it is another eternity. The gate itself is 'the moment' dividing the eternal past from the eternal future. As he explains this thought in more detail to the dwarf (who grumpily comments that 'Everything straight is a lie... all truth is bent, time itself is a circle') he himself speaks more and more quietly, 'for I was afraid in the face of my own thoughts and the thoughts underlying them'. Distracted from these fearful thoughts by the howling of a dog he turns to see an extraordinary sight. A young shepherd is lying on the ground. A long black serpent has crawled into his mouth and has bitten into his gullet. Never has Zarathustra seen such a repulsive sight. Unable to pull the snake out he calls on the shepherd to bite. He does so and, spitting out the snake's head, springs to his feet, laughing with a laughter that consumes Zarathustra and awakens in him a thirst and a longing that, he says, can never be stilled.[58] Is this, in Michalski's terms, the longing for an eternity that will give him the time that the dwarf's interpretation of the circle of time seems to deny?

To answer this question, let us hear what happens later, when, as Zarathustra is recovering from a sickness, his animals visit him and attempt to comfort him with the doctrine of eternal recurrence:

> Everything breaks apart, everything is put together anew; eternally the self-same house of being is being built. Everything parts, everything returns with greetings again; the ring of being remains eternally true to itself.
>
> Being begins in every Now, for every Here there rolls the ball of There. The middle is everywhere. The path of eternity is bent.[59]

---

[57] Assuming that the *Übermensch* is not to be understood as a historical expectation of either a redeemer figure or a new kind of human being but as a model for what every individual capable of willing eternal recurrence might be.

[58] See Nietzsche, *Zarathustra*, 406–10.   [59] Nietzsche, *Zarathustra*, 737.

But it is just at this point that Zarathustra, laughing, interrupts them, playfully calling them 'mischievous rascals' and 'barrel-organs'. He recalls the vision of the shepherd, applying the story to himself, and saying how 'that monster crept into my throat and choked me! But I bit off its head and spat it away from me'. And now, he accuses the animals, you too are making this into a barrel-organ melody. Even if the narrative differentiates between the hostile figure of the dwarf and the well-meaning animals, it therefore seems as if the latter too are missing something crucial about eternal recurrence, namely, that it is not a view or theory of life to be simply repeated or taught but a condition to be lived and a challenge to be responded to.[60] But, for their part, the animals seem also to have something positive to teach Zarathustra when, in response to his reproaches, they tell him that in order to enter his destiny as the teacher of eternal recurrence he must stop talking and *sing*. And this is just what happens when, in the next section, 'Of the great longing', Zarathustra himself addresses his soul and urges it to sing:

> O, my soul, now I gave you everything and also my last, and I have emptied all my hands for you: that I might sing you, that was my last!

> That I might sing you, speak now, speak: which of us should now be the one to give thanks? But better yet: sing to me, sing, O, my soul! And let me give thanks.[61]

This invocation mutates into an encounter between Zarathustra and life itself, culminating in his song-like echoing of the midnight chimes that mark the moment of passage from one day to the next:

> One!
> O, man! Pay heed!
> Two!
> What says the deep midnight?
> Three!
> 'I was sleeping, I was sleeping...
> Four!
> I have awoken from this dream...
> Five!
> The world is deep,
> Six!
> And deeper than the day had thought.
> Seven!

---

[60] That the movement between the two visions, of the gate and the shepherd, marks a move from theory to action is also an important element of Vattimo's interpretation.

[61] Nietzsche, *Zarathustra*, 743. In *Ecce Homo* Nietzsche comments that 'One might perhaps be permitted to categorize all of Zarathustra as music' (*Ecce Homo*, 574).

Deep is its woe…
Eight!
Pleasure—deeper yet than heartfelt sorrow;
Nine!
Woe says: Pass away!
Ten!
Yet every pleasure wants eternity…
Eleven!
…wants deep, deep eternity!'
Twelve![62]

This 'song' confirms Michalski's analysis, since the intense pleasure ('*Lust*') that longs for eternity is simultaneously a readiness to welcome the coming new day. The point is reinforced in Zarathustra's song of 'The Seven Seals (or: The Song of Yea and Amen)',[63] which concludes the third part of the book. The title alludes to the seven seals of the Book of Revelation, thus positioning Zarathustra's own teaching in relation to the apocalyptic vision of a world in which 'there is no more time'. The song intones Zarathustra's longing for eternity, a longing that has made it impossible for him to live an earthly life, loving eternity more than he could love any woman (Augustine, surely, could have said the same). The point is underlined by the way in which each verse concludes: 'For I love you, O eternity'. In the seventh and final verse, Zarathustra recalls the 'wisdom of birds' that 'there is no "above", no "below". Cast yourself around, forward, back, light as you are! Sing! Speak no more! Are not all words for those bound by gravity! Don't the light ones find all words to be lies? Sing! Speak no more!'[64]

For the one for whom the task is no longer the theoretical contemplation of the human condition *sub specie aeternitatis* but the active willing of time as expressive of its own eternal possibility, words, the medium of articulated knowledge, are no longer adequate. Perhaps they are not even relevant. In his affirmation of will as the eternal source of temporal life, Nietzsche repeats and, it may be said, consummates the trajectory of German Idealism. In his embrace of song he repeats the impulse of Romantic poetics to rise above the structured limitations of language into the infinite liquidity of song, supremely instantiated for many of the poets in the freedom of birdsong—the skylark and the nightingale. Is this, as Nietzsche may have hoped, an overcoming of philosophy or is it perhaps simply a breaking-loose from philosophy and an abandoning of philosophy for an eternity that philosophy can never think? An eternity that can only be sung, not said? But is this then a reversion (or even a

---

[62] Nietzsche, *Zarathustra*, 746–7.

[63] In early editions was the conclusion of the whole work before the addition of a 'Fourth and Final Book' that has, however, often been seen as semi-detached from the rest of the work.

[64] Nietzsche, *Zarathustra*, 750.

regression) to what lies below the threshold of philosophy? Is this not perhaps evidence for what Löwith incisively calls Nietzsche's lack of 'middle and measure' in his account of the relationship between human beings and their world, an account marked by a rhetoric of persistent excess (note the characteristic Nietzschean prefix *Über-*!), culminating in a 'midday-abyss' that is at once the 'highest being and deepest nothing'?[65]

That this loss of measure, this opening of an immeasurable rift in the flow of time, might connect to what, in Chapter 1, we saw to be arguably central to a religious interest in God's timelessness is more than possible.[66] But Nietzsche's recourse to song at just this point in Zarathustra's story also raises problems regarding the form of religious consciousness appropriate to a time-experience that seems to disclose an immeasurable 'other' time. Particularly, it sharpens the question as to just what we might *say* about such a time-experience.

With primary reference to Wordsworth, Simon Jarvis has defended the possibility of 'a philosophic song'.[67] But if song can, in its way, *respond* to philosophical questions, it scarcely seems to *answer* them. If I speak to you and you respond by singing or whistling, it may be—if you are not simply registering your total indifference to me and to what I have just said—that you are subtly (or not so subtly) indicating the pomposity or emptiness of my words or summoning me to go with you in a movement beyond language. But that is still not an answer. Yes, maybe there is a point at which philosophy just has to stop and acknowledge that it has arrived at the things of which we cannot usefully speak. And eternity might have a good prima facie claim to be one of the things. But if, as this reading of Nietzsche might suggest, philosophy simply has to give up talking about eternity and transmute its discourse into song, what is its relation to religious and theological discourse about eternity? The monotheistic religions, at least, have strong traditions arguing for the limitations of philosophy with regard to fundamental religious truths, including what it means for God to be 'the Eternal', and they also have a strong sense of the religious value of various kinds of hymnody, from lamentation to praise. Yet theologies of this kind are also likely to insist on the primacy of 'the Word' and on revelation as having the character of verbal address: 'the Lord spoke'. Revelation is not the communication of a wordless feeling, but the communication of a word of command, promise, or love.[68] If eternity marks a limit to the possibilities of

---

[65] Löwith, *Nietzsches Philosophie*, 179ff.

[66] See 'Closing Reflections' in Chapter 1 of this volume.

[67] See Simon Jarvis, *Wordsworth's Philosophic Song* (Cambridge: Cambridge University Press, 2007).

[68] See for example the discussion of Rosenzweig's category of vocativity in Chapter 9 of this volume; however, this is not simply set up by him as an antithesis to song and the hymn of the community is also given a crucial role in his account of the relationship between the eternal God and His eternal people.

unaided human reason, even—perhaps especially—at that limit, we remain in need of a word.

These comments indicate a tangle of issues that would become salient in the philosophy and theology of the early twentieth century. For all their differences (and I shall not be minimizing them), voices as diverse as Karl Barth and Martin Heidegger would, as it were, 'call time' on philosophical attempts to interpret time through its relation to eternity and to see temporality as the manifestation of the eternal. Both also make critical reference back to German Idealism and to Nietzsche and both also insist—albeit in their very different ways—on the word, *logos*, as the eminent mode in which time, eternity, and their difference are made humanly understandable.

However, before we move on to consider those like Barth and Heidegger who, for diverse reasons, criticized what they saw as the confusion of time and eternity represented by the thinkers we have been surveying, we should note one further form in which the unity of time and eternity has found expression in modern thought. Anticipated in the literature and philosophy of Romanticism, the form in which it took shape in the twentieth century was also influenced by a certain reading of Nietzsche. This I shall call the mysticism of the eternal now or of the timeless moment.

THE ETERNAL NOW

In the last section, I briefly hinted that Nietzsche's abandonment of 'middle and measure' in his invocation of eternal recurrence might offer a bridge towards the soteriologically-inflected religious meaning of divine timelessness. We might gloss such a moment of abandonment in terms of the 'timeless moment' or 'eternal now' found in a number of accounts of religious experience. Where such accounts typically differ from Nietzsche, however, is that they do not involve any act of will. They are events in which the world and, more importantly, the self are disclosed as eternal or essentially timeless and that occur as ontologically and psychologically unexpected, gratuitous, and entirely surprising. That there may, nevertheless, be a connection back to Nietzsche relates to that moment in the Nietzschean itinerary of the self in which the adult becomes a child, given over to the 'free play of becoming'.[69] It is probably no coincidence that this kind of mysticism would become especially prominent

---

[69] This can be related to Löwith's description of the tension between freedom and nature in Nietzsche's thought. The—in his view unsustainable—paradox is that what we are to will in willing eternal recurrence is precisely a state that is, in a sense, beyond the will, a return to nature and the Dionysian life of the cosmos.

in the period after Nietzsche, even when it was articulated (as it often was and is) in terms of Asian rather than Western philosophy. Yet perhaps the most cited and concise statement of what it is about is in lines from William Blake's 'Auguries of Innocence': 'To see a World in a Grain of Sand, | And a Heaven in a Wild Flower, | Hold Infinity in the palm of your hand, | And Eternity in an hour'.[70]

But, to repeat, the kind of vision or experience I am trying to talk about here is not something we are to strive 'to see' or in any way to bring about by an act of will. Indeed, actively willing it might be one of the chief obstacles to it happening; if it happens at all it must happen involuntarily.

An extreme case of such an involuntary rapture is the epileptic vision described by Dostoevsky, whose eponymous 'idiot', Prince Myshkin, speaks of 'flashes and glimpses of a higher self-sense and self-awareness' immediately preceding his epileptic seizures in which, as he says, ' "I was somehow able to understand the extraordinary phrase that time shall be no more. Probably," he added, smiling, "It's the same second in which the jug of water overturned by the epileptic Muhammad did not have time to spill, while he had time during the same second to survey all the dwellings of Allah" '.[71]

However, if Prince Myshkin's visions have the required character of involuntariness, the particular form of 'the eternal now' I am wanting to focus on here is not a rapture that takes us out of the world of time, but the revelation of the world of time as itself manifesting the truth and presence of the eternal. Perhaps a closer approach is that of the case of a 'Canadian psychiatrist', R. M. Bucke, cited in William James's lectures on *The Varieties of Religious Experience*. Bucke describes how, on the way home from an evening discussing poetry and philosophy with friends, 'All at once, without warning of any kind, I found myself wrapped in a flame-colored cloud...the fire was within myself'. And, as the experience continues, 'I became conscious in myself of eternal life. It was not a conviction that I would have eternal life, but a consciousness that I possessed eternal life then; I saw that all men are immortal...The vision lasted a few seconds and was gone but the memory of it and the sense of the reality of what it taught have remained'.[72] However, even if this vision is described as coming 'without warning' and therefore as involuntary, it is probably no coincidence that it followed on from an evening discussing poetry and philosophy.[73] Perhaps the friends had been discussing Emerson,

---

[70] William Blake, 'Auguries of Innocence', in *Poetry and Prose of William Blake*, ed. Geoffrey Keynes (London: Nonesuch, 1967), 118.

[71] Fyodor Dostoevsky, *The Idiot*, tr. Richard Pevear and Larissa Volokhonsky (London: Granta, 2003), 227.

[72] William James, *The Varieties of Religious Experience* (London: Collins, 1960), 385.

[73] This is not the only example cited by James that includes a reference to the eternal although, interestingly, he himself does not strongly thematize the time-consciousness of the religious experiences he discusses.

whose essay on 'Immortality' affirms that 'Future state is an illusion for the ever-present state. It is not length of life but depth of life. It is not duration, but a taking of the soul out of time... when we are living in the sentiments we ask no questions about time. The spiritual world takes place;—that which is always the same'.[74] And perhaps behind Emerson (himself a significant influence on Nietzsche) is also Schleiermacher, who, in his *Speeches on Religion* of 1799, complained of the misunderstanding of immortality amongst religious believers: 'In search of an immortality, which is none, and over which they are not masters, they lose the immortality that they could have... But try to yield up your life out of love for the universe. Strive here already to annihilate your individuality and to live in the one and all'. And, as he sums up, 'To be one with the infinite in the midst of the finite and to be eternal in a moment, that is the immortality of religion'.[75]

Yet these Spinozan visions do not quite engage the quality of the lived moment itself as eternal which I am seeking to focus on here. Consider Aldous Huxley's influential account of his first mescalin experience:

> ...along with indifference to space, there went an even completer indifference to time.
>
> 'There seems to be plenty of it,' was all I would answer when the investigator asked me to say what I felt about time.
>
> Plenty of it, but exactly how much was entirely irrelevant. I could, of course, have looked at my watch; but my watch, as I knew, was in another universe. My actual experience had been, was still, of an indefinite duration or alternatively of a perpetual present made up of one continually changing apocalypse.[76]

Like Huxley, sometime Anglican priest Alan Watts became what Watts's surviving partner called a 'guru of the flower children'. In his *Behold the Spirit*, sub-titled 'A Study in the Necessity of Mystical Religion' and written as a work of Christian apologetics, Watts already identified the sense of an eternal now that he would later articulate in terms of Daoist philosophy. 'God', he stated, 'is the most obvious thing in the world—the simplest, clearest and closest reality of life and consciousness'.[77] Union with God is simply what reality, at its most real IS (typically capitalized by Watts). From our point of view 'Any attempt to grasp it, by action or by inaction, suggests that it is not absolutely present'.[78] But it is—all we have to do is to see that it is.

[74] Ralph Waldo Emerson, 'Immortality', in *The Complete Works of Ralph Waldo Emerson*, VIII: *Letters and Social Aims* (Boston, MA: Houghton, Mifflin, 1883), 329.

[75] Friedrich Schleiermacher, *On Religion: Speeches to its Cultured Despisers*, tr. and ed. Richard Crouter (Cambridge: Cambridge University Press, 1988), 139–40.

[76] Aldous Huxley, *The Doors of Perception* and *Heaven and Hell* (London: Granada, 1977), 16.

[77] Alan Watts, *Behold The Spirit: A Study in the Necessity of Mystical Religion* (New York: Vintage, 1971), 91.

[78] Watts, *Behold The Spirit*, 94.

The focal point of Reality is now—this present moment, this elusive image of eternity, so small that it has no temporal length and yet so long that we can never escape from it. Here in this present moment life is most lively; here alone do we really exist…In this moment we live and move and have our being, and nowhere else. What we have to realize, therefore, is not the getting of union with God, but the not being able to get away from it. It is in, it is this Eternal Now, wherein God so lovingly holds us.[79]

But the point is 'not to be still and gaze into the Eternal Now with love', as in classic versions of Christian contemplation, but 'to give oneself wholeheartedly to the work which the present moment brings; it is to go straight ahead with the Eternal Now, accepting its changing forms with one's entire being but not grasping hold of any of them'.[80] This is not so much a Nietzschean act of willing the eternity of the now, however, but simply, living it with both active and passive capacities—with the whole of the human reality that we are (even now! right now!), in play.

Moving away from Christianity, Watts became a leading popularizer and expositor of East Asian thought, notably Zen Buddhism and Daoism. In the same period a similar view was being expounded by the Zen scholar D. T. Suzuki, also extensively read in Western theological, philosophical, and literary traditions. In his essay 'Living in the Light of Eternity', Suzuki cites Raymond B. Blakney's translation of Meister Eckhart to the effect that 'The now-moment in which God made the first man, and the now-moment in which the last man will disappear, and the now-moment in which I am speaking are all one in God, in whom there is only one now'.[81] He then interprets this saying in a Zen perspective: 'I have been reading all day, confined to my room, and I feel tired. I raise the screen and face the broad daylight. I move the chair on the veranda and look at the blue mountains. I draw a long breath, fill my lungs with fresh air and feel entirely refreshed. I make tea and drink a cup or two of it. Who would say that I am not living in the light of eternity?'[82]

In the twentieth century, Meister Eckhart would be taken by both Christian and non-Christian commentators on mysticism as a salient representative of the mysticism of the eternal now. He would also play some background—or perhaps not so background—role in shaping Heidegger's attempt to think his way through to a new way of envisaging time that would be freed from precisely the kind of reifying concepts and language that had troubled Nietzsche. The position at which Heidegger seems eventually to arrive is indicated by a

---

[79] Watts, *Behold The Spirit*, 94–5.   [80] Watts, *Behold The Spirit*, 240.

[81] D. T. Suzuki, *Mysticism: Christian and Buddhist* (London: George Allen and Unwin, 1957), 111.

[82] Suzuki, *Mysticism*, 111. Like Reiner Schürmann, I refrain from pursuing the question as to whether or to what extent Suzuki's Zen interpretation of Eckhart is historically justifiable. See Reiner Schürmann, 'Appendix: Meister Eckhart and Zen Buddhism', in *Wandering Joy: Meister Eckhart's Mystical Philosophy* (Great Barrington, MA: Lindisfarne Books, 2001), 217–22.

consciously Eckhartian word, *Gelassenheit*, which could be translated as 'equanimity' but also implying the idea of 'letting-be'. It is thus, for Heidegger, a way
of being that has let go the ambitions of the Nietzschean will-to-power and lives
with simple and fundamental openness to the unfolding possibilities of being
and, as we shall see (in Chapters 4 and 9), this might (just) allow for speaking
of a kind of presence of the eternal in time—despite Heidegger's general avoid-
ance of what he sees as the inherently metaphysical and dualistic associations
of the concept of eternity.[83] Although it is arguable that Heidegger essentially
strips the Eckhartian term of its religious significance,[84] and would certainly
resist the kind of direct and seemingly unproblematic use of 'presence' found
for example in Watts or Suzuki, Reiner Schürmann has argued for seeing
Heideggerian *Gelassenheit*, which he translates as 'releasement', as offering an
appropriate philosophical appropriation of Eckhart's thought. Heidegger also
reflects a complex of Eckhartian themes of which, in addition to *Gelassenheit*,
Schürmann lists 'energetic identity', 'being as process and appropriation', 'presence under the mode of nothingness', 'word', 'without a why'. These combine in
a philosophical position that, under the perhaps Nietzschean rubric 'wandering identity', Schürmann sees as a 'meditative' thinking, a 'letting-be [that] is
the condition for mystery to manifest itself'.[85] And, we may add, to manifest
itself in, with, and under the moving pattern of our human time-experience.

We shall return to the vexed question as to whether Heidegger's focus on
temporality allows for time to be experienced as manifesting the eternal in
any meaningful sense. But he certainly shares at least some of the concerns of
those who, in the wake of Nietzsche, have sought to re-envisage the relation
to the eternal as allowing for the affirmation of human time-experience and,
at the same time, as involving a moment that is beyond the Nietzschean will
to power. Indeed, we may say that it is very much part of the function of the
term 'eternal' in this context that it indicates a limit to what can be willed. That
what is perceived or felt is perceived and felt as 'eternal' means that it does not
wait upon our mental activity in order to bring it into being; we ourselves do
not construct it, but can relate to it only under the mode of givenness, as revelation, or, as Watts might put it, it just IS: 'This is It', as the title of one of his
books proclaims.[86]

---

[83] Interestingly, as Suzuki looks from Zen to Eckhart, Heidegger for his part also looked from
Eckhart to Daoism, and if Reinhard May perhaps exaggerates in speaking of East Asian thought
as a 'hidden source' of Heidegger's work, he does seem to have sensed significant analogy and
at one point considered engaging in collaborating with a Sinologist colleague on translation Lao
Tse. See Reinhard May, *Heidegger's Hidden Sources: East Asian Influences on his Work*, tr. Graham
Parkes (London: Routledge, 1996).

[84] This is the argument of John D. Caputo in his *The Mystical Element in Heidegger's Thought*
(New York: Fordham University Press, 1986).

[85] Schürmann, *Wandering Joy*, 196.

[86] Alan Watts, *This is It and Other Essays on Zen and Spiritual Experience* (New York: Vintage,
1960).

The re-reading of medieval mystical sources, especially Eckhart, alongside some versions of Zen Buddhism and aspects of the later Heidegger's exploratory proposals for a more meditative kind of philosophical mindfulness, both reflected and influenced widespread cultural approaches to religion in the twentieth century. The motif of 'the eternal now' also appears in, for example, the popular sermons of Paul Tillich[87]and T. S. Eliot's *Four Quartets*,[88] as well as in the renewed popularity of the kind of 'mysticism' found in Thomas Traherne (quoted by Watts, Suzuki, and Huxley) or amongst Buber's Hasidim (finding philosophical articulation in the Third Part of *I and Thou*: 'Through every single You the basic word addresses the eternal You'[89])—and, of course in William Blake, with whose 'Auguries of Innocence' this section began. There are many fairly straightforward aspects of our cultural situation that make this kind of mysticism attractive: it reflects a disenchantment with the rationalism and compartmentalization of modern life and intellectual training, it offers an ecumenical approach that makes the resources of a range of religious traditions available for contemporary self-understanding, and it answers a felt need for ecstatic experience—as well, of course, as articulating what many of its proponents believe to be something that has actually happened to them, beyond language and cultural construction. Here, however, the point is that it also offers a development—and perhaps a certain culmination, a *ne plus ultra*—of a long-standing philosophical trajectory that, in the modern world, runs from Spinoza to Nietzsche and that, in a succession of ways, attempts to think the unity of time and experience. That movement, I am suggesting, runs from the ambition of a fully intellectual love of God in which the world is seen for what and as it is *sub specie aeternitatis*, through a sequence of views that attempt progressively to incorporate into such a view the manifold and changing forms of life lived in time. When this theoretical enterprise reaches a point at which the tension between timeless contemplation and lived time can no longer be sustained, the project mutates into the Schellingian–Nietzschean attempts to conceive the unity as a manifestation of will rather than intellectual contemplation. But, as developed by Nietzsche, this valorization of a willed unity of time and eternity brings us, in the end, to a point at which there is nothing more to say and words break out into song. But whether the pure lyricism of such a philosophic song can still be spoken of as an act of *will* seems questionable. Perhaps, as some Romantic poets themselves claimed, it is the voice of nature itself, of which the poet is only the mouthpiece. Prepared by the Nietzschean will to will eternal recurrence, then, the twentieth century is repeatedly drawn

---

[87] For further discussion of Tillich, although in another context, see Chapter 6.

[88] I have discussed Eliot's version of this theme elsewhere, and will therefore not repeat myself here. However, were the reader to pause at this point and re-read *The Four Quartets* themselves before proceeding, it would not be time wasted. See my *God and Being* (Oxford: Oxford University Press, 2011), 133–5.

[89] Martin Buber, *I and Thou*, tr. Walter Kaufman (Edinburgh: T. & T. Clark, 1970), 123.

to experiencing the presence of eternity in the sheer happening of the world, of life, of the self. This is it—where 'is' is not a static copula marking the presence of a timeless eternity, but the active happening of time: 'is' as always and ceaselessly, even 'now', becoming.

Whether or not this brief summary works as the historical narrative of a certain line of modern philosophy, what is more important is that the positions that are represented in it show both the promise but also the challenges facing our attempts to think the human condition in the mirror of eternity. Of course, another line of modern thought had long since given up all interest in the eternal, looking instead to the deliverances of the positive sciences.[90] But for those interested in preserving, interpreting, and even continuing the enterprise of classical philosophy, from the Greeks through to the early modern world, the question of the eternal would be both as necessary and as seemingly resistant to being thought as, in Chapter 1, we found to be the case in modern analytic thought. In this respect, perhaps, the two worlds of 'Anglo-American' and 'Continental' philosophy are not so far apart as they may seem. Positively, we also saw at least a possibility that a Nietzschean experience of the immeasurable depth of time might have some correlation to the immeasurability of time that, I argued, did the essential religious work of philosophical appeals to divine timelessness. In the development of a certain strand of modern mysticism, this relates to the experience of an eternal now, a moment in which the conventionally linear construction of human time-experience breaks down or goes into suspension—for a moment!—allowing for the sense of another time, the time of the eternal in the now or the timeless moment, to take hold of and re-orientate the subject's understanding of time.

But whilst the idea of an eternal now will be a continuing point of reference at subsequent stages of this study, it is by no means a sufficient answer to the question as to what eternity might mean in a world encompassed by the horizon of temporality. Furthermore, if Nietzsche did, in his unique way, contribute to the development of a new kind of mysticism, his transposing of philosophy into song could also be seen as a way of recognizing that it had become impossible to *think* time and eternity together in the manner of a Spinozan or a Hegelian and that his 'solution' was, therefore, no solution—at least, not for philosophy. With Nietzsche as marking a certain historical watershed, we therefore look now to a succession of interrelated approaches that, against German Idealism, insist on the necessary disjunction and difference between time and eternity.

---

[90] That this too could (e.g. in Neo-Kantianism) also come back to the question of the eternal would be another story again, although, once more, it would connect to the particular problematic engaging the Heideggerian development of Husserlian phenomenology.

4

Time or Eternity

We have considered a sequence of attempts from Spinoza through to Nietzsche and beyond that have sought to think the unity of time and eternity. In this chapter we turn to the opposite tendency, that is, to those who have insisted on their disjunction and even their incommensurability. This could be seen as merely the obvious next step in running through the spectrum of possible theoretical positions. However—and this is already true of the movement from Spinoza to Nietzsche—what is at issue is more than just a matter of setting out some intellectual choices, since a certain experience of history (that is, of time) comes ever more clearly into view as internal to the theoretical position being advanced. Ricoeur spoke of the French Revolution as marking a rupture within the continuum of historical time and, as such, placing a continuing and standing challenge to philosophical reflection. Similar claims could perhaps be made for other historical events—from the Jewish exile in Babylon, through the Sack of Rome and the Protestant Reformation to the World Wars and genocides of the twentieth century.[1] However, the point is surely that any such contingent historical event becomes significant for theological and philosophical reflection when, over and above its unquestionable impact on those who experience it at first hand, it is also seen as revealing a basic structural element in human experience. It is at this point that the experience of history moves us to reflect on the meaning and scope of time as the condition of historical life and therefore also on the possible relation of time to eternity. In relation to the specific modern experience of history as catastrophe this then means paying increasing attention to the possibility that time and eternity are, quite simply, incommensurable and irreconcilable.

As is well known, Nietzsche saw his mature philosophy as inverting the thought of his own 'educator', Arthur Schopenhauer, and I shall therefore begin this chapter with a brief glance at Schopenhauer's radical alternative to

---

[1] In this respect, we might consider Theunissen's discussion of Pindar as indicating a response to an analogous rupture, namely the rupture between the mythical world of archaic time and the emergence of history itself (see Chapter 7).

German Idealism that, as we shall see, also sought to deny what Nietzsche sought to affirm: the eternity of temporal becoming. Turning then to a more explicitly theological source—the early theology of Karl Barth—we shall certainly see what we might expect to see: a significant and deliberate attempt to draw a clear line of demarcation between human and worldly temporality and divine eternity. However, whilst acknowledging and emphasizing the primarily negative thrust of Barth's commentary on Paul's Letter to the Romans, we shall also have occasion to see—which seems not to have been so clear to most of its first readers—that it is also a work that, in its own way and in a new key, repeats themes of the nineteenth century philosophical discussion. From Barth we shall look back to Kierkegaard, one of his acknowledged influences, and, specifically, to Kierkegaard's attack on what he regarded as the illusory nature of Hegelianism's attempt to see human life *sub specie aeternitatis*. Like Barth's commentary on Paul, Heidegger's early philosophy of *Existenz* was also shaped in part by the reading of Kierkegaard. But even as he draws on key elements of Kierkegaard's understanding of time, Heidegger sees Kierkegaard's insistence on relating human life to the eternal as a tell-tale sign that the Danish thinker has misconceived the nature of time, even if his investigations are also deemed to have been amongst the most 'penetrating'.[2] However, as we shall see in later chapters, each of Heidegger and Kierkegaard is too complex simply to be categorized as 'for' or 'against' the unity of time and eternity, not least because—as Heidegger points out—how we think of eternity is in part dependent on how we think of time: we cannot therefore rule out in advance the possibility that a rethinking of time might provide the occasion also to rethink what we mean by eternity, and to do so in such a way that the distortions consequent upon the older metaphysical views are avoided.[3]

A WORLD OF PAIN

In beginning this part of our investigation with Schopenhauer, we begin with a philosopher who, both in his own eyes and in that of many historians of philosophy, stood outside the mainstream of modern European philosophy.[4] Yet

---

[2] On Heidegger's distinctive use of the term 'penetrating' see my article 'Existence, Anxiety and the Moment of Vision: Fundamental Ontology and Existentiell Faith Revisited', in Anthony Paul Smith and Daniel Whistler (eds), *After the Postsecular and the Postmodern: New Essays in Continental Philosophy of Religion* (Newcastle upon Tyne: Cambridge Scholars Publishing, 2010), 128–51.

[3] This comment is, indeed, central to the impulse behind this present study in its entirety.

[4] To the extent that he exerted a wider influence on academic study in the nineteenth century this was, on the whole, in relation to the nascent field of Oriental Studies. See Christopher Ryan, *Schopenhauer's Philosophy of Religion: The Death of God and the Oriental Renaissance* (Leuven: Peeters, 2010), esp. 226ff. However, as Ryan also points out, Schopenhauer did exert quite a considerable influence on artists and writers.

he would provide a decisive stimulus to the philosophical development of the young Nietzsche, even if Nietzsche would then go on to invert Schopenhauer's own negative evaluation of worldly life.[5] And if he also consciously (and often provocatively) looked to Indian philosophy and religion for key elements of his arguments, dismissing conventional European assumptions with haughty disdain, he nevertheless also saw himself as developing the essential insights of Kant, which, he thought, Kant himself had misconstrued.

Kant's great discovery, according to Schopenhauer, was 'the distinction of the phenomenon from the thing-in-itself'.[6] What this conveyed to Schopenhauer was that the world of human experience is what and as it is as a result of our own activity in representing it thus. Whether it is a matter of objects out there or inner mental states, everything we can experience and know is only a form of representation. However, Schopenhauer also maintained that Kant erred in supposing the thing-in-itself to be the causal ground of phenomena.[7] This, he asserted, is to introduce a distinction into the nature of reality that is without any real basis. There is no difference between the immanence of phenomena and the transcendence of the thing-in-itself; there is only the one reality, only immanence—but appearing in two different ways, as the represented world of phenomena and as the thing-in-itself, which is not and can never be made into an object of representation (since it is the unrepresentable reality of the world that the manifold of representation can only ever represent). But this also means that there is no fundamental distinction in reality itself between the variety of phenomena, between a mountain, a tree, an insect, or a human being. What differentiates them is only ever temporary, provisional, and local.

So how might we speak of this one reality that, being unrepresentable, cannot really be known or spoken about? It is what Schopenhauer calls the will, the one constant, omnipresent will, 'which lies outside time and space, outside the *principium individuationis*'.[8] In an image used by Schopenhauer himself, the relationship between will and phenomena resembles the working of a magic lantern: it is one and the same light that causes all the different images to appear. But if all things are in this way projections of the one universal unrepresentable will, why should the will 'want' to reproduce itself in the manifold forms of representation? Schopenhauer's answer is, essentially, that it doesn't, since it has no 'wants' or 'wishes' and harbours no rational or even irrational purpose in pursuing its own endless self-externalization. It simply is the blind, purposeless urge to self-perpetuation and self-manifestation.

---

[5] It is as pointing towards Nietzsche that Rosenzweig identified Schopenhauer as inaugurating the 'new thinking' that he saw himself as representing.

[6] Arthur Schopenhauer, *The World as Will and Representation*, tr. E. F. J. Payne, 2 vols (New York: Dover, 1969), I. 417.

[7] Schopenhauer, *World*, I. 252ff.   [8] Schopenhauer, *World*, I. 113.

In these terms, however, the will behaves uncannily like the Platonic good that seeks to create as many possible beings as can bear some measure of being and of the goodness that inheres in all being—only for Schopenhauer there can be no talk of 'good'. On the contrary, the world produced by the will is inevitably a world of incessant conflict and suffering. Why? Because since each of the appearances through which the one will manifests itself is conscious of itself as a separate, individual entity, it will spontaneously perceive itself as being in opposition to other individual entities that it sees as similarly striving to assert and maintain their own claim to life. By virtue of the process of individuation that is necessary for the appearance of a manifold and differentiated world, the will becomes divided against itself in a motiveless and endless agony of conflicted becoming.

But does it 'become'? Again, Schopenhauer takes inspiration from Kant and Kant's view (as he understands it) that time is not a feature of the real world but is an ideal form deployed by the human mind to order the chaos of the manifestations of will. In other words, in the relationship between the will and the world of representations, time belongs to representation and not to the will. As Schopenhauer says, 'Time is merely the spread-out and piecemeal view that an individual being has of the Ideas. These are outside time, and consequently *eternal*.'[9] However, although he follows this by citing Plato's definition of time as a moving image of eternity, Schopenhauer does not regard Ideas or Forms as having any independent ontological basis. Although they offer an intermediate stage between the sheer force of the will and the manifold of appearances, they too are only forms of manifestation and do not directly reveal the immediacy of will. Only the will is really eternal, although the contemplation of ideas can give a relative respite from the otherwise ceaseless torment caused by the struggle of all against all. Such respite is also found in art, which, as Schopenhauer put it, 'stops the wheel of time'.[10] But it stops it only for a moment and the relief that philosophy and art offer is only temporary. The only radical solution is for the individual will to deny its own will to assert the existence over against others, that is, to deny its own will to live—to cease to be, to become nothing.

This is not, however, a doctrine of suicide. On the contrary, those who believe that the torment of existence can be solved by suicide have entirely failed to see the point that their individuation as suffering and wretched human individuals is an illusion. Rather than kill themselves, they have only to step outside themselves and see that really they are already nothing, just one manifestation amongst trillions of a single life-force. It is only the

---

[9] Schopenhauer, *World*, I. 176.

[10] Schopenhauer, *World*, I. 185. This is especially true of music. Thus, Nietzsche's understanding of song as the eminent mode in which to express the temporality of eternity once more contradicts the basic movement of Schopenhauer's metaphysics.

intellect, representing the individual as an individual possessed of its own temporal history, that erroneously believes that biological life and death bring about any fundamental ontological change. Consequently, Schopenhauer affirms Spinoza's assertion that 'in the depths of his self-consciousness, man feels himself to be eternal and indestructible'—only this apparent affirmation of immortality is immediately qualified by the statement that since both memory and expectation of the future are bound to the forms of representation that bestow our sense of individuation in time and space 'nevertheless he can have no memory... beyond the duration of this life'.[11] Judaeo-Christian ideas of individual immortality are therefore entirely misplaced, since an individual life has a beginning in time and what has a beginning in time cannot be immortal. True immortality is the eternity of that which never came into existence and never passes out of it. In this regard, Indian teaching about the eternity of the soul as identical with the prime unchanging substance of Brahman is more logical than the Christian view, although, alluding to the absence of belief in immortality in the Old Testament, he concedes that the latter is at least consistent in realizing that those who are created out of nothing must return to nothing.[12]

In an image that recalls the Platonist model of a line that without deviating from its own course is touched at various points by a sequence of temporal vectors,[13] Schopenhauer pictures time as a rolling sphere that touches an unmoving tangent that, in itself, has no temporal duration. From the standpoint of the one will there is no passage of time. 'Past and future contain mere concepts and phantasies; hence the present is the essential form of the phenomenon of the will, and is inseparable from that form. The present alone is that which always exists and stands firm and immoveable.'[14] In this sense, then, Schopenhauer affirms what he calls the 'scholastic *Nunc stans*', although, where this is associated in scholasticism with God, Schopenhauer's eternal present is constituted solely by and as the blind impulse of universal will. Rather than offering itself in a mystical flash of insight into the transcendent divine wisdom, this *nunc stans* is figured as a rock on which the stream of time breaks, or an everlasting midday whose heat is unrelenting.[15]

With unyielding consistency Schopenhauer therefore rejects any view that sees temporal becoming as affecting either the one abiding thing-in-itself or our relation to it. In dismissing those (such as Hegel) who take such an approach he does not attempt to conceal his contempt: 'anyone who imagines that the inner nature of the world can be historically comprehended, however glossed over it may be, is still infinitely far from a philosophical knowledge of the world. But this is the case as soon as

---

[11] Schopenhauer, *World*, II. 271.   [12] Schopenhauer, *World*, II. 487.
[13] See the section 'Can a timeless God know temporal beings?' in Chapter 1 of this volume.
[14] Schopenhauer, *World*, I. 279.   [15] Schopenhauer, *World*, I. 280.

a becoming, or a having-been, or a will-become enters into his view of the inner nature of the world; whenever an earlier or a later has the least significance'.[16]

We can now see why Nietzsche, reversing what he saw as Schopenhauer's 'Buddhist' negative evaluation of life and calling on his readers to embrace the pain and agony of the world of becoming as a necessary and integral part of joyous self- and life-affirmation, should also see this affirmation as, in some sense, 'eternal'. In a sense, nothing has changed from Schopenhauer's view of the 'Now' as an eternal present, an unending noon-day (an image that recurs also in Nietzsche)—except that where Schopenhauer says 'No!' Nietzsche calls upon us to say 'Yes!' The infinite and unqualified affirmation that Schopenhauer reserved for the will is now lavished by Nietzsche on what Schopenhauer could only see as the cause of endless misery and therefore as what needed to be renounced, namely, life in time. Where the one (Schopenhauer) sees eternity as the great alternative to time, the other (Nietzsche) depicts time itself as (in some sense) eternal.

Schopenhauer claimed that despite his opposition to theism his philosophy served to rescue the genuine impulses of Christian asceticism and world-renunciation. In this respect Kierkegaard, whom we shall consider later in this chapter, could see a certain parallel to his own protest against Hegelian theology's conflation of time and eternity.[17] Karl Barth, however, saw Schopenhauer more as a modern version of the Marcionite heresy.[18] In any case, his thought undoubtedly encouraged the gradually increasing circle of those in the West experimenting also with Asian ascetic practices. In an age largely dominated by the attempt to think historically and therefore to project the possibility of a common horizon shared by both time and eternity, Schopenhauer put down a reminder that, leaving aside any purely philosophical considerations, the soteriological impulse of religion might demand satisfaction from something more than or other than time and history. Strange as his conception of the eternal was, it provoked the thought that the eternity of the philosophers might be far from satisfying for the religious soul's longing for an eternal salvation. As we shall shortly see, Kierkegaard and Barth are just two prominent voices from within the modern Christian tradition itself who agree and who similarly suppose that eternity may have a negative and even hostile relation to time, at least to the kind of time focused exclusively on such fulfilments as time itself can offer.

---

[16] Schopenhauer, *World*, I. 273.

[17] For the Kierkegaard–Schopenhauer relationship see Niels Jørgen Cappelørn, Lore Hühn, Søren R. Fauth, and Philipp Schwab (eds), *Schopenhauer – Kierkegaard: Von der Metaphysik des Willens zur Philosophie der Existenz* (Berlin: De Gruyter, 2012).

[18] See Karl Barth, *Church Dogmatics*, ed. G. W. Bromiley and T. F. Torrance (London: T. & T. Clark, 2004), III.1. 334–40.

We have seen that in a certain—if surprising—sense, one aspect of Nietzsche's reversal of Schopenhauer is that time itself becomes invested with the value of what, in previous philosophical and religious thought, had been construed as eternity. But the easier (and probably prevalent) reading of Nietzsche is that he simply extended Schopenhauer's critique by eliminating the realm of eternal truths and values entirely. On this reading, 'eternity' is one amongst other attributes of what Nietzsche calls the 'true' world proposed by idealist philosophers and Christian theologians, that is, a world unaffected by the changes and chances of life in time, of which Nietzsche writes that it is 'an idea for which there is no more use and no longer even one we are obliged to hold... We have got rid of the real world'.[19] Whether this 'true' world is conceived in terms of a level of causality higher than the chain of observable causes that operate within the continuum of worldly being or as a realm of unchanging moral values, Nietzsche judges it to be an illusion that is no longer fit for purpose. Even if it might in the past have served to elevate human beings above brutishness and to train them in the subtleties of thinking, it can do so no longer. Insight into its fictive character makes it impossible to use it in good conscience, not even as an instructive myth. Having been shown by science and philosophy to be no longer a possible object of knowledge—to which religious references to the incomprehensible or mysterious nature of God also indirectly testify—we cannot evade the reality that this 'true' world is in fact only the projection of a disordered state of the worldly body. Belief in such a 'true' (or 'other' or transcendent or perfect or eternal) world is so far from providing genuine insight into the order of things as to be rather the product of decadence, a failure on the part of the will to live and to embrace life in its manifold contradictions and endless changeableness. It is therefore very easy to take Nietzsche's reversal of metaphysics as simply wiping the horizon clean of all metaphysical ideas, including eternity. But whilst this is not false, it is clear that it is the following features of metaphysical ideas that are the particular objects of his attention: that they are possible objects of knowledge, that they are timeless, and that they are valued above the actions and passions of embodied life. That Nietzsche has to be opposed to any version of metaphysics that supposes the possibility of knowing a timeless God or of knowing the world *sub specie æternitatis* is therefore undeniable and it is therefore entirely correct to take Nietzsche as a critic of divine eternity in this sense. However, as we have also seen, he is also prepared to transfer the attribute of eternity to time itself—an eternity that is no longer knowable but that can be willed and sung, as expressive of the infinite value of temporal life.

[19] Friedrich Nietzsche, *Götzen-Dämmerung* in *Werke in drei Bänden III*, ed. K. Schechta, (Frankfurt am Main: Ullstein, 1972), 409.

'IF I HAVE A SYSTEM ...'

In the Preface to the second edition of his commentary on Paul's letter to the Romans, Karl Barth famously wrote that 'if I have a system it is limited to a recognition of what Kierkegaard called the "infinite qualitative distinction" between time and eternity, and to my regarding this as possessing negative as well as positive significance: "God is in heaven, and thou art on earth".'[20] What is implicit in this programmatic statement and becomes explicit at many subsequent points is that 'eternity' functions for Barth as an especially significant name of God, as when he speaks of 'the hidden line, intersecting time and eternity, concrete occurrence and primal origin, men and God' (ER, 29).[21] Of course, these two citations already highlight a distinctive tension in Barth's conception of the relations of time and eternity. For if the words from the Preface speak of their distinction, separateness, and distinction, the second quotation speaks of 'intersection'. However, this is not a simple or naive contradiction since it is of the essence of Barth's overall position that it is precisely the God who, qua Eternal, is eternally infinitely and qualitatively distinct from human beings, is also paradoxically present to them in that 'moment' of intersection that is the event of Jesus Christ and, more specifically, his resurrection. The name 'Jesus Christ' is the place where 'two worlds'—God's eternal world and human beings' temporal world—'meet and go apart' (ER, 29).

Much of the impact of Barth's work was—and to new readers surely remains—the expressionistic verve with which he drives home the negativity of this relationship, piling metaphor upon metaphor so as to obstruct any attempt to think this point of intersection as constituting an event in time that could be assimilated to the general continuum of human history. Variously referring to this event as the Word or Gospel or Power of God, as the name Jesus Christ, as the Resurrection, or as the divine Judgement on human existence, it is in each of its forms a crater left by an exploding shell, a void (ER, 29), it comes 'vertically from above' (ER, 30), a tangent that both touches and doesn't touch the circle of history (ER, 30), a frontier and a new world (ER, 30), 'not an event in history at all' (ER, 30), 'the KRISIS of all power' (ER, 36), a barrier and an exit (ER, 38), a 'contradiction' (ER, 38), 'the divine incognito' (ER, 39), 'the hidden abyss' (ER, 46), 'a fiery miracle' that leaves behind only a pile of burnt-out clinkers (ER, 65)—and so on. As Barth sums up the temporal significance of the human condition apart from the gospel, 'time is nothing when measured by the standard of eternity [and] all things are semblance when measured by their origin and by their end, [and] that we are sinners, and

---

[20] Karl Barth, *The Epistle to the Romans*, tr. Edwyn C. Hoskyns (Oxford: Oxford University Press, 1933), 10. (Further references are given in the text as 'ER'.)

[21] 'Primal origin' too is a term that Barth uses at a number of points as a name of God. As we shall see, this is not without significance for his conception of the relations of time and eternity.

that we must die' (ER 43). Or, in brief, 'We confound time with eternity. That is our unrighteousness' (ER, 44).

Barth knew himself to be taking up and developing the late nineteenth century's rediscovery of eschatology in the origins of early Christianity. Writing about Johannes Weiss's book *The Preaching of Jesus Concerning the Kingdom of God* (1892), Albert Schweitzer stated that it was as important as Strauss's *The Life of Jesus Critically Examined* (1835) in that it too confronted theology with a basic decision about the life of Jesus. As Schweitzer put it: '*either* eschatological *or* non-eschatological!'[22] Schweitzer himself chose the former option, arguing that whilst it was possible to construct a plausible life of Jesus on the basis of the historical materials provided by the New Testament, the Jesus thus revealed was an entire stranger to the world-view of modern times. He lived and died entirely within the horizons of a set of apocalyptic expectations about the coming Kingdom of God that modern people just can't share. Therefore, he concluded, the real historical Jesus 'will be to our time a stranger and an enigma...He passes by our time and returns to His own'.[23]

Barth could certainly agree to this last, even if he did not go along with the emphasis placed on historical-critical methods by Weiss, Schweitzer, and others. In a way that he would subsequently see as 'one-sided', the commentary on Romans insisted on the eschatological futurity of God's relation to time.[24] The Christian's relation to divine eternity is a relation to 'the "Moment"—which is no moment in time—when the last trumpet shall sound and men stand naked before God' (ER, 109). But although judgement is an ineluctable element of this relation, Barth is also clear that the Christian attitude is essentially one of hope: the 'No' that reveals human beings' exclusion from eternity is the other side of the 'Yes' that invites them to the new world of eternal life; 'the barrier marks the frontier of a new country' and 'We have therefore, in the power of God, a look-out, a door, a hope' (ER, 38); thus, in an allusion to the prophet Habbakuk, 'The prisoner becomes a watchman. Bound to his post as firmly as a prisoner in his cell, he watches for the dawning of the day' (ER, 38). As Barth puts it in commenting on Romans 8.15, the world as seen in the light of God's will to save—the world to which God says 'Yes'—is 'the world of life...in which nothing that is limited and trivial and passing to corruption exists as it is, but IS what it signifies. Related to its final origin and purpose, to its final

---

[22] A. Schweitzer, *The Quest of the Historical Jesus*, tr. William Montgomery (London: A. & C. Black, 1954), 237.

[23] Schweitzer, *Quest*, 397.

[24] See Karl Barth, *Church Dogmatics*, II.1: *The Doctrine of God*, ed. G. W. Bromiley and T. F. Torrance (London: T. & T. Clark, 2004), 634–5. In addition to the scholarly 'discovery' of eschatology by Weiss and others, Barth also draws attention to the less academic and more popular apocalyptic spirituality of Christoph Blumhardt. The presence of a strong current of apocalyptic thought in late nineteenth century German Pietism is also illustrated in Gerhardt Hauptmann's extraordinary novel *Joseph Quint: Der Narr in Christo* (*Joseph Quint: The Fool in Christ*).

meaning and reality, the course of this world becomes, even in its questioning, full of answer; even in its transitoriness, full of eternity', and we ourselves become 'new men standing on the threshold of this new world' (ER, 178).

But this prioritization of the future and the accompanying imperative of hope is not intended by Barth to be a back-door to reintroducing continuity between time and eternity. As we have just seen, even though the 'moment' is the 'moment' of the Last Judgement, it—like the Resurrection—'is no moment in time'. Consequently, it is not part of a historical time series. As 'the creation of the divine righteousness in us and in the world' it is

> a new creation; it is not a mere new eruption, or extension, or unfolding, of that old 'creative evolution' of which we form a part and shall remain a part, till our lives' end. Between the old and the new creation is set always the end of this man and of this world. The 'Something' which the Word of God creates is of an eternal order, wholly distinct from every 'something' which we know otherwise. It neither emerges from what we know, nor is it a development of it. Compared with our 'something' it is and remains always—nothing. (ER, 102)

Barth's early theology was often referred to as 'dialectical theology' and we can see why: every human 'Yes' is countered with a divine 'No': every divine 'Yes' also has as its corollary a human 'No'. As Barth continues on to Romans 8:24 he says it again. In Jesus Christ 'there is that which horrifies: the dissolution of history in history, the destruction of the structure of events within their known structure, the end of time in the order of time' (ER, 103)—but, as he adds, Christ's life is also and at the same time 'history pregnant with meaning; it is concreteness which displays the Beginning and the Ending; it is time awakened to the memory of Eternity; it is humanity filled with the Voice of God' (ER, 103–4).

There are some striking phrases here, not least the reference to the event of Jesus Christ—the intersection—as 'time awakened to the memory of Eternity'. That time might contain a memory of eternity is scarcely a possibility that Barth's accumulated negatives might have allowed us to suspect. However—and apart from the fact that he acknowledges Plato as having offered some inspiration in his work on the second edition of the commentary—it is a possibility he has flagged early on in the context of expounding 'the night' of human existence without God (with reference to Romans 1:3): 'our life in this world has meaning only in its relation to the true God. But this relation can be re-established only through the—clearly seen—memory of eternity breaking in upon our minds and hearts' (ER, 48). Or, shortly before that: 'He is the hidden abyss; but He is also the hidden home at the beginning and end of all our journeyings' (ER, 46). And, perhaps with the image of history as 'pregnant with meaning' we might ponder the words of some verses cited by Barth in the preface to the fifth edition of his work:

God needs MEN, not creatures
Full of noisy, catchy phrases.
Dogs he ask for, who their noses
Deeply thrust into—To-day,
And there scent Eternity.
Should it lie too deeply buried,
They go on, and fiercely burrow,
Excavate until—To-morrow. (ER, 24)

This may not be great poetry, but it moves Barth to declare his wish to be such a 'Hound of God' (a 'Domini canis', as he adds with a touch of lugubrious humour). But after all that we have heard of the 'essential, sharp, acid, and disintegrating ultimate significance' (ER, 49) of the distance between God and history it seems strange to be faced with images that might even be seen as suggesting a quasi-Hegelian (or perhaps Nietzschean) suggestion that eternity is the true meaning of temporality. At the very least, it is clear that however original his solution is going to be, the question with which Barth is struggling is essentially related to the question that also gripped German philosophy from Hegel to Nietzsche, namely, how a thoroughly temporalized and historicized world can manifest or be interpreted through a relation to the eternal.

In the *Church Dogmatics*, Barth would parse divine eternity in terms of what he called pre-temporality, supra-temporality, and post-temporality. Theologically this distinction could—and he suggests did—come to expression in, respectively, the Reformers' preoccupation with the activity of the predestining divine will 'from before the foundation of the world', the instantaneousness of time and eternity in Romantic religiosity, and Barth's own early eschatological emphasis, which, as we have heard, he came to regard as one-sided. True divine eternity is not one dimension of eternity more than another. It is all three, connectedly. However, Barth's self-criticism is perhaps unfair on what is actually going on in the text of the commentary on Romans, since there too, the pre-temporality of God is repeatedly emphasized. 'The Gospel is the Word of the Primal Origin of all things' (ER, 28), whilst 'the world of the Father' is set in apposition to both 'the primal Creation' and 'the final Redemption' (ER, 29). What God effects in 'the End' is a new creation, but it is also, simply, 'creation', bringing into being the world as he wills it to be. Similarly, images of intersection and tangent or the emphasis on the divine event not being an event in history point to God's supra-temporality and to the possibility of his being present to any moment in the historical process.

Why this threefold qualification of God's relation to time is important in the argument of the *Church Dogmatics* is because what Barth sees as being at issue is not God's metaphysical qualities but his constancy in willing to save and in showing his gracious favour towards human beings. This, as he sees it, would be only inadequately and misleadingly expressed by insisting on divine timelessness. Eternal duration, he suggests, does better religious work than timelessness. As he puts it 'Because and as God has and is this duration, eternity, He can and will be true to himself, and we can and may put our trust in Him'.[25] Perhaps surprisingly, he offers a rather warm commendation of Boethius' definition, taking Boethius—as against his scholastic interpreters—to be describing duration, not timelessness. For whilst the human experience of the *nunc* (the 'now') is of an empty and transient moment, this is not true of Boethius' divine 'now', precisely because this is a now that involves the perfect possession of eternal life. 'But we know eternity,' he writes

> primarily and properly, not by the negation of the concept of time, but by the knowledge of God as the *possessor interminabilis vitae*. It is He who is the *nunc*, the pure present. He would be this even if there were no such thing as time. He is before and beyond all time and equally before and beyond all non-temporality. He is this *nunc* as the possessor of life completely, simultaneously and perfectly, and therefore to the inclusion and not the exclusion of the various times, beginning, succession and end.[26]

Barth consequently endorses the interpretation of divine eternity as *sempiternitas* (although probably not intending the connotation of having a beginning in time), commenting also that 'the statement that God is eternal tells us what God is, not what He is not'.[27] However, if this might be read as arguing for a kind of temporalizing of God, Barth's fundamental thesis concerning the separation and distinctness of time and eternity remains in place. As these lines themselves indicate, God's eternity is not merely distinct from temporality—it is not dependent on or definable in relation to it in any compelling way. God is eternal, with or without time. God's duration, as Barth also says, 'is just the duration which is lacking to time'.[28] In fact, 'Time can have nothing to do with God'.[29] Yet once again the dialectic of Yes and No makes its presence known, since if we once allow God's complete transcendence over time, it also follows that God is not excluded from acting in time or from making time what Barth calls 'the formal principle of his activity outwards'.[30] But God can act in time, as the Bible bears witness, although He does not Himself become temporal in so doing.[31] He acts in time as the eternal, eternally constant and self-same in love and faithfulness, which, as Barth goes on to explain, is what it means for God to manifest his glory. Nevertheless, this clearly does not cancel the dialectic and presupposes rather than qualifies what Barth still sees as the decisive distinction between time and eternity.

---

[25] Barth, *Church Dogmatics*, II.1. 609.   [26] Barth, *Church Dogmatics*, II.1. 611.
[27] Barth, *Church Dogmatics*, II.1. 613.   [28] Barth, *Church Dogmatics*, II.1. 608.
[29] Barth, *Church Dogmatics*, II.1. 608.   [30] Barth, *Church Dogmatics*, II.1. 609.
[31] A different reading is offered by Eberhard Jüngel, *God's Being Is in Becoming: The Trinitarian Being of God in the Theology of Karl Barth*, tr. John Webster (Edinburgh: T. & T. Clark, 2001).

And yet—again!—if, despite this distinction, faith is able to see time as pregnant with eternity or to sniff out the eternity that is hidden in time, what is
Barth saying that really goes beyond the repeated efforts of German philosophy to think the unity of time and eternity? The answer is, surely, the simplest
and most obvious answer, that for Barth this is not accomplished as theory
(as for Hegelianism), nor as an act of supra-temporal will (as in Schelling),
not yet as willing the eternal recurrence of time (Nietzsche) but precisely as
faith. And faith, for Barth, is not in the first instance an act of human will. It is
dependent—entirely dependent—on 'the faithfulness of God', that is, on God's
constant will to be faithful to his original act of creation and to bring even
fallen humanity into the sphere of the love that that act itself enacted. Even
with regard to the Yes in which God both wills and reveals the unity of time
and eternity, then, the difference between temporally extended human life
and eternally enduring divine life is repeated and underlined. Why? Because
of ourselves we can neither see this unity nor will it into being: it must be
revealed to us if we are to become capable of believing it. In time there is nothing more we can do.

From the Barthian perspective, the error of the nineteenth century trajectory is that it entirely presupposes the competence of human beings to resolve
the infinite qualitative difference between what belongs to God and what
belongs to the creature. Putting it otherwise, Barth's claim is that the Christian
view presupposes a fundamental passivity or limitation on the part of human
beings. This is not a matter of simply promoting heteronomy over autonomy
or of drawing attention to the dependence of human life and mental processes
on, say, their biological or social conditions. It is rather a matter of indicating
a limitation we just can't do anything about or with or to. The eschatological
'Moment' is not 'a supreme human experience' we can cultivate or enjoy (ER,
109): it is simply the reality of our limitation vis-à-vis God. Being as we are,
we are thoroughly and through and through conditioned by temporality. Yet
we are neither the initiators nor the maintainers of our own time. Our time,
as Barth several times remarks (alluding to Goethe), is a parable, and only a
parable of eternity (ER, 50, 107). This is what we persistently forget. And it is
precisely when, still in this state of forgetfulness, we try to think or to will the
unity of time and the eternal that we become guilty of what he calls 'confound-
ing' time and eternity, rather than truly living their truly divine unity.

But what we also forget when we thus confound time and eternity is that our
own life is handed over to time in such a way and to such a degree that we too
are as transient as all other temporal things and that our life too can and must
come to its end in death: 'the operation of this forgetfulness and the end of
our wandering in the Night is—Death' (ER, 54). Barth places a question-mark
here and, as we might assume, it is a question that is answered very differently
according as to whether we are approaching it from a human and temporal
point of view or from a divine and eternal point of view. But within a human

perspective that has not been illuminated by faith, there is no doubt that the answer is indeed: death. 'Our life is confronted with a steep precipice, towering above us, hemming us in on every side, and on it are hewn the words: *All things come to an end*' (ER, 170). In the perspective of faith, everything is changed in the light of Christ's resurrection, but in 'the night' of a life in time that has no graced vision of eternity all things—ourselves included—come to an end. If eternity opens the possibility of eternal life, time, without eternity, is—to anticipate what we will read in Heidegger—thrownness towards death.

All of this raises many questions. As a dogmatic theologian, Barth is arguably true to his metier in asserting rather than arguing for the durational character of divine eternity or the impossibility of thinking the unity of time and eternity without reference to God's saving action. Of course, these are not merely arbitrary assertions but appeal to an established tradition of interpreting the text of the Christian Bible. Nor is Barth ignorant of their distinction from the manifold of philosophical positions that his contemporaries are taking to the same complex of questions. This is not naive dogmatism and Barth is not to be reproached for not answering questions that he consciously rejected as inappropriate. But if we do allow the legitimacy of human questioning in these matters, there remain large and unresolved issues, not least relating to the possible cognitive content of the 'knowledge of God' given in faith. If faith is faith in an event (the Resurrection) that is not an event in time and if the possibility of any human cognitive act is dependent on temporal experience or (as we considered in Chapter 1) on temporally inflected propositional utterances, what can the content of Christian knowledge possibly be? What do we *know* concerning the relationship between time and eternity when we know Jesus Christ as the point of intersection of the two planes or two worlds of divine eternity and creaturely temporality? And how, more concretely, might such knowledge transform our lives as we continue to live them in time?

These are questions that have haunted the commentary on Romans since its first publication, as Barth's own comments in the prefaces to later editions acknowledge. And for many readers they are not 'solved' by the *Church Dogmatics*. Here too we are confronted with the same fundamental aporia. Let us say that it is good that we are thus confronted and that forcing the issue in the way that he does is one of Barth's theological merits—a merit that is also beneficial to those who come at the questions from a more philosophical angle. But this does not mean that we are obliged to give up on our own attempts to break through 'the barrier' or to cross 'the polar zone' of the distinction of time and eternity as expounded by Barth. Nor need we hold back from pressing further with the question as to what, in the light of this distinction, the thought of eternity could nevertheless mean for the thoroughly temporal beings that we ourselves are.

As we have seen, Barth acknowledged Kierkegaard as an influential companion in his reading of Romans, and we shall see that there are many points at which there is a deep congruence between aspects of their thinking of time and eternity, not least with regard to their criticisms of speculative attempts to think life in time *sub specie aeternitatis* and their shared recognition of a fundamental 'passivity before passivity' in our relation to our own temporality. Yet, and as Barth himself would become aware, there are also significant differences in tone and approach, not least with regard to those elements in Kierkegaard that made him a source for Heidegger's philosophy of *Existenz*—a remark that is no less in place if we also acknowledge the possibility that this philosophy also significantly *mis*read Kierkegaard. But before we can consider that possibility we must attempt to set out just what it is that these two— Kierkegaard and Heidegger—did in fact have to say on the relation of time and eternity.

GOD IS IN HEAVEN, WE ARE ON EARTH

Barth declared the second edition of his commentary on Romans to be significantly in accord with Kierkegaard's insistence on the infinite qualitative difference between God and humanity. As we shall see, Kierkegaard's position contains conflicting tendencies and cannot be reduced to a simple formula.[32] Nevertheless, it is the case that there is a strong current in Kierkegaard's thought that, with Barth, emphasizes the *difference* of time and eternity. This finds expression both in his pseudonymous work as well as in some of the religious writings published under his own name.

Perhaps the best-known discussion is found in the *Concluding Unscientific Postscript*, in the context of Kierkegaard's critique of speculative idealism. It is central to Kierkegaard's argument here that the fundamental error of the speculative philosophers (and more particularly of the Hegelians) is precisely to have forgotten their own ineluctable temporality and to imagine that they can comprehend the whole of existence in a single unified system '*sub specie aeterni*', as he puts it.[33] This objection is developed with particular reference to a series of theses 'possibly or actually' attributable to Lessing: that the subjectively existing thinker is attentive to the dialectic of communication; that the subjectively existing thinker is as existentially negative as positive, is as comic as expressive of deep pathos, and is constantly in a process of becoming; that historic events can never provide proof for eternal truths and that the transition from historical existence to eternal blessedness can only occur by virtue of a leap; and that if God held all truth in his right hand and the unceasing search

---

[32] We shall explore the more positive aspects to Kierkegaard's treatment of time and eternity in Chapter 8.

[33] SKS7, 81/CUP, 81.

for truth in his left, it would always be preferable to choose the left. Kierkegaard makes much of the fact that his readers will in all likelihood regard the appeal to Lessing as rather outdated, but in fact it goes to the heart of a set of debates integral to the history of German Idealism itself, especially Idealism's relation to Spinoza and the question of pantheism.

After introductory reflections on the nature of subjective communication—which also have significant implications for attempts to present truth systematically or objectively—Kierkegaard comes to what he calls the negativity of the subjective individual. What does he mean by this? His starting-point is that the subject exists only in a process of becoming or, as he also puts it, in a dialectic of being and non-being. It is precisely the error of speculation that, on this basis, it nevertheless believes it to be possible to pass from the deliverances of sense certainty and to proceed through historical knowledge so as to arrive at a speculative result. But this, Kierkegaard says, is precisely to have forgotten the conditions under which an actually existing thinker thinks. All knowledge is only ever an approximation and the difference between being and non-being could only ever be resolved in infinity. Now, as he also says, the human person is 'existing infinite spirit' (SKS7, 81/CUP, 82) but this formula involves giving equal weight to existence and to infinity as well as to the difference between them. The paradox of existing subjectivity, then, is 'this monstrous contradiction, that the eternal becomes, that it comes into existence' (SKS7, 81/CUP, 82). That the individual existing subject is characterized as 'infinite' does not mean that it has the attribute of infinity or is capable of infinitely transparent self-consciousness however. Rather, it points to the negative condition that no actual finite manifestation of the existing subject's life is ever going to encompass the whole—and if the subject pretends to a positive infinite knowledge, then it becomes merely comical. The finite and infinite poles between which human existence is poised—or being and non-being—can only be related in an unresolved tension, never mediated. As Kierkegaard puts it

> The subjectively existing thinker, who has infinity in his soul, has it all the time, and for this reason his form is only ever negative... [H]is positivity consists in continuously internalizing [this condition] in which he is conscious of this negativity... He knows about the infinite negativity of his existence and constantly keeps the wound of this negativity open, which can sometimes prove salvific (whilst those who let the wound heal over and who become positive are deceived) (SKS7, 84/CUP, 84–5).

In other words, those who bear this existential condition in mind will never forget that they do not exist otherwise than in a process of becoming and that being a self is therefore not a fixed quantity but a matter of constantly striving to become and to remain who we are or choose to be.

But this also means that our relation to the prospect of an eternal blessedness (which Kierkegaard takes to be a basic feature of the religious view of life) cannot mean affirming that the human soul is immortal, since this would be merely an item of objective or speculative 'knowledge'. Instead, it is a matter of how we strive and choose to be in life. Conventional human wisdom sees eternal blessedness as 'an eternal presupposition located behind one, in immanence, [the same] for every individual. Qua eternal each one is higher than time and therefore has his eternal blessedness behind him' (SKS7, 93/CUP, 95)—in other words, as an attribute or qualification 'in' the soul from its inception. This 'pastness' correlates to the objectivity of this supposed knowledge. But, says Kierkegaard, Christianity teaches that eternal blessedness is only one possible outcome of human existence and we can never allow ourselves to forget that its opposite—eternal misery—is also possible. And whether it is the one or the other that awaits us is not a matter of some attribute applicable to human beings independently of their life in time but of how, in time, we choose to be. Another way of putting this is that the passage from time to eternity is not an immanent development along a continuum, but a leap. And this is also to say—again!—that it is not a relationship that can ever be regarded as complete or finished, which is precisely the illusion entertained by those who strive for and even claim to have attained systematic completeness. In brief, then, those who aspire to a speculative view of the whole of existence have forgotten the conditions under which they are human. As Kierkegaard remarks, 'Such a one wants to be speculation, pure speculation and I must therefore give up talking to him since, in that moment, he ceases to be visible to me or to any weak, mortal human eye' (SKS7, 105/CUP, 109).[34]

This leads Kierkegaard to the following distinction: that a logical system is possible but an existential system is impossible. Again, he emphasizes the role of time and movement. In agreement with Trendelenburg's *Logical Investigations* he dismisses Hegel's claim to have developed movement on the sole basis of logical relations. Movement can only be introduced into the system, he insists, if it is already presupposed or, less politely, has been smuggled into it on the basis of unacknowledged empirical experience. Consequently, anyone who believes they have succeeded in bringing the manifold of existence into a unified speculative system is self-deluding and has given up the basic condition of being human. The system cannot be completed by any earthbound individual. Humorously, he suggests it might find readers on the moon and, more seriously, that whilst it is impossible for human beings it might be possible for God: 'The one who is himself outside of existence and yet in existence, who, in his eternity, is eternally complete and yet comprises existence within himself' (SKS7, 115/CUP, 119)—but such a one is and can only be God.

---

[34] This whole section of the *Postscript* is amongst the most satirical parts of Kierkegaard's authorship, and alongside his logical objections to Hegelianism he makes plentiful use of mockery. On the philosophical justification for this procedure see John Lippitt, *Humour and Irony in Kierkegaard's Thought* (Basingstoke: Macmillan, 2000).

Kierkegaard returns to these themes in a later section of the *Postscript* where he is analysing what it means to be ethical or to be actually subjective and describing one he calls 'the subjective thinker'. As he states at the outset of this section 'abstract thought is *sub specie aeterni*' and therefore 'disregards the concrete, temporality, the becoming of existence, and the need that the existing person has to combine the eternal and the temporal in existence' (SKS7, 274/CUP, 301). This could seem to recall Hegel's own criticism of Spinoza, but—once more with reference to Trendelenburg— Kierkegaard asserts that, precisely because it has recourse to characteristics of experience that, on its own premises, should not be available to it, the Hegelian 'solution' is merely specious. Hegel, he says, 'is entirely and absolutely right in saying that, in an eternal perspective, *sub specie aeterni*, in the language of abstraction, in pure thinking and pure being there is no *aut—aut* [either—or]' (SKS7, 277–8/CUP, 305). But it is a different matter again when this is meant to be a point of view available to concretely existing individuals, and Kierkegaard sarcastically asks whether an existing individual is *sub specie aeterni* when he 'is sleeping, eating, blowing his nose and whatever else a person does' (SKS7, 278/CUP, 306). For the existing person eternity can only be present as a future possibility that will only ever become actual (or not) in relation to that individual's concrete and temporally-situated *choices*.

> Where everything is in a process of becoming...where eternity relates as *à-venir* to becoming, there we have the absolute disjunction. That is, when I juxtapose eternity and becoming I do not get rest but *à-venir*. This is one good reason why Christianity has proclaimed eternity as what is to come, because it is proclaimed to existing individuals and therefore supposes that there is an absolute *aut—aut*' (SKS7, 280/CUP, 307).[35]

If thinking always tends towards a timeless eternity, abstracting from the concrete and local interests of time-bound entities, existential thinking cannot but be interested in the thinker's own existential predicament and it is 'interested' also in the technical sense that it reflects a kind of *inter-esse*,

---

[35] I am freely translating Kierkegaard's '*Tilkommelse*' as à-venir, a play on words made familiar by Lévinas and which is in fact etymologically more akin to the Danish word used by Kierkegaard (literally 'coming-towards') than is the English 'future'. One might use a variation on 'advent', although in the present passage, where Kierkegaard uses both *Tilkommelse* and *Tilkommende* it might then be more appropriate to use both 'advent' and 'advenient', but apart from sounding excessively ecclesiastical this would also seem more forced. Using the French form suggests— deliberately—that what Kierkegaard is saying here anticipates the kind of point that Lévinas makes by punning on *à venir* ('to come') and *avenir* (future). In other words, the eternal is a future that is always coming but that never arrives in time. A further anticipation of Lévinas might be seen in the role of the infinite in the constitution of the self, discussed at the start of this section. However, I am not suggesting—and there seems little evidence for the view—that Lévinas in any way 'derived' these views from Kierkegaard.

a being-between being and non-being, between possibility and actuality, between time and eternity, living a tension that can never, in life, be resolved.

If, in these ways, Kierkegaard develops the idea of the 'absolute disjunction' between time and eternity in the context of a philosophical critique of some of the basic assumptions of Hegelian philosophy, he elsewhere develops the same idea in more directly religious terms. In 1847, a year after the *Postscript*, he published a collection of religious discourses entitled *Upbuilding Discourses in Various Spirits*, the third part of which, 'The Gospel of Sufferings', contained a discourse on the theme that 'the school of sufferings educates us for eternity'. The discourse is based on the text of the Letter to the Hebrews 5:8: 'nevertheless His Son [i.e. God's son, Christ] learned obedience from what He suffered'. Although, as Kierkegaard puts it, 'His will was in agreement with the Father's from eternity and his free resolve was the Father's will' (SKS8, 351/UDVS, 253), the conditions of incarnation meant that he had to 'learn' obedience and to do so through suffering. As becomes apparent, Kierkegaard does not regard these sufferings as primarily physical but rather as connected with Christ's being misunderstood, ignored, or rejected by those he loved and came to save. This suffering culminates in the prayer in Gethsemane, prayed in relation to his imminent crucifixion: 'if possible, Father, let this cup pass from me, but not my will but thy will be done' (SKS8, 353/UDVS, 255). Here we can see that the connection between suffering and obedience is by no means immediate or automatic. As Kierkegaard points out, suffering also exposes a person to despair and loss of faith, unless they are willing to let themselves be helped by God. This, as he puts it, 'is the highest risk for the highest gain, and the highest gain is the eternal' (SKS8, 354/UDVS, 256).[36] We can learn much by going out into the world and finding out all sorts of things, but suffering turns us in upon ourselves and it is through such an inward turn that we learn obedience and, through obedience, gain the eternal. 'When we suffer and want to learn from what we suffer, then we will only ever learn about ourselves and about our relation to God and this is a sign that we are being educated for eternity' (SKS8, 355/UDVS, 257).

Suffering brings us up against the limits of our autonomous powers, confronting us with what we are unable to do or to control. Here is the limit to what we can will, and if obedience is not a response to externally imposed suffering, then, Kierkegaard suggests, it might after all be a case of self-will—a masochistic preference for self-abnegation. Suffering that is not self-inflicted, however, detaches us from reliance on our own will-power and, if it is then

---

[36] It is striking that in these religious discourses, Kierkegaard shifts from talking about 'eternity' (as in the *Postcript*) to 'the eternal', although (as we shall see) he continues to use 'eternity' as well. I suggest that the intention here is to emphasize that 'the eternal' is not just a metaphysical quality but a name of God, whereas the 'eternity' of the *Postscript* was an abstract concept that could comprise both God and other non-divine eternal truths, such as the eternal truths of logical or mathematical relationships.

brought into relation to God, it teaches obedience and, specifically, teaches us 'to let God rule, let God reign'. 'And,' Kierkegaard asks, 'what is any eternal truth if not: that God reigns; and what is obedience but letting God reign, and what other connection and agreement is possible between the temporal and the eternal than: that God reigns and letting God reign!' (SKS8, 355/UDVS, 257). And this is also why, where the world is in a state of perpetual motion and restlessness (i.e. of becoming), the God-relationship can establish a basis for rest and peace. Therefore, as Kierkegaard adds, 'to find rest is to be edu-cated for the eternal' (SKS8, 356/UDVS, 258). Why? Because, once more, it involves the surrender of self-will.

Kierkegaard concludes with two final points. Firstly, he suggests that the highest qualifications usually require the longest schooling and that, by anal-ogy, what educates us for eternity must last as long as temporal existence itself. The relation to the eternal that is the goal of learning obedience through suf-fering is not an 'aim' that can be achieved in time but is a qualification of the (unattainable) totality of temporal existence. If we take all this to heart in the spirit of repentance it will, secondly, prove to be a joyful thought that restores spiritual youth.[37] And so Kierkegaard ends by again emphasizing the intercon-nectedness of suffering, obedience, faith, and eternity: 'Only suffering educates us for eternity: for eternity is in faith, but faith is obedience, but obedience is in suffering. Obedience is therefore not external to suffering, faith not external to obedience, eternity not external to faith. In suffering, obedience is obedient; in obedience, faith is faith; in faith, eternity is eternity' (SKS8360/UDVS, 263). However, whilst this discourse might seem to suggest that suffering provides some kind of bridge from temporal existence to eternity, it is clear that this only happens under certain significant conditions: that the relationship is nei-ther one of knowledge nor one of will (as in Hegel and Schelling respectively) and that it has an essentially negative relation to what is generally regarded as the positive content of life. In terms that Kierkegaard will develop more extensively in other later religious writings, it is an 'inverse' relationship.[38] The relation to eternity does not give us knowledge of or mastery over time, but, in time, detaches us from reliance on and identification with time whilst simul-taneously, qua suffering and qua obedience, guarding against the self-forgetful hubris of identifying ourselves with the eternal and absolute knowledge that is proper solely to God.

An especially clear application of this 'inverse dialectic' to the relations of time and eternity is found in the second section of the *Christian Discourses*

---

[37] The idea is not really spelt out, but is presumably that a young person, in Kierkegaard's time, is assumed to be a person whose life-choices are subject to guidance by another and who therefore does not have to bear the weight of sole and entire responsibility for their existence.

[38] The idea of an 'inverse dialectic' is a major theme in Sylvia Walsh's interpretation of Kierkegaard's Christian writings. See Sylvia Walsh, *Living Christianly: Kierkegaard's Dialectic of Christian Existence* (University Park, PA: Pennsylvania State University Press, 2005).

from the following year 'Joyful Notes in the Strife of Suffering', and is already indicated by the titles of the first and fifth discourses: 'The Joy in the fact that one only suffers once but triumphs eternally' and 'The Joy in the fact that what you lose in temporal terms you gain eternally'.

In the first of these discourses Kierkegaard explains that what is at issue is not just a matter of how to deal with particular cases of suffering that might befall a person from time to time in their lives but our relation to temporality as such and as a whole. As he remarks near the start of the discourse, the 'once' of temporality is the 'once' of the 'moment' and 'once' is only ever 'once', 'even if it lasts seventy years...or seven times seventy times', whilst 'The temporal itself, in its entirety, is the moment; understood eternally temporality is the moment and eternally understood the moment is only once' (SKS10, 110/CD, 97–8). It is, he suggests, like a killing on stage. It seems to the audience that the victim has been run through but we all know that afterwards the actor gets up and goes home, just as Daniel walked safely out of the lions' den or Shadrach, Meshach, and Abednego out of the burning fiery furnace. 'For all suffering in time is an illusion and death itself, eternally understood, is a comedian' (SKS10, 113/CD, 101–2). He also refers his readers to an altarpiece in the Church of the Saviour in Christianshavn, which depicts the angel offering the cup of suffering to Christ in Gethsemane. If we were to sit and contemplate this picture hour after hour, year after year, with the pious aim of deepening our sense of Christ's suffering, there would come a moment when we would have to admit that this was not, in fact, the end of the story: he took the cup, he suffered, and he triumphed. But the opposite is not true. If we were similarly to contemplate an image of the Resurrection, we would never get to the point of seeing this merely as a transitional moment. The suffering is for once, the victory is eternal. That is, of course, a logic that makes sense for those who have accepted the Christian faith in Christ's resurrection and not for others—this is from a set of specifically *Christian* discourses, after all—but, from another angle, it also illustrates again the argument of the *Postscript*: that there is no mediation of time and eternity outside the movement of faith that precisely focuses on their *difference*. Time and eternity are essentially incommensurable and, at best, can only ever be related in the mode of suffering, that is, inversely. The one is what the other is not and vice versa. On this basis, Kierkegaard can argue both philosophically against speculative philosophy's view that time and eternity can be thought together and also religiously against what he sees as his contemporaries' absorption in the things of time.

Looking at the beginnings of the modern entertainment industry in the contemporary phenomenon of Tivoli, Kierkegaard comments that lovers of Tivoli are not lovers of eternity (SKS7, 261/CUP, 286). Similarly, he will comment that the growing enthusiasm for firework displays is likewise illustrative of a kind of absorption in time that has lost touch with eternity: such displays have to become ever more spectacular with ever-diminishing intervals between the successive explosions if they are to maintain the ever-decreasing attention span of the public (SKS8, 282–3/UDVS, 184–5). To have time is to lose eternity. To gain eternity, we must be ready to sacrifice time and all that belongs to time. Superficial as they are—and precisely in their superficiality— Kierkegaard sees such phenomena of a nascent entertainment culture as signifying an epochal shift in contemporary society. This shift finds its extreme and perhaps decisive manifestation in what he calls politics, meaning the movement for and absorption in the democratic changes that transformed Danish society from the late 1840s onwards. In Kierkegaard's view, this kind of politics meant turning one's back on eternity in order to become preoccupied with the affairs of time, the *saeculum*, the secular present age. As such it also meant turning away from what properly concerns the existing single individual and losing oneself in 'the crowd'.

It is in these terms that one of the decisive lines of confrontation in Kierkegaard's later writings is drawn between 'the crowd' and eternity. From the late 1840s onwards, Kierkegaard thought endlessly about the meaning of his own authorship and in a 'Supplement' to his posthumously published *The Point of View for my Work as an Author* entitled 'The Single Individual' he set out the essential issue of the authorship as being the choice between the crowd (which, as he repeatedly says, 'is untruth') and the Eternal. Starting with the words 'In these times everything is [about] politics' (SKS16, 83/PV, 103), the first part of the Supplement ends by appealing to 'the eternal truth: the individual' (SKS16, 92/PV, 112).[39] By virtue of a logic already operative in the *Postscript*, this is because society—the crowd—lives by the principle of the numerical, which means reducing human identity and behaviour to what is statistically computable, whereas the self-relation of the individual is grounded on a relation to the Infinite, in the qualitative or eminent sense of the term.[40] Succinctly, Kierkegaard asserts that 'Eternally, divinely, Christianly, Paul's saying that "Only one reaches the goal" is valid… [and God] the great examiner, he too says, "Only one reaches the goal", which is to say that everybody can and everybody should become this "one", but only one reaches the goal' (SKS16, 86–7/PV, 106–7). The 'distance' between the sublime and infinite elevation of God and the individual life cannot be traversed by the quantitative measure of 'politics' but only—paradoxically—at the heart of individual existence itself, seemingly the most fragile and transient manifestation of human life.

[39] This distinction, between the crowd and politics on the one hand and the individual and eternity on the other is also paralleled elsewhere in Kierkegaard's writing, such as in his frequent characterization of Judaism and Christianity in the 1850s journals as, respectively, devoted to enjoyment of life in this world and renunciation of this world for the sake of eternity.

[40] This also relates to his view of sin as grounded in the principle of comparison, i.e. the substitution of the self's proper self-identity through the God-relationship with an identity drawn from a mimetically derived social 'self'. For further discussion see my *Kierkegaard and the Theology of the Nineteenth Century* (Cambridge: Cambridge University Press, 2012), 141–5.

Yet Kierkegaard also wants to insist that his objections to political activity do not amount to a misanthropic neglect of human striving for fulfilment. 'By truth,' he writes, 'I always understand "eternal truth". But politics and such like has no business with "eternal truth". A politics that understood "eternal truth" and that seriously sought to introduce "eternal truth" into the world as it actually is would in that same instant show itself as the most unpolitical thing that could be imagined' (SKS16, 89–90/PV, 109–10). And yet, although the religious is said to be '*toto caelo*' different from politics, 'religion's eternity is the transfigured re-edition of politics' most beautiful dream' (SKS16, 83/PV, 103). Why? Because it is the religious that leads to genuine human fulfilment and is premised on the genuine equality of human individuals, namely, equality vis-à-vis the infinite elevation and claim of the eternal. We are and cannot be otherwise than in time, i.e. temporal, and, as such, exposed to the intrinsic limitations and the perpetual attrition of powers and knowledge that belong to life in time. Nevertheless—or for this very reason—the key to identity, to becoming a self in the fullest sense, is not to be found by attending to the things of time but by turning from the temporal to the eternal, and it is precisely this turning that suffering is especially suited to help bring about.

NOTHING BUT TIME

Heidegger's 1927 work *Being and Time* had as dramatic an effect on the development of modern European philosophy as Karl Barth's commentary on Romans did on theology. Superficially the two works could not be more different, however. Barth's is written in a polemical and expressionistic prose that piles metaphor upon metaphor and image upon image and that argues for and on the basis of the authority of a received text so as to assert the rights of eternity against the claims of time. Heidegger's breakthrough work is written for the most part in the rigorously 'scientific' style of Husserlian phenomenology, suspending all claims to privilege any particular approach to the matter at issue, '*die Sache selbst*'—and where Barth speaks, as it were, for eternity, Heidegger sets out to demonstrate that time is, as he puts it, 'the horizon for the meaning of Being' so that the being of human existence and of all human reflection on existence becomes inseparable from its defining temporality.[41]

It is not to deny the reality of these differences to add that there are also some significant commonalities. For although Heidegger's language is more scholarly than Barth's, his project of pursuing a fundamental ontology with the

---

[41] Heidegger, Martin, *Gesamtausgabe*, II: *Sein und Zeit* (Frankfurt am Main: Klostermann, 1977), 18; Eng.. *Being and Time*, tr. John Mcquarrie and Edward Robinson (Oxford: Blackwell, 1962), 39.

tools of scientific phenomenology is also polemical in relation to what he calls 'a hardened tradition'[42] of interpreting the question of being—just as Barth sees access to the original force of the Pauline message as having been blocked out by the accumulation of human traditions. And although Heidegger's philosophical and Catholic background led him to focus more on the role of scholasticism in this hardening of tradition, he too, like Barth, was specifically contesting the approach of contemporary German historicism, namely the view that historical research could deliver a solid factual basis for interpretation and that interpretation itself could yield general laws and deliver universally binding conclusions. For theologians such as Adolf von Harnack and Ernst Troeltsch (also amongst the specific targets of Barth's polemics and the most significant representatives of liberal Protestantism at the start of the twentieth century), the potential outcomes of historical research included establishing the facts of the life of Jesus of Nazareth and justifying Christian claims to be the 'absolute' religion by means of a universal science of religion.

Although *Being and Time* itself shows few direct traces of this particular debate, Heidegger had expressly opposed just these two figures (and especially Troeltsch) in lectures on the philosophy of religion (winter semester 1920/21) and on Augustine (summer semester 1921). The approach of the theological liberals was avowedly 'historical'; nevertheless, Heidegger saw it as a kind of residual Platonism, manifesting a perennial tendency on the part of the intellectual tradition to consider time and history in terms of essentially timeless structures of understanding.[43] As an alternative to this approach, Heidegger proposed returning to the original texts of early Christianity—the letters of Paul—and, rather than attempting to reconstruct Paul's world-view or classify Paul according to some taxonomy of religion, seeking instead to elicit what he calls Paul's *'faktische Lebenserfahrung'* ('factical life-experience'). Neither God nor religion are to be treated as objects of a value- or interpretation-free science, since it is solely on the ground of human experience that they become (or fail to become) meaningful for us. As a salient instance of historical experience, Paul's experience can only be understood if we are ready to place ourselves inside it and, as Heidegger will say 'write the letter with him'.[44] And if Heidegger does not forget that what he is about is nevertheless an interpretation and, as such, cannot be satisfied with simply repeating Paul's words after him, we are to let Paul's self-understanding be determinative for our interpretation and not vice versa. Quite specifically, this means taking Paul's proclamation of the gospel as our starting-point.[45] Possibilities of convergence with Barth are not hard

---

[42] Heidegger, *Sein und Zeit*, 22 (Eng. *Being and Time*, 44).

[43] Martin Heidegger, *Gesamtausgabe*, LX: *Phänomenologie des religiösen Lebens* (Frankfurt am Main: Klostermann, 1995), 38–50.

[44] Heidegger, *Phänomenologie*, 87.   [45] Heidegger, *Phänomenologie*, 80–1.

to see, including a hermeneutic optimism regarding the possibility of entering into a direct relation to Paul's text. Nevertheless, where Barth will articulate the content of the proclamation in terms of Jesus Christ, Heidegger will look more to Paul's own personal and religious experience.[46] Against nineteenth-century historicism his enquiry is not led by 'the ideal of a theoretical construction' but seeks 'the originality (*Ursprünglichkeit*) of what is absolutely historical in its absolute unrepeatability'.[47]

What Heidegger then finds is that the proclamation is indissociable from the entirety of Paul's existence. It is not the theoretical content yielded by a momentary conversion experience but it 'encounters a person at a particular moment' but it then 'accompanies them through life'.[48] And, for Paul, the proclamation is lived out in a life entirely turned towards the Parousia, the expected return of Christ. This, Heidegger says, is more important than any momentary ecstatic experience, nor is it to be confused with the attitude of waiting for some future event on the plane of standard historical time, i.e. clock-time. Paul himself dismisses his readers' worries about the day and the hour of Christ's return because the point is precisely not the day or the hour but the total attitude towards God involved in the proclamation. 'It is not a matter of the meaningfulness of some future content [i.e. what will happen in history at some as yet unknown date] but of God. The meaning of temporality is defined by the basic relation to God, and in such a way that eternity is only to be understood in relation to what is lived in temporality as the medium of its accomplishment'.[49] As throughout these lectures, we sense Heidegger struggling to find the vocabulary in which to say what he wants to say, but whilst the result is often rather contorted the broad thrust of what he is attempting seems clear.[50] This is that the factical life-experience of the early Christian community, as exemplified by Paul, is indissociable from its experience of living through the time between the Crucifixion and the Second Coming, a time of suffering expectation.[51] As Heidegger puts it, 'The meaning of the "when" of the time in

---

[46] In these terms Heidegger is closer to his Marburg colleague Rudolf Bultmann and we may note that it is precisely around this issue that Barth and Bultmann will themselves ultimately diverge. Where Barth insists on the need to interpret Christian proclamation as teaching about Jesus Christ qua revelation of God, Bultmann looks instead to interpret it in terms of human self-experience and of the possibilities offered by kerygmatic proclamation for a renewed self-commitment. In Barth's terms, this means that Bultmann has, with Heidegger, turned the theological content of the gospel into an anthropology. See, e.g. Karl Barth, 'Rudolf Bultmann— An Attempt to Understand Him', in H. W. Bartsch (ed.), *Kerygma and Myth: A Theological Debate*, tr. R. H. Fuller, 2 vols (London: SPCK, 1972), II. 83–132.

[47] Heidegger, *Phänomenologie*, 88.   [48] Heidegger, *Phänomenologie*, 117.

[49] Heidegger, *Phänomenologie*, 117.

[50] Of course, this is not to say that it is successful or persuasive.

[51] It is theologically interesting that Heidegger interprets Paul's experience of this time as primarily an experience of tribulation, a time abandoned to itself as a result of the rejection of Christ and not yet taken back into the fullness of the divine purpose. However, Heidegger seems to omit what many theologians might say is the more important theme for Paul: that it is the time between, not the Crucifixion but, the Resurrection of Christ and his coming again. Seen in this

which the Christian lives has a quite particular character. Previously we gave a formal characterization of this as "Christian religiosity lives temporality". It is a time without its own order or fixed points, etc. It is impossible to gauge such temporality in the perspective of any objective concept of time'.[52] It is, he says, time that can only be known in 'the Spirit', in bondage and expectation.[53]

Naturally, not all of the seeds of the conception of time that comes to expression in *Being and Time* can be found in this early work on Paul—not least because the later work also contains what Heidegger had in the intervening time gleaned from Kierkegaard and from his continuing work on classical sources (Plato, Aristotle, and Plotinus) and on Kant, German Idealism, and Dilthey. But what is already clear and what will remain as a basic element of his subsequent work is that time is not to be understood in terms of its relation to eternity (as in the Platonic 'time is a moving image of eternity') but must be understood from itself, in terms of the temporal experience of the time-bound existent that we ourselves are. In this context, even the God-relationship is more fundamentally a variant of human time-experience than an experience of eternity outside of time.

These points are restated and developed in a 1924 paper to the Marburg Theological Society on 'The Concept of Time'. Here, having posed the question 'What is time?' Heidegger immediately comments that 'If time finds its meaning in eternity, then it must be understood starting from eternity'.[54] But if 'eternity' is simply a name of God and if faith is the only means by which human beings can have knowledge of God, then only theologians would be competent to say what time is. Heidegger concedes that since it is necessarily concerned with the temporal being of human existence in its relation to eternity and with an event that took place once in time (the Incarnation), theology is properly concerned with time. But, he adds, 'the philosopher does not believe. If the philosopher asks about time, then he has resolved *to understand time in terms of time* or in terms of the *aeí*, which looks like eternity but proves to be a mere derivative of being temporal'.[55] Although his own approach is not that of a theologian (and is also said to be *pre-* rather than *actually* philosophical) it can nevertheless serve theology—precisely by 'making the question concerning eternity more difficult'.[56]

---

light it is a time in which, even now and even in the situation of the absence of Christ, the dominant note of Christian existence is one of joy.

[52] Heidegger, *Phänomenologie*, 104.   [53] Heidegger, *Phänomenologie*, 123.

[54] Martin Heidegger, *Gesamtausgabe*, LXIV: *Der Begriff der Zeit* (Frankfurt am Main: Klostermann, 2004), 107; Eng. *The Concept of Time*, tr. W. McNeill (Oxford: Blackwell, 1992), 1. The emphasis on the question of eternity seems likely to be connected with the fact that this was a lecture to a theological rather than a philosophical audience.

[55] Heidegger, *Der Begriff der Zeit*, 107 (Eng. tr., 2).

[56] Heidegger, *Der Begriff der Zeit*, 108 (Eng. tr., 2).

Briefly noting the views of Aristotle and of more recent physics, he gives particular emphasis to Augustine's discussion of time in Book 11 of *The Confessions*, paraphrasing the closing passage in which Augustine states that it is the mind itself that, in measuring time, makes time what it is. Strikingly, Heidegger glosses Augustine's 'mind' (*mens*) as 'spirit' (*Geist*), a concept he will eschew in *Being and Time*. At the same time, he anticipates the later work by describing this act of measuring as occurring in and as dependent on a 'disposition' (*Befindlichkeit*) of Dasein—the characteristic term for the human existent that will take centre-stage in *Being and Time*. But *Befindlichkeit* too is one of the basic terms of *Being and Time*. Translated by Robinson and Macquarrie as 'state-of-mind' it is said by Heidegger to constitute one of the three basic and interdependent forms of Dasein's self-conscious life: state-of-mind, understanding, and discourse. That is to say, Dasein never exists otherwise than finding itself in a certain state of mind (say, feeling 'alright'), which involves a certain understanding of itself (that it is able to feel alright or not alright), and that is articulated as logos or discourse (being able to say to others or to ourselves 'I'm feeling alright' as the way in which we *know* we're feeling alright).

Time, then is not a secondary attribute that qualifies human being, as if we might say that 'the human being is in its essence a rational animal and has the further attribute of temporality' (to which we might add sociability, belligerence, religiosity, etc., so that temporality becomes one attribute amongst others). As Heidegger will say in the conclusion of his paper, 'Dasein always is in a manner of its possible temporal being. Dasein is time, time is temporality. Dasein is not time, but temporality'.[57] When, at the start of *Being and Time*, he will say that the best place at which to begin an enquiry into the meaning of being is with ourselves and when he then goes on to assert that time is in fact the horizon for the meaning of being, these are not two arbitrarily connected statements. If we commence our enquiry into the meaning of being with Dasein, the being that we ourselves are, then, since Dasein proves to be a being that is its temporality, it more or less inevitably follows that the meaning of being will prove to be temporality. And what is also crucial for the question as to the relation of time and eternity is not only that, in a negative sense, this

---

[57] Heidegger, *Der Begriff der Zeit*, 123 (Eng. tr., 20). I take it that this sequence of four statements—1) Dasein is time; 2) time is temporality; 3) Dasein is not time; 4) Dasein is temporality—but 'Dasein is time … Dasein is not time'—is not an exercise in blatant self-contradiction but that the qualification of time as temporality points away from 'time' in the sense of the passage of clock-time to something like time-ness as the quality of being essentially temporal and that this is eminently the case with regard to Dasein. It 'is' time, but not in the sense of having a life-span that might be measured in terms of the passage of time. Rather, it is essentially temporal since, in a sense close to that of Kierkegaard, we can only become who we are, fully personal human beings, by choosing and acting in time. In this sense Dasein is different from, say, a piece of wood, that has time as one attribute amongst others (it is five years old, for example) but is not constituted by its reflexive relation to its own time.

seems to leave no role for eternity in the conception of time, but that time assumes the positive role ascribed to eternity in traditional Christian theology. In the Introduction to *Being and Time* Heidegger will write 'Being and the structure of Being lie beyond every entity and every possible character which an entity may possess. *Being is the transcendens pure and simple*'.[58] But if Being itself is grounded in temporality, then Heidegger has effectively replaced the transcendence of the eternal God over time with the transcendence of temporality over all possible beings—including any supposedly 'eternal' God.[59]

Already in the paper on 'The Concept of Time' Heidegger offers a preliminary survey of how this works out, identifying a number of the key ideas of *Being and Time*: being-in-the-world; being-with-one-another; discourse; speaking as the eminent mode of self-disclosure; mineness; the 'one'; care for our own being; being disposed in a certain way;[60] the ineluctability of average everydayness in Dasein's self-experience and self-understanding; the possibility of an authentic relation to being; the grounding of this possibility in a relation to death that is characterized as running on ahead of itself; this primary and authentic relation to the future as grounding an authentic relation to the past and to the present; historicity as made possible through the repetition of what qua temporal is irreversible.[61]

As even this brief list indicates, Heidegger's account is complex and multi-faceted and I cannot do more here than highlight a few key points that have particular bearing on the relations of time and eternity.[62]

Pivotal to Heidegger's account of Dasein is what he sees as the ubiquitous phenomenon of *care*. Dasein is a being that is always ahead of itself in care for itself. In other words, our actions do not occur in a timeless present nor do they occur as carried along by a flow of causal events running from the past to the future. Instead, our relation to our world and to ourselves is determined primarily by the way in which we are always ahead of ourselves. We can see how this might be so just by considering some simple everyday situations: I get up to open the door to admit my visitor, I think about what to buy

---

[58] Heidegger, *Sein und Zeit*, 38 (Eng. *Being and Time*, 62).

[59] Although Nietzsche is scarcely mentioned in *Being and Time* this might seem to complement what we have considered as Nietzsche's affirmation of time itself as eternal. However, as opposed to the idea of eternal recurrence, Heidegger's formulation leaves no obvious chink by which to reintroduce eternity into a thoroughly temporalized philosophy.

[60] See the discussion of *Befindlichkeit* above.

[61] These are the terms in which these points are considered in the paper. They vary slightly from those of *Being and Time* itself, but will be recognizable to readers of the latter work.

[62] Nothing, of course, can substitute for the study of *Being and Time* itself, which is in any case a prerequisite for any serious engagement with twentieth century European philosophy. There are, however, many helpful introductions to this work, such as Stephen Mulhall, *Routledge Philosophy Guidebook to Heidegger and Being and Time* (London: Routledge, 1996). My own view is developed at greater length in my *Heidegger on Death: A Critical Theological Essay* (Farnham: Ashgate, 2013).

for dinner tonight, I read an article in order to prepare my lecture tomorrow. But this is not enough to provide Heidegger with a sufficient basis on which to account for the meaning of being, since such examples only show relatively localized occurrences in the overall configuration of a human life. In such cases we only run on ahead of ourselves a relatively short way. And even if we can look many years into the future, planning retirement or anniversary parties, such prospects are never able to take the whole of our actual being into account. As Heidegger puts it in 'The Concept of Time', echoing an ancient Epicurean saying, 'how is this entity to be apprehended in its Being before it has reached its end? After all, I am still underway with my Dasein. It is still something that is not yet at an end. When it has reached the end, it precisely no longer is. Prior to this end, it never authentically is what it can be; and if it is the latter, then it no longer is'.[63] In other words, we could only be assured of having got the whole of Dasein—the phenomenon to be investigated—into view if we could also take into view its death. Now, of course, we *can* do this if we treat Dasein as the object of biological or medical or historical science. We can describe and define the processes of death and we know all about the deaths of Socrates, Jesus, and other great figures of the past. But that is to forget the character of 'mineness' that Heidegger says clings to every investigation into Dasein, since Dasein is not just any old being but the being that we ourselves always are (yes, including you the reader and I the writer of these words). I can see and conceive the deaths of others, but this doesn't tell me what it means for me to have to die—and, mostly, under the influence of the average everyday and evasive way of talking about these things characteristic of 'one'— that is, humanity in general—I either avoid thinking about this entirely or else think about it in such a way as to tranquillize (as Heidegger puts it) the shocking reality that, one day, I will certainly cease to exist: I, not just a bit of me, not just my body, but all of me. To be thrown into existence is to be thrown towards death, towards the entire annihilation of everything I am, ever have been or ever will be. Now Heidegger will later reject the view that this was a 'philosophy of death' and assert that death is not offered as the sole focus of the transition to authenticity.[64] In *Being and Time* itself he claims that his interpretation is neutral as regards the possibility of some future life.[65] Nevertheless, a straightforward reading of *Being and Time* would seem to

---

[63] Heidegger, *Der Begriff der Zeit*, 115 (Eng. tr., 10).

[64] See the opening pages of Martin Heidegger, *Gesamtausgabe*, XLIX: *Die Metaphysik des deutschen Idealismus. Zur erneuten Auslegung von Schelling: Philosophische Untersuchungen über das Wesen der menschlichen Freiheit und die damit zusammenhängenden Gegenstände (1809)* (Frankfurt am Main: Klostermann, 2006). Recent philosophical approaches to Heidegger have similarly argued against taking the thematization of death in *Being and Time* at face-value, suggesting that it functions more as a placeholder for a general point about human finitude. For discussion see my *Heidegger on Death*, 8–10.

[65] Heidegger, *Sein und Zeit*, 247–8 (Eng. *Being and Time*, 292).

make death a defining feature of human existence and to make our relation to death the hinge that enables us to make a transition from an inauthentic to an authentic relation to being.

And how might we make that transition? Knowledge of our own future decease is not available to us, but if cognition is not possible this does not exclude an existential comportment that Heidegger describes as 'running-towards' (rendered by Macquarrie and Robinson as 'anticipation') in which I grasp the 'impossible possibility' of my having to die as the basis of my existence as a whole—and because this comportment is grounded in the threefold structure of self-consciousness as *Befindlichkeit*, understanding, and discourse it therefore also becomes possible to analyse it and talk about it. This, then, provides the ground for an interpretation that is able to grasp Dasein's being as a whole. But, as Heidegger will further explicate, this means grasping Dasein's being as temporal through-and-through. Having to die and being pure temporality are two sides of the same coin. If there were an eternal life 'beyond' death it is hard to see how the thought of such a life could be entertained without entirely dissolving the structure of Heidegger's thought at this point.

Having established the possibility of running towards death and authentically grasping our being as a whole, Heidegger proceeds to show how this future-orientated act of self-choice also lays the ground for our understanding of both past and present. Where earlier philosophies of time had sought to ground our sense of time in the immediacy of the present or in a retrospective view of the past, it is the future that provides Heidegger with the axis on which to develop the relationship between the three time 'ecstases' as he refers to them. Adopting the Kierkegaardian concept of 'the moment of vision' (*Augenblick*),[66] Heidegger sees these ecstases as susceptible of unification in and through an act of resolute self-commitment. Those who resolutely run towards their death and make this the basis of their understanding of their lives as a whole will also be fully open to all the possibilities of their present and of their past. There is, as it were, nothing from which they might have to hide, and the tension of their futural self-projection will hold them back from slipping into a mere absorption in the present that is satisfied with doing what everyone else ('one') is doing or thinking or feeling.[67] But this does not involve appealing to anything outside our own human time-experience: the unification of time occurs not—as Augustine and other theological thinkers would have thought necessary—by reference to the abiding unity of the eternal (timeless or durational) but solely by a certain development or modification of temporal experience itself. Indeed, as Heidegger also says, time itself will be different for us according as to whether we relate authentically or inauthentically to our death. It is not a matter of a given temporality that we can live

---

[66] See Chapter 8.
[67] Heidegger, *Sein und Zeit*, 337–8 (Eng. *Being and Time*, 387–8).

either authentically or inauthentically, but how we experience time is dependent on how we live it.[68]

This development does not, however, mean moving away from or recanting the basic thesis that our time-experience is what and as it is by reference to our thrownness towards death. If there is any possibility of consistency or coherence in our lives at all, it is only within the parameters of our being towards death and—using another Kierkegaardian concept—this possibility is established in and through the repetition of resolute self-choice. There is no outside help. Time itself is the only transcendence there is. Berdyaev may have been excessive in describing Heidegger's position as one of pessimism, but perhaps he is not far off the mark when he says that *Being and Time* suggests, in effect, a victory of time over eternity and being towards death as precisely articulating a view of history that has renounced all reference to the eternal.

In the light of this summary we can perhaps see the thrust of a much-cited footnote in which Heidegger acknowledges Kierkegaard's contribution to the idea of a 'moment of vision' before going on to emphasize its limitations. Heidegger states that Kierkegaard's 'penetration' in seeing 'the existentiell phenomenon of the moment of vision' is not matched by his subsequent interpretation of it, since 'He clings to the ordinary conception of time, and defines the "moment of vision" with the help of "now" and "eternity". When Kierkegaard speaks of "temporality", what he has in mind is man's "Being-in-time". Time as within-time-ness knows only the "now"; it never knows a moment of vision'.[69] Although these highly condensed comments call for more comment than I shall give them here (not least because we have not yet explored Kierkegaard's own account of the moment of vision), it is clear that Heidegger is troubled by what he sees as Kierkegaard's view that the moment of vision is constituted by the manifestation of 'the eternal' in a pure 'now' of experience. Axiomatic for Heidegger—to say it once more—is that beneath, behind, or beyond the repetition of the moment of vision there is *no* eternal being or power, only the transcendence of temporality itself. In the following section (though not with reference to the passage from *The Concept of Anxiety* where Kierkegaard discusses the moment of vision) and again in Chapter 5, we shall question whether Heidegger is entirely doing justice to Kierkegaard at this point and whether, in fact, the situation may be less of an either/or than is presented in *Being and Time*.

But is this Heidegger's last word on the question of eternity? In lectures on Hölderlin's poem *Germanien* and the poet's own appeal to eternity, Heidegger will say that 'there is no intrinsic definition of eternity but the idea and concept of what we call eternity are always defined in accordance with the ruling idea of time'.[70] Both the sempiternal duration and the *nunc stans* of Boethian

---

[68] See, e.g. Heidegger, *Sein und Zeit*, 304 (Eng. *Being and Time*, 351–2).

[69] Heidegger, *Sein und Zeit*, 338 (Eng. *Being and Time*, 392).

[70] Martin Heidegger, *Gesamtausgabe*, XXXIX: *Hölderlins Hymnen 'Germanien' und 'Der Rhein'* (Frankfurt am Main: Klostermann, 1989), 54–5.

timelessness, Heidegger says, are derived from a common idea of time that runs through Christian theology and on into the philosophy of Hegel. This idea is that of time as the endless vanishing of one moment into the next or the persistence of an all-encompassing now. But neither of these do justice to Hölderlin's experience of time and they inhibit us from thinking eternity on the basis of this other time-experience revealed in Hölderlin's poetic work. What if the relation of eternity and time were not contrastive (as Heidegger supposes is the case with Kierkegaard) but, perhaps as he is hinting is the case with regard to Hölderlin, eternity and time were somehow mutually implicating?

Heidegger never thematically pursues what such a revised conception of eternity might be, although some of his later writings offer hints. One example is his popular lecture on the seventeenth century preacher Abraham à Sancta Clara, whom he admires as, in his own way, a poetic thinker.[71] In the lecture he focuses on an image that he describes as 'perhaps the most astonishing and beautiful poetic word-picture' left us by Abraham: 'Come hither, you silver-white swans, who, with your snow-defying wings, row round and about upon the water'. Heidegger juxtaposes this with a further 'poetic word-picture': 'Do you not know that human life is like snow and clover, neither of which abide'.[72] Reading these together, he says, suggests that whilst snow melting on water offers an image of the transience of temporal life, the snow-white plumage of the swan offers a counter-image of transient life being counter-intuitively maintained in being ('snow-defying') as it moves on the surface of the water. In Heidegger's words, 'The movement of the white swans on the water is an image for what does not pass away in the midst of what passes away'.[73] Could this be an opening to some trans-temporal experience of duration that could in turn open the way to a re-conceptualization of the eternal? I have, in a preliminary way, explored this elsewhere and now only raise the possibility that, even on Heidegger's own terms, the acceptance of time as the horizon of being may not exclude an appropriately modified rethinking of the eternal.[74]

If Barth's distinction of time and eternity left eternity as the only true measure of time, Heidegger's way seems, on the whole, to have led to time as the sole true measure of eternity. Both are equal and opposite possibilities of the radical separation of time and eternity. As we have seen, Heidegger will later let fall some hints (though scarcely more than hints) that this is not the end of the story. Nevertheless, the impact of his thought—whether intended or not—was to further the mood engendered by a certain way of reading Kierkegaard and

---

[71] The occasion of the lecture is the fact that Abraham, like Heidegger, came from the Black Forest town of Meßkirch.

[72] Martin Heidegger, *Gesamtausgabe*, XVI: *Reden und andere Zeugnisse eines Lebensweges 1910–1976* (Frankfurt am Main: Klostermann, 2000), 607.

[73] Heidegger, *Reden*, 607.   [74] See my *Heidegger on Death*, 145–53.

Nietzsche that led to the conclusion that human life was handed over without reserve to sheer flux and final oblivion. But, whether or not this is a correct reading of Heidegger himself, is it the only conclusion that can be drawn from investigating human temporality solely from within the parameters of human time-experience itself? In other words, is it phenomenologically adequate to the matter at issue? Helped by three post-Heideggerian philosophers, Vladimir Jankélévitch, Jean-Louis Chrétien, and Emmanuel Lévinas, we shall in the next chapter start to sketch how a certain kind of time-experience might itself open a path towards the transcending of time, in time, through time, and for the sake of life in time. The chapter will conclude with a reading of *The Brothers Karamazov* as a literary staging of these issues in terms of the struggle between oblivion and hope.

5

Oblivion, Memory, and Hope

REMEMBRANCE IS THE SECRET OF REDEMPTION

If time is nothing but time, then it seems that, apart from the paradoxical or impossible possibility of intervention by a supra-temporal eternal God, then temporal beings such as we are must accept that, like Hölderlin's *Hyperion*, we exist as thrown towards an ineluctable annihilation and final oblivion. And even if we accept the Nietzschean and Heideggerian challenge to say an unrestricted 'Yes!' to this temporal thrownness, and even when this 'Yes!' is actually lived as a mystical moment of ecstatic bliss, there seems to be no scope for a relation to the eternal in the terms presupposed by Augustine, Boethius, and subsequent Christian philosophers and mystics. But isn't it also the case that if time bears us all towards annihilation it simultaneously bears us towards new tasks, new responsibilities, new relationships and, in doing so, bringing us into ever-new variations of companionship with our fellow mortals? It is only thanks to time that we experience our own identity as bound up with our relations to lovers, family, friends, enemies, and the practically indefinable multitude of all whose lives coincide in time—in some sense and measure—with our own. Time is full of things to do and to discover, friendships to nurture, new people to know. Who could ever grow weary of being in time?

Yet even if we do experience time as enriching rather than diminishing our stake in being, can we escape the fact that the multiform landscape of interwoven personal and common life is criss-crossed by fissures that mark loss and limitation? Isn't it the case that the more we love, the more we must, in the end, lose, so that to commit ourselves to love is to commit ourselves to loss?[1] In traditional wedding services the couple vowed to be together 'till death us do part', indicating that even the moment of maximum fulfilment was inseparable from anticipated loss. Whereas a rash lover declares 'I can't live without her', the love that finds expression in the marriage vow acknowledges that bereavement and loss

---

[1] See my *'The Heart could Never Speak': Existentialism and Faith in a Poem of Edwin Muir* (Eugene, OR: Cascade Books, 2013), 40–9.

are inseparable from love. And, when it comes to the wedding reception, one of the most time-honoured toasts is to 'absent friends', usually and primarily those—parents or grandparents perhaps—who died before they could see the happy day.

If such reflections might seem morbid to some, they may nevertheless help us move beyond the Heideggerian account of human being as thrownness towards death and do so in one crucial respect. When the annihilating power of time is brought into relation to the complex of our defining loves, it is no longer a matter of individuals left to confront their ownmost possibility of annihilation in solitary authenticity. In the perspective of love, the remembrance of death passes from self to other. But can the faithful remembrance to which love commits itself do more than push back the time of final oblivion in which there is, literally 'after all', no one left to remember? Surely, no matter how rich our lives may be in love and friendship, the fact remains that any story we tell about ourselves, our lives, and our loves, will be a story of what time has taken or will take away. The tale—even the best-told tale—is a mere shadow of life's fullness. We remember times past and we retell how it was: but although what we remember and what we retell is not nothing it is somehow less than it was. It has become a shadow world. And, turning to the future, we see something similar. If what gives purpose and direction to our lives consists in the plans, hopes, dreams, and expectations that we have for the future, this future itself is only ever 'real' as and when it becomes present and, when it does so, it is often not the future we expected: the beloved loves another, the train is late, the vote goes against us. But if memories are, in the end, but shadows, and if hopes are uncertain, what sense can we make of a present that is abstracted from a shadowy past and an uncertain future and that, as so many have attested, vanishes in the moment it appears? Each of past, present, and future plays into who we are and who we are is inseparable from who we have been and who we hope to be. Yet what weight can be put on a self that is constructed of such fragile elements? Time as constitutive of the life-time of the unique individual and of the complex societies brought about by the concrescence of such individuals is threatened and uncertain in all directions. Is it not as true today as it was in the days of Solomon, that everything passes away into forgetfulness?

The eighteenth-century Hasidic teacher the Baal Shem Tov is reported as having said that 'remembrance is the secret of redemption; forgetfulness leads to exile' and the possibility of temporality becoming capable of sustaining a fully meaningful life has often been linked to the challenge of counteracting the entropic force of forgetfulness. Perhaps all things must indeed pass—but if only we could sustain a memory of what had been, then we would have an assurance that it had not been in vain.[2]

---

[2] This logic is widely manifested in popular discourse of mourning, as mentioned at the start of the Introduction to this volume, cf. the saying 'To live in the hearts of those we love is not to die' or (especially of the war dead) 'We shall never forget them'.

Much that has been said in the preceding chapters may be read as supplying some philosophers' responses to this challenge, with Hegel perhaps offering the most robust defence of the view that the retrospective recollection of all that has happened in time is sufficient to yield a blissful and fulfilled self-knowledge on the part of Spirit. Both the power and the limitations of such a view are effectively summed up in his well-known dictum that it is only at dusk that the Owl of Minerva (the goddess of wisdom) spreads her wings, which is to say that it is only when an epoch of history has come to an end that its story can be told, its telos known, and philosophy is able to paint its 'grey on grey'.[3] But, as Hegel's many critics from Kierkegaard on have made clear, such confidence is by no means universally shared. If Anselm countered his critics with the assertion that 'you have not yet considered the extreme gravity of sin', we may say that the Hegelian view has not yet considered the extreme power of forgetfulness and the concomitant frailty of memory. How far can recollection really take us in relation to the more radically temporalized intellectual landscape of post-Nietzschean thought? Can it yield grounds for hope? And, more particularly, can it yield grounds for hope that might make it once more meaningful to speak of a God-relationship that was also and as such a relation to the eternal?

In addressing these questions, I begin with a philosopher, Vladimir Jankélévitch, who has offered modest and cautious responses that certainly fall short of and do not claim to ground any claims regarding the eternal. Their virtue, however, is precisely in this same modesty and caution, and it is sufficient, I think, to take us a small but significant step forward in our enquiry. I shall mainly discuss his study *The Irreversible and Nostalgia*, but will also cross-refer to other writings.

ETERNALLY IRREVOCABLE

Temporality is one of the recurring threads of Jankélévitch's uniquely lyrical philosophical explorations. Like many post-Kierkegaardian and post-Heideggerian thinkers, Jankélévitch is committed to thinking the temporality of time from within human time-experience itself. To this extent (and although his style sets him at a significant remove from more formally phenomenological philosophers) he may be regarded as contributing to a phenomenology of time and eternity. It should immediately be said that Jankélévitch has no interest

---

[3] G. W. F. Hegel, *Werke*, VII: *Grundlinien der Philosophie des Rechts* (Frankfurt am Main: Suhrkamp, 1970), 28.

in defending an idea of God as timelessly eternal nor even in establishing the metaphysical existence of some kind of sempiternal God. Although, as we shall see, he does employ the idea of sempiternity, this is in a somewhat idiosyncratic sense and is not, in the first instance, applied to God. But here, as at many other points in Jankélévitch's work, a marked silence concerning God does not mean that what he says is without significance for theological reflection.

From the beginning, Jankélévitch is clear that irreversibility 'is the very temporality of time' in that it defines 'the entirety and the essence of temporality, and of temporality alone': 'Time is irreversible in the same way that the human being is free: essentially and totally'.[4] Nor is the mention of freedom merely illustrative. Irreversible time and freedom are not just attributes of human being, they are together constitutive of it. Both are comprised in the assertion that human being is being that becomes (*devient*). As Jankélévitch puts it in a succession of verbs that are based on *venir* (to come), '[The human being] becomes (arrives [*advient*], appears [*survient*], sometimes even recollects himself [*se souvient*]), but he never comes back [*revient*]' (IN, 8). In this regard time is very different from space. I can revisit a location in space, but I can never once revisit the past. Time's irreversibility is irresistible and is more inflexible than even the laws of logic (IN, 12). The nature of time is futurition, endlessly and uninterruptedly pouring itself forth into the future and in such a way that the past can now only ever be present as memory, whilst the future cannot be anticipated 'in [any] development that is in the process of becoming there is always a dimension of non-achievement that escapes us' (IN, 280). If it were otherwise and if we could come and go in time as we do in space we would be as gods![5]

In fact—and this is a central element in the argument that follows, repeated in ever varied form—even the return or attempted return to the past (which Jankélévitch calls preterition) is unable actually to reverse the flow of time. For all such attempts are themselves temporal. If the would-be restorers of the Ancien Régime were to have succeeded or ever would succeed, they would not return the world to 1788 but to 1832 or 1932 or whenever the attempt was being made. However, Jankélévitch does not intend these observations in the

---

[4] V. Jankélévitch, *L'irréversible et la nostalgie* (Paris: Flammarion, 1974), 7. (Further references are given in the text as 'IN'.)

[5] Jankélévitch does not reference Kierkegaard's *Repetition* in this work, but these comments provide a way of understanding the necessary failure of the pseudonym Constantin Constantius' attempt to establish the possibility of repetition by returning to Berlin and trying to relive the experiences he had enjoyed there the year before. As Constantin poses the question of repetition at the start of the work it is a question of temporal repetition, which he opposes as such to the Platonic recollection. However, since his 'experiment' takes the form of a journey in space it makes what Kierkegaard would call a 'category error'. All of which I assume to be both intentional and integral to the comic dimension of this little work.

spirit of 'vanity of vanities'. Always leaving the past behind and always moving on into a future that never fully or finally comes, time is not empty but, in words Jankélévitch cites from Unamuno, *Plenitudo plenitudinum et omnia plenitudo*, time is itself fully the fullness of time. In the same spirit, time does not, as for the idealist, cause the depreciation of the eternal fullness of meaning; on the contrary, time and the possibility of meaning (*sens*) are interconnected. This does not mean that time itself can be grasped as an essence, a 'what', or a quiddity, but, in one of Jankélévitch's key expressions, the fact of futurition is the elementary quoddity, the 'that' of time that never yields itself to adequate theorization.[6] But, by virtue of a disjunction familiar from both Kierkegaard and Heidegger, how we live that quoddity is indeed an issue for freedom, the freedom that we ourselves are.

One consequence of these basic principles is that each instant of time has what Jankélévitch calls the character of 'primultimacy', meaning that each instant is both first and last; it has never been before and it can never be repeated. Reaching almost to the threshold of existence, the moment is, as he puts it, 'superannuated' by the following moment, dwindling away to the point of almost vanishing, so that we can find ourselves wondering whether it really happened, or whether it really happened *to me*, or whether I was maybe dreaming (IN, 49). It is almost as if it never happened—and yet it did. Elusive as it is, the quoddity of each primultimate moment is irreducible: just as the 'that' of this moment now occurring (the moment of my writing and your reading, each primultimately!) just is real (is all the reality there is, in fact), so by the same logic of quoddity 'the fact of having once existed is irrevocable and interminable' (IN, 51). However, Jankélévitch insists that this is not said by way of metaphysical consolation for the passing of time—although it is perhaps a point from which he will develop something like a metaphysical or quasi-metaphysical consolation.

In the light of these definitions, the past can only exist for us as a 'pure and naked pastness' (IN, 59), that Jankélévitch will also call 'passeity'. Again he insists that even if something like a repetition were possible, the 'second time' would really be a new 'first time'. It could not be a second first time on account of its own newness in the occurring of the now and the having-been of the original first time. Even if two events are in all other respects identical, the passage of time marks them out as irreducibly different: 'the fact of having acted or having been changes everything', he writes (IN, 60). Amongst the other things that it changes is that the primultimacy of the now in its unceasing futurition cannot be construed as empty or vanishing, a pure now without reference backwards or forwards. The past cannot be recalled into life—it has been and no longer is—but a purely momentary consciousness

---

[6] The connection to Bergson (see Chapter 1) seems clear.

would be the consciousness of a mollusc and not a human being. We can remember and we can, similarly, anticipate. Each primultimate moment is therefore also pro-retrospective. Every adieu is also an au revoir—and vice versa (IN, 66). 'The miraculous transformation [by which the flow of time is lived as full of meaning] occurs in the context of a continuous transformation' (IN, 66). The paradox of time is thus that '[it] is at once open-ended futurition and the secreting of an eternal, an irrevocable having-been which it deposits behind itself...that which has been has been and can no longer not have been' (IN, 72).

However, if this seems like the beginnings of a metaphysical consolation that Jankélévitch has forsworn, he now takes a series of dramatic steps to rule out what he regards as some of the most-familiar attempts in this direction. There can be no reversal of time, he insists, and any attempt to do so ends in a fiasco. We cannot be rejuvenated or resurrected (and even Lazarus' resurrection is no real reversal or undoing of time itself), whilst a final resurrection of all things—a dawn for which no vigil can prepare us (S. Bulgakov, cited on p. 99)—is a chimera. Eternal recurrence, whether in its Greek or Nietzschean versions, fares no better and perhaps worse. Perhaps worse, since it relies on a sub-personal logic of the interdependence of polarities (Plato) or the spatial finitude and temporal infinity of the world (Nietzsche[7]), and neither version considers Bergson's insight that 'The moments of becoming do not exist in act before the unforeseeable process that causes them to arise' (IN, 109).[8] In any case, it would still remain true that the 'second time' of recurrence would itself be a first time, separated forever from the first time precisely by the passage of time. For similar reasons, Jankélévitch rejects the option of an eternal present or a simple, perhaps Parmenidean, eternity in which there is no becoming. Such 'an eternally eternal eternity' would never even have allowed the possibility of time and if we are capable of envisaging a time when time and death will be no more, it will then still have been the case that time and death have been. We can work on whether time, our time, passes more or less slowly or more or less quickly (i.e. its quiddity), but we can do nothing about the fact of its passing (its quoddity). We can, to a limited degree, 'effect time in time but not the time of time' (IN, 131), through memory and anticipation, but our chronometry, the measure we bring to time, has no purchase on chronology, the time of time itself (IN, 132). We can develop techniques for getting a job done in ever-diminishing units of time, but 'an hour is always an hour' (IN, 143). The irreversibility of time 'has neither resemblance nor common measure with speed' (IN, 149).

---

[7] Whether this is the heart of Nietzsche's theory is debatable. See the discussion in the section 'Eternal Recurrence' in Chapter 3.

[8] This could be taken as what Nietzsche himself meant, if we follow the interpretations of Stambaugh and Michalski. Again, see the section 'Eternal Recurrence' in Chapter 3.

It seems, then, that we have no alternative but to accept our lives as borne ever forward by time's ceaseless primultimate futurition, from the cradle to the grave. But this is not so easy. The impossibility of a return to the past engenders a certain resentment of time, 'a sentiment of impotence which is the subjective corollary of an objective impossibility' (IN, 154). This can lead us, like Ecclesiastes, to a 'bilateral despair', a despair in which both past and future are seen as equally futile and inaccessible to human purposes—the scenario we considered at the start of this chapter. Some may entertain a bilateral hope, involving hope for the past and the future. Perhaps this could comprise something like the Russian visionary Fedorov's utopian project of 'raising the ancestors' as a necessary part of creating a truly ideal future.[9] In any case, Jankélévitch regards any such hope as, bluntly, 'an incomprehensible absurdity' (IN, 157).[10] But can we at least entertain what he calls a 'unilateral' hope, that is, hope for the future?

We *can* hope, of course—who's to stop us?—but hope does not exist independently of what it looks to overcome. This is regret, which Jankélévitch calls the organ-obstacle of hope. It is the organ of hope because what we regret—such as the prospect of our future death—is what stirs us (stirs even the condemned man, perhaps (IN, 162)) to hope. Consequently, hope always presupposes a tacit despair (IN, 164) and will always have the quality of a sweet-bitter melancholy (IN, 166–7). Similarly, hope is ambivalent in relation to will, to actually setting about doing something to change our lives. It is more militant than merely wishing for something, yet in comparison with willing it has an optative quality, a sense of 'may be' (IN, 171).

We shall return to this optative quality of hope at several points in our subsequent enquiries, but more immediately problematic for hope as offering a possible way towards the eternal is precisely its entanglement in regret. What we hope for, even if it does not yet exist, can become real, but what we regret is locked in the past, in 'the necropolis of the impossible' (IN, 180). Jankélévitch, a musicologist as well as a philosopher, points to the evidence of such regret in musical works devoted to 'souvenirs', and if hope is always sweet-bitter, regret is always bitter-sweet. It is *too late*, the musical expression of regret seems to say, and Jankélévitch invites us to feel the full force of the French '*hélas*'. Yet, he asserts, '*hélas*' is not the expression of pure pessimism since it also offers 'the inexhaustible delights of reminiscence' (IN, 192), such as we experience them in the musical adagio (IN, 210). Consolation and desolation are intertwined at every step on the way (IN, 193). But the object of regret is not really this or that

---

[9] See Nikolai Fedorovich Fedorov, *What Was Man Created For? The Philosophy of the Common Task: Selected Works*, tr. and ed. Elisabeth Koutiassov and Marilyn Minto (Lausanne: Honeyglen/ L'Age d'Homme, 1990).

[10] Whilst Fedorov's science-fiction-like project might seem to be fairly easily discounted, contemporary posthumanism appears to be nurturing a not dissimilar vision.

singular event but passeity itself, the fact of having-been, *fuisse*. Nevertheless, by having-been an event has, as we have seen, become—in a sense—eternal. Although it is as if an unknown slave who died in 500 BC never existed, it is, Jankélévitch asserts, *only as if* he never existed. He *did* exist and his existence can never, in all time, be undone. Not eternally eternal, like timelessly eternal truths, his life is 'eternalized' as an instance of sempiternity (IN, 200). 'The advent of the event has retrospectively become an eternal preterite' like 'the mysterious *néant* of a little girl exterminated in a German camp, vanished for ever' who is nevertheless an eternal moment in history (IN, 201).[11]

We should pause to consider what is being said here. We are close not just to the traumatic epicentre of Jankélévitch's thought, but to a point at which anything said or to be said in philosophy, theology, or poetry must surely begin to hesitate, to stumble, to be lost for words... The death, the crime, will never be undone. Never. And yet...

We resume.

We resume, as we always must, finding ourselves carried along by the irresistible tide of futurition. And why can we not break loose from all our regrets and just say 'Yes' to its ceaseless flow? If our misery lies in our resistance to time's irreversibility, wouldn't such a 'Yes' issue in joy—'a very simple word for a very rare thing, a monosyllable for an instant' (IN, 222).[12] But such a 'Yes' would not be, or would not have to be, a mere denial of the past. If the irrevocable past can be the occasion of regret, surely it is also integral to the possibility of meaningful life. Is a declaration of love made just once really a declaration of love, asks Jankélévitch rhetorically (IN, 226). Isn't the 'second time' (ad infinitum...) a basic condition of being (IN, 227)? Yet won't that almost inevitably carry us via the temptations of repeatability and reversibility away from the infinitesimal point that is the actuality of the primultimate moment towards one or other kind of objectification? The 'joyous consent to irreversibility', then must have 'the same unilaterality as hope' since 'Everything a man can do, he can do in the direction of the future' (IN, 235)—and only in the direction of the future, we might add. So, we are not free whether to go future-wards or not, but we are free as to how we might relate to the future-ward movement of our lives. The free person 'makes as a gift a voluntary donation of the tax that

---

[11] I have preserved the French *néant*, often translated as 'nothingness', to acknowledge Jankélévitch's distinction between this term, which has the form of a present participle and, as such, participates in the primultimate plenitude of the moment, and *rien*, nothing that is simply nothing at all. Even reduced to an infinitesimal *néant* the girl was not and never will be a *rien*. And that despite the intention of a Nazi ideology that willed her to be nameless and nothing and in doing so (in Jankélévitch's view) reflected and enacted its 'metaphysics of nothingness'. But, as he says, 'one can annihilate being but not having been' (IN, 201).

[12] Although Jankélévitch himself has little time for the idea of eternal recurrence, what he is discussing here seems close to at least some of the interpretations of Nietzsche's eternal recurrence considered in Chapter 3.

destiny demands of him and that he will, in any case, be bound to pay' (IN, 239). Nietzschean *amor fati* is modulated into the courage that 'inscribes itself in futurition' (IN, 241). And this is always particular and local. It is not wishing for the impossible but looking to what is possible—'the impossible is easy, but it is the possible which is difficult and dangerous, which demands audacity and courage' (IN, 241).

In examining what this involves, Jankélévitch lays particular emphasis on gratitude. In being grateful for the destiny or fate that comes to us, we honour the act of donation over the gift and a readiness for 'inexhaustible fidelity' towards what has (irrevocably) been given. As he will say later on, it is the *datio* not the *donum* (IN, 356). Like the cry *'hélas',* the word of thanks, *merci*, indicates an excess: in the one case it is the excess of inconsolability over the passage of time, in the other the excess of gratitude over the gift. In thus looking to the act of giving rather than the gift, gratitude distils the possibility of love. Love, as Jankélévitch points out, is, of course directed towards the other person, the beloved, and not towards one or other mode of time, but love nevertheless demands—or, at least, characteristically involves—a certain attitude towards time: faithfulness to what has been given, gratitude to the giver, and hope for the future. As such, the moment of love is eminently primultimate: '"Hic incipit vita nova." This is the beginning of spring and the recommencement of eternal youth' (IN, 248).

But this is too easy. Think back to the examples Jankélévitch gave of irrevocable events, eternalized in their passeity. These are not the kind of events one can just swing over one's shoulder as one jauntily strides out on a spring morning. If regret allows for the bitter-sweet adagio of nostalgic remembrance, there is the harder case of remorse, the despair of never being able to revoke the irrevocable. The silence of remorse in its despairing contemplation of the past is not 'the apophatic silence that envelops love and temporality but the purely negative muteness in which death encloses us' (IN, 273). If irreversibility has an ineffable quality that we find expressed in the music of Debussy and Fauré (two composers often cited by Jankélévitch), what is irrevocable relates more closely to what is simply unspeakable, *'indicible'*— and we recall that it was precisely the substantive *L'Indicible* that Jankélévitch used to name the attempted annihilation of European Jewry by the Nazis.[13] Remorse—as opposed to regret—is and must be 'anti-poetry' par excellence. Even if, as we have seen, forgetfulness may make it seem as if the slave never existed or the girl never died, it can only ever make it 'quasi non-existent', as if it hadn't happened—but it can never bring it about that it didn't happen (IN, 289). Even if everyone forgets it, it still exists in what Jankélévitch calls 'a kind of meta-empirical firmament' (IN, 294). Only the grace of pardon can

---

[13] See Vladimir Jankélévitch, *L'imprescriptible: Pardonner? Dans l'honneur et la dignité* (Paris: Seuil, 1986).

do away with the irrevocable misdeed, an event, he says, that would be 'an intimate and total conversion of being' (IN, 292). A God might change the past but we cannot—and this is the real sting of what is irrevocable: that we can do nothing at all about it.

In *On Forgiveness* Jankélévitch speaks of the impossible possibility of forgiveness in such a way as to invite belief in the possibility of such an impossible event.[14] But even if, in some utopian sense—or even some practical sense—we may hope for forgiveness, it can only ever be as an event outside all rational calculation and all economy of reciprocity. The whole point of forgiveness being forgiveness is that it's not a matter of time having healed or (with the help of forgetfulness) having tempered the bitterness of memory, nor is it a reasonable response to a practical programme of rehabilitation, nor is it the recognition that really there's nothing to forgive (there is). Forgiveness—as Christian theology has always insisted—is not a matter of merit or desert and cannot be gained as a reward for quantifiable acts of contrition or paid off with prescribed penances.[15] That both giving and receiving forgiveness are a matter of the heart does not mean they can be willed but that they exceed will and emotion as surely as they exceed rational calculation. Forgiveness is always and only possible as transcendence, breaking in on immanence as sheer grace.

The discussion of time in *Irreversibility and Nostalgia* does not lead directly to the question of forgiveness but turns instead to the phenomenon of nostalgia. This might at first seem to be a phenomenon of space as much as of time, since the most obvious instance is, perhaps, nostalgia for one's native land in the pathos of exile. And there is a good reason for this, since real, geographical space is not interchangeable as geometrical space is. Yet nostalgia for place is always also a longing for another time: Ulysses longing for Ithaca is longing for his own youth on Ithaca. And, ultimately, nostalgia is an expression of our own self-relation. There's nothing—or there doesn't have to be anything—special about my childhood, my home, or even my beloved—except that it or he or she is mine. It is not the beauty of the place that elicits nostalgia but the nostalgia that creates the beauty (IN, 355). The nostalgist doesn't have to have been this or that: she or he just has to have been, since it is the passeity as such, the quoddity of time in its irreversibility and irrevocability that is the object of nostalgic melancholy (IN, 357). But, as we have learned, that past doesn't exist except as irrevocably past. The

---

[14] See Vladimir Jankélévitch, *Forgiveness*, tr. Andrew Kelley (Chicago: University of Chicago Press, 2005). As of the time of writing, this is one of the few works by Jankélévitch to be translated into English. Curiously, it seems to be out of print in French.

[15] It is, of course, unfortunately the case that historically the Church has made it look as if this is just what we can do. And if that is a charge that seems especially pertinent to the Roman Catholic Church, we may add that Protestantism has developed its own ways of 'earning' forgiveness, as in ritual performances of repentant emotion.

native land sought by nostalgia is 'another world' that exists 'nowhere' (IN, 361). As Plotinus—and the Bible—recognized, the true native land is eschatological, the Fatherland or the heavenly Jerusalem (IN, 362). This being so, there is, in time, no final word and every word is only ever pen-ultimate. Having returned, Ulysses must set off again—as in Kazantzakis's modern sequel to Homer's epic. 'It is to find you again that I am leaving' (IN, 367). This is at once a recognition of the irreversibility of time and the past for ever being past but also a summons to live the moment, the quoddity of futurition in its living plenitude. And so we hear such a modern Ulysses call out '*Holà! Compagnons de voyage, à vos rames!*'[16] as he sets course for an ocean without limits (IN, 383).

It might seem as if, by taking the modern Ulysses' part, Jankélévitch has in these final pages left unresolved the burning question of irrevocable deaths and unspeakable crimes. As he ponders the power of music, especially in the modern neo-Romantic tradition, to express the deepest sighs of our metaphysical nostalgia, we are left to wonder whether this is it. Faced with all that is irrevocable and unspeakable in history and with the possibility of forgiveness hovering in the background as an impossible possibility, is settling down of an evening, turning up the music, and yielding to the bitter-sweet consolation of some gentle music all that we can do? Perhaps—whatever the period or genre of music, poetry, and art we individually favour—this is precisely how many of us deal with the stress and distress of existence. We cannot solve all the world's problems and especially cannot bring back all the innocent victims of all of history's many and continuing unspeakable events, some perpetrated by fellow citizens (and all of them, really, perpetrated by brothers and sisters on brothers and sisters). Who can blame us if we seek to lose ourselves for an hour a day in the possibility of things having been otherwise and so, so better? Is this not our post-religious vespers, our post-modern compline? And, if critics such as Kierkegaard suspect this will enervate the moral will, who's to say that such evenings might not, in fact, sufficiently renew and reinvigorate us so as to enable us to take up the tasks towards which futurition so relentlessly carries us? It is, at least, a humane vision, not without tenderness, not without a certain amiability, and not without a sense for the unfathomable mixture of tragedy and joy that is so integral to who we are. But is it all? Is the eternity of having-been towards which Fauré's nostalgic strings sweep us, conjuring forth a distant, unrealized and unrealizable intimation of that 'other country', all the eternity there is?

---

[16] 'Ahoy, fellow travellers, to your oars.' Many readers will perhaps think of how Eliot's *Four Quartets* insist on the inseparability of endings and beginnings and the need for even old men to become, again, 'explorers'.

ETERNALLY UNHOPED-FOR

Jean-Louis Chrétien's approach to questions of time, pastness, forgiveness, and the eternal is at many points similar to but also significantly different from that of Jankélévitch. Jankélévitch approaches time as a self-constituting continuum, refusing any solace other than the solace that time itself can give us. If he gestures towards an 'impossible' transcendence from which the grace of pardon for the wrongs of an irrevocable past may yet appear, this would have to be of such an all-encompassing and world-transforming character— 'apocalyptic'—that it is scarcely allowable as an object of hope (which, according to Jankélévitch, must always be hope for something possible). Chrétien, by way of contrast, will go further and suggest that this impossible, not-to-be-hoped-for transcendence may, however indirectly, however elusively, enter into the weave of our human experience.

I shall focus here on Chrétien's study *The Unforgettable and the Unhoped-for*, although it might more accurately have been entitled *The Immemorial, The Unforgettable, and the Unhoped-for*, since the question of the immemorial, of what precedes and forever eludes memory, is the focus of much of the book. All three topics are, of course, interconnected in complex ways around the key question as to the limits of what can be experienced, known, and expected in time. Chrétien's approach is, in Michael Theunissen's sense, 'negative'.[17] That is, he looks to draw the contours of what cannot and ought not to be forgotten by an analysis of the phenomenon of forgetting. As he states at the beginning, the book concerns loss and, more particularly, the form of loss experienced in forgetfulness. However, as long as we still remember what it is we have lost (as in memories of childhood), Chrétien suggests that it is still not entirely lost. Yet loss is needed if we are to become separated from our origin and 'become ourselves in truth'. Why? Because as long as we go on clinging to the image of ourselves that is reflected in memories from the past we never become who we are in the actuality of the present.[18] We have to forget who and what we were so as to become who we are now becoming. But is such radical forgetting possible?

'Forgetting' Chrétien says, has in fact accompanied Western philosophy from its beginnings, as that from which Platonic anamnesis seeks to deliver us. However, modern ways of interpreting this—Leibniz, Kant, Hegel, and Natorp are cited—typically emphasize that the forgetting that is reversed by the act of anamnesis was never really total. As Hegel's understanding of the

---

[17] See the section 'A Negative Theology of Time' in Chapter 7 of this volume.
[18] Jean-Louis Chrétien, *L'inoubliable et l'inéspéré* (2nd edn, Paris: Desclée de Brouwer, 2000), 9–10. Further references will be given in the text as II. The point may be compared with Lévinas's notion of the necessary 'atheism' of human existence, i.e. that fully to be who we are we must separate out from our point of origin and do so entirely and without reserve.

German *Erinnerung* ('memory', but also 'internalization') makes especially clear, anamnesis is a form of recollecting an a priori knowledge that was never truly lost. Forgetting therefore shows itself to be merely a temporary and partial condition. Such interpretations strip away the mythical elements in which Plato presents his teaching and explain it in terms of the a priori structures of self-consciousness. But if this is so, then, as far as truth is concerned, there is 'no past, no forgetting, no immemorial' (II, 24) and all that has been is fully revealed to the light of consciousness. Later, Chrétien will say that this kind of approach is already anticipated in Plotinus, since Plotinus taught that there was one part of the soul that never did fall into the world of time and embodiment and that had therefore never entirely forgotten itself (II, 44).[19]

Chrétien asks us to consider that, in fact, the mythical aspect of Plato's teaching may be vital to the whole and that, as he puts it, rather than reading Plato through Kant we should perhaps try reading Kant through Plato (II, 24).[20] In other words, the temporality of the myth is crucial to what it is teaching and the 're-' of re-collection is at the very heart of the theory of anamnesis. But if what has been forgotten and must be re-collected also belongs—as Plato seems to say—to the essential constitution of the self (and is not, as Natorp suggests, always 'really' present to the self), we are led towards the idea of a kind of split in time, to a past 'that is other than any past in which I am already human, a past that is other than any past in this incarnate life' (II, 27–8). Chrétien is, of course, conscious of the apparently paradoxical nature of this situation and has no intention of minimizing the paradox. Forgetting the immemorial past—the past of contemplating timeless eternal truths—is precisely what we have to do in order to become fully human. But what, then, is the purpose of re-collection? Is it an attempt to return to what we might call the metaphysical womb and revert to some pre-human state? No. Deploying a logic that recalls Jankélévitch's insistence that preterition never escapes futurition, Chrétien proposes that thinking our way towards this immemorial past is, paradoxically, an exercise in anticipation (II, 28–30). For if the entire structure of knowledge were to be a priori and timelessly present, time and history would lose all significance; but because knowledge is founded in what can never entirely be recalled, time, and therefore the future, become vitally important to who we are and what we can know. Consequently, the quest for the immemorial past becomes interdependent with the quest for what we can hope for, since it is only by seeking knowledge in time or seeking the future perfection of such knowledge as we already have that we could ever approximate to the immemorial.[21]

---

[19] The reference is to *Ennead* V.I.1.   [20] Schelling would surely have agreed.

[21] The interconnection between knowledge and hope is already established in Kant's summary of the three basic questions of philosophy: What can I know? What should I do? and What can I hope for?

The first beginning is irretrievable—immemorial—and philosophy can only ever begin with a second, derived, or historically-acquired beginning.[22] Drawing on Plotinus, Proclus, and Philo, Chrétien argues that, nevertheless—and certainly paradoxically—the good that we desire to attain through recollection is always with us, though immemorially, as what we have always forgotten. But this forgetfulness is itself 'the measure of the excess of the good over us' (II, 62). In other words, we are always so much more than we are ever able to know. Philosophy, however, has for the most part interpreted this forgetfulness in terms of a truth that is hidden rather than lost and therefore a truth that, in principle, is retrievable. Everything essential is preserved, and if forgetting is a human act—*my* act—then what is forgotten must, in principle, be recuperable by an equal and opposite human act. Nietzsche and Valéry are amongst those cited in support of this view, whilst in the case of Bergson memories are so little 'lost' as to be independently active in the subconscious, striving, as it were, to make themselves heard (II, 76–8). Likewise, Ravaisson maintains what we might regard as the Hegelian point that pure spirit would forget nothing and would see everything 'in the form of eternity' and that it is only the veil of material life that obscures such knowledge. But, as Chrétien comments, this is no longer really what we mean by memory (II, 80–1). As Chrétien says of the personalist idealist I. H. Fichte, this 'reabsorbs time into eternity' (cited II, 95). In such philosophies, which Chrétien takes as representing the mainstream of Western thought, 'memory is the place where we free ourselves from time; it surpasses the past, abolishing the relation to the past as such in order to attain the atemporal presence of essence' (II, 95).

Augustine too argues for the possibility of total recall. However, in his case, there is a significant difference from the Platonic tradition, ancient and modern. This is that such recall is precisely *not* possible for human beings. It is God and God alone who remembers my life in its entirety (II, 82). And, whatever may be the case with God, it is certainly true that I am not, in fact, the only one who remembers my life since others also do. Perhaps, then, whether with regard to God or just other people, I am not myself the site of what is immemorially basic to my life (II, 83–4)?

This thought is further developed through the theme of promise, introduced by Chrétien with reference to Kierkegaard's *Either/Or* and the distinction between the aesthete of Part I and the ethical point of view represented in Part II. The aesthete deliberately lives without making any firm or lasting commitment, neither to friends, nor to a beloved, nor to any social task. By

---

[22] The reference to a 'second beginning' is presumably implicitly alluding to Heidegger's idea that the cycle of philosophical thinking that began with the first beginning of Greek philosophy is being consummated in the advent of global technology and that we are therefore called to prepare for a 'second' or 'other' beginning. I take Chrétien's point (arguably contra Heidegger) to be that, strictly, there never was a 'first' beginning and that the 'first' beginning of philosophy was itself already a 'second' beginning.

contrast, the ethicist bases his life-view on the possibility of making a prom-ise—primarily, though not necessarily exclusively, the marriage vow—and it is this promise, a commitment to the future, that provides a fixed point to which, as it were, the chain of memories can be attached and therefore acquire sufficient shape to become knowable. 'Nothing is irreversible except by means of a promise' (II, 108). But a promise can only be spoken in a world in which I am not alone. Minimally, it must also involve the one to whom the prom-ise is made. Consequently, 'If there is a parousia'—that is, an experience of the plenitude of recollected meaning—'I am not it and I cannot be the place for it. Its place is exodus' (II, 109)—the place at which I am led out beyond myself and my present possibilities. Kierkegaard is again referred to in order to extend the point that memory of pardon, or, more precisely, memory of the *promise* of pardon, is the condition of forgetting sin (II, 112). In a word, it is love that establishes both the possibility of forgetting and the possibility of memory. Turning this time to St John of the Cross, Chrétien writes that 'Wounded by the forgetfulness that dispossesses [the soul], it is by love that it is wounded. We are not unforgettable. Only the immemorial is properly unforgettable since it gives itself to hope and it gives hope by means of and in forgetting' (II, 113–14).[23]

And so we come to what is properly unforgettable. This is not, and could not be, some empirical fact that we were never able to erase from our memo-ries, since any particular memory is, at least in principle, forgettable. What we cannot forget is, paradoxically, what we have never been able to remember or, to be more precise, what we cannot forget is the memory of the loss of what we have never been able to remember. If we were to think of it as a past event, it would have to be the kind of past event that could never become entirely past—perhaps like the grief of a deep personal loss (II, 115–16), the presence of an absence, a loss that is still with us (II, 117): what we don't have and can't therefore forget. As such it therefore coincides or at least converges with the immemorial. It is not—for Augustine at least—the unforgettable memory of paradise lost since paradise lost can't be remembered and can only be known, if it can be known at all, by faith (II, 126–7). That is to say that, as in the case of the immemorial, we ourselves are not the site of what is unforgettable.[24]

But nor—despite the negative prefix— is the unforgettable a primarily nega-tive phenomenon. For William of St-Thierry (writing in the Augustinian tra-dition), what is truly unforgettable is the eternal (II, 127), which is not just an event of the past but also the goal of all our striving (II, 128). If it is God it is not God as somehow internal to the soul (as some Christian mystical traditions

---

[23] For further discussion of the theme of promise see the sections 'A Theology of Promise?' and 'A Certain Messianicity' in Chapter 6.

[24] A powerful poetic statement of this insight is Edwin Muir's poem 'The Absent'. See my '*The Heart Could Never Speak*', 40–9.

might suggest) but, in Chrétien's view, God 'comes to memory so as to wound it with a wound of love that not even eternity can cover over again' (II, 129). God, present to us as the positive meaning of what is truly unforgettable, is unforgettable precisely not as a fact that we could ever know as the object of an intentional act of consciousness: 'God is only present to us by comprehending us and exceeding us in all our ways' (II, 131). Such presence can only be eschatological in the twofold sense of occurring only at the extreme limits of human possibilities, at the end or boundary of our human way of being (II, 131). Forgetfulness, or, more precisely, self-forgetfulness, the forgetting of my human reality—is therefore integral to any faithful response to such a presence, as Paul speaks of forgetting what lies behind for the sake of what is ahead (II, 132).[25] But we must even forget what God has done in us, since we are not the place where this is to be seen for what and as it is (II, 133).[26] An extreme and exemplary instance of this is Jesus' own cry of dereliction on the cross (II, 134–6). In this we see what it means that 'The alterity of God inscribes itself unforgettably at the heart of our intimate being' (II, 137)—a statement that could be taken as expressing the heart of Chrétien's vision.[27]

The excess of the divine presence as what is truly unforgettable is further specified in terms of Chrétien's final category: the unhoped-for. Again, this is not to be seen as a negative phenomenon but as the revelation of a power other than our own (II, 143). In early Greek tragedy the idea of events that were unexpected often had a negative import and led to uncertainty rather than to hope (II, 144ff.).[28] Even within the Christian dispensation, we might have to say that we shouldn't hope for the unhoped-for (II, 148–9) since—by definition—it is what exceeds all anticipation and is that to which no known paths can lead (II, 151). A paradigm of how we might relate to the advent of such an unhoped-for event is in Philo's commentary on Abraham's response to the divine promise: he fell to the earth as an expression of his humility and his incapacity for receiving such a promise, and he laughed because it was of God (II, 154–5). Again, this underlines the notion of promise as the mode in which the immemorial, the unforgettable, and the unhoped-for become a reality in human life, a view that, Chrétien says, 'is at the heart of all biblical theology' (II, 162).

Perhaps curiously, it is Kant who at this point provides Chrétien with a point of reference that supports his analysis yet from which he also decisively

---

[25] Cf. Philippians 3:13–14.

[26] This might be taken as a significant warning to certain Christian cultures in which the recitation of the signs of God's favour in the believer's life are a well-established feature of pious practice.

[27] These remarks can also be compared with Kierkegaard's attack on the 'forgetfulness' of Hegelian philosophy, a forgetfulness that he sees precisely in terms of its forgetting the embodied and therefore mortal nature of human existence.

[28] See Chapter 7.

distinguishes his own position. In his *Anthropology in Pragmatic Perspective*, Kant speaks of the foundation of moral character as occurring in and as a rebirth, an unrepeatable event inaugurating a new epoch in personal life that would have a good claim to be unforgettable (II, 162). Yet, against Kant, Chrétien comments that 'this solitary baptism of autonomy transposes the features of actual baptism to a place within the limits of mere ethical reason; but actual baptism is always received from another. A purely autonomous self-baptism, however, would not be able to lead to a strong sense of the unhoped-for' (II, 163–4). An example of such a self-baptism would be Kant's view that we are ourselves originators of the moral law, but, on this view 'We cannot and will not be able to receive anything...It would [therefore] be absurd and immoral to hope for what is unhoped-for, that is, grace'. The Kantian God 'beholds, and gives nothing' (II, 164). Against Kant, then, I do not make my own promises: with regard to what is promised me, I am addressed or called and the promise comes as if suddenly, unexpectedly, from the place of its immemorial genesis. That I can nevertheless lay hold of this promise and make it the focus of my life and self-understanding is because it becomes actual and unforgettable as the divine word. And, for Chrétien, it is important that Scripture, understood as the Word of God, begins precisely with the story of creation, since this means that the Word and the promise articulated in the Word are to be understood as foundational to my life as a whole. In terms of Philo's allegorical interpretation of the promised land the vine and the olive that epitomize the blessings of the land are to be understood as joy (the vine) and the twofold blessing of nourishment and light (the olive), which also indicates how the olive might become a symbol of the Word that illuminates the meaning of human life (II, 167).

It is probable that Chrétien, who is unabashed in saying that he writes both as a philosopher and as a theologian, will by this point have lost many of his secular-minded readers. Yet I think it plausible to interpret everything said here—including what is said of God—within the parameters of a phenomenological approach that does not presume upon assent to any metaphysical claims about divine existence. For the point is not to define God or to defend a particular view of the divine nature and attributes but, simply, to show how a religious understanding of God might be configured within a certain 'how' of human experience. Whether or not God 'exists' is almost beside the point. Indeed, the whole thrust of the argument is that whatever God 'is' for us cannot be forgotten because he can never be either remembered or anticipated in any intentional act of knowledge. He is (precisely) immemorial and, only in that sense, is he able to be also unforgettable and unhoped-for. So, the book concludes with the most indirect—some might say 'weakest'—of claims: 'For those who hope because they remember and remember because they hope, who keep watch through silent nights, there is always already and always still to be heard far off the light and supple whispering wind that, however, wounds us in the most secret intimacy [of our being] and that, as it passes, blesses the olive and brings it to flower' (II, 168).

The imagery evokes John of the Cross. And if there are deep structural commonalities with Jankélévitch's account of the impossibility of remembering or anticipating what belongs to the sheer unknowable quoddity of time's mystery, the mood—can we say 'experience'?—at which it hints is significantly different. God is not metaphysically affirmed, but the hope and the demand of the divine promise (that is also, note, a promise of forgiveness and a demand to attend to the needs and hopes of others) says more than what could be seen as the ultimately aesthetic, musical solace offered by Jankélévitch. I have not wished to belittle the latter, but Chrétien sets before us a different possibility: the possibility of faith. But does he do so too easily? Is he too swift in closing the gap between the immemorial and the experienceable? If his 'mysticism' will appeal to some, is it, in the end, a philosophical short-cut? We recall Heidegger's comment that the theologian always comes too soon. Is Chrétien too previous in all that he reads into the stirring of the night wind? Though how much exactly does he read into—or out of—it?

These questions are not merely rhetorical but are asked out of concern for the issue that lies at the heart of Chrétien's own project, namely, whether our time-experience can be open to another time, a time that he calls the time of the immemorial, unforgettable, and unhoped-for, and a time that we might also, within the limits of phenomenological investigation, call the eternal. But can the presence of the eternal, of a time not our time, an 'other' time be more than an object constituted by the kind of nostalgic or mystical intimations of which Jankélévitch and Chrétien speak? Can it indicate more than the possibility of longing for the advent of such a time? Over and above the consolations of music or the stirrings of the night wind, could it ever take shape in a community that would truly be a community of remembrance and hope? Must we accept that time, in all the thick complexity of its historical unfolding, cannot offer anything more substantial than this?

ETERNALLY TO COME

An alternative approach that nevertheless shares much common ground with Jankélévitch and Chrétien but which more extensively addresses the question of temporality and community is that of Emmanuel Lévinas, for whom the obligation of care for the other, summed up in the biblical injunction to seek justice for the stranger in the land, the widow, and the orphan, is the only genuine measure of the God-relationship that we have. But, as we shall now see, this is also fundamentally connected with the character of human temporality.

A key idea in Lévinas is that of the infinite. Following Descartes, he sees the idea of infinity as intrinsic to the very possibility of thinking—yet it also bursts open the limits of pure thought. The infinite carries us beyond any possible 'interest' we might have in being, dis-interesting us in the active sense of dissolving all attempts to grasp or anticipate being: if being is truly infinite, it is, quite simply, beyond us; but if, at the same time, the infinite is inseparable from the possibility of thought and therefore self-consciousness, it follows that we are assigned to an attitude of 'pure patience' and 'deference' in the double sense of having to be humble before what we are waiting for and always being subject to deferral. The relation to the infinite is 'as irreversible as time', it is 'older' than consciousness and yet it is also 'the most profound thought of what is now'.[29] It is 'the very dia-chrony of time...a way of "being dedicated" that precedes any conscious act...which is precisely not intentionality in the compound of the noetic [the thinking] and the noematic [the thought]'.[30] In other words, it is not a thought that could ever be intuited in a single moment of vision in which the meaning of all time could ever be given.

This seems to register a view comparable to what we saw in some analytic philosophers' concern that a timeless God could not be known or indeed interact with time at any moment, not even for the most glancing, evanescent 'now'. But could we say that although we cannot *know* the infinite, we can relate to it in the manner of desire, as in the Platonic *eros*, an infinite longing at the heart of our being? Maybe, but if desire is truly infinite, it will never coincide with what it desires—although without the pull of the infinite, love will degenerate into eroticism in the vulgar sense and even sink down to the level of pornography. But how is such a transcendence of the desirable beyond interest and eroticism possible?

Lévinas's answer is to propose that the desired but unattainable God may be *near* in the mode of holiness. Never fully present, He is nevertheless 'proximate'. Refusing to be the object of our enquiring gaze, God is never our 'Thou' but only 'He'. We, however, do become *his* object: I becomes 'me' and subject becomes accusative as the divine command directs my love towards what was never the object of my desire, namely, towards the needs of others, that is, towards the neighbour, the stranger, the widow, and the orphan, whose claim on my love can never benefit me and, apart from the divine command, could never be the object of my (self-)interest. The good that we desire is in this sense a good beyond being.[31] The ultimate passivity of the self is not simply a passivity vis-à-vis being but is now glossed as an obligation to the other issuing from an unrepresentable past that was never present and as a responsibility that is prior to any law or social contract. I encounter the other 'in his "without recourse"[32] that is to be heard as a crying-out to God with neither

---

[29] Emmanuel Lévinas, *De Dieu qui vient à l'idée* (Paris: Vrin, 2004), 9–10.

[30] Lévinas, *De Dieu qui vient*, 12.   [31] Lévinas, *De Dieu qui vient*, 113–14.

[32] That is, without recourse to any justice other than what I am now called upon to effect on his behalf.

voice, nor thematization [i.e. articulated content]. Assuredly, the resonance of silence—"*Geläut der Stille*"—resounds there'.[33] Our responsibility in relation to the other is 'limitless', summed up in a citation from Dostoevsky's *The Brothers Karamazov* that is a kind of leitmotif of Levinasian thought: 'We are all guilty before everyone and for everything and I most of all'.[34] Others thus become the measure of the dia-chrony, the time marked by the immemorial presence of the infinite, the God-in-us whom we cannot come to know through reflection on our own interiority but only through being extraverted out towards the call of the neighbour. With regard to eternity, this certainly dispenses with any effectual notion of a non-temporal eternal God, an 'intellectual God' as Lévinas says—but, by the same measure, time should not be regarded as a 'degradation of eternity'. It is the 'always' of the non-assimilable, non-knowable, because never adequate relation to the absolutely other.[35]

Like a number of other critics, Lévinas sees Heidegger's focus on death as necessarily diminishing the significance of this relation to others and as misrepresenting the essence of human temporality. He is, of course, well aware that Heidegger adduces *Mitsein* (being-with) as an intrinsic feature of Dasein's being-in-the-world. However, he does not see being-with the other as sufficient to account for the call to become responsible-for the other. Since Heidegger also makes the relation to death paradigmatic for human beings' relation to their own being and sees death itself in terms of the annihilation of the existing individual the priority of the ethical is further undermined in favour of the supposed authenticity of bearing the guilt of being thrown towards nothingness. But, for Lévinas, death is not primarily annihilation but a mystery, and the duality announced by death is better parsed in terms of the relationship between time and others.[36]

An important aspect of Lévinas's position concerns the material character of human existence. Materiality is not a prison into which the soul falls but the proper accompaniment of all subjectivity. For Lévinas we have to eat to stay alive and it is therefore eating in which our self-relationship first becomes manifest, as the joy or pleasure of being able to satisfy our hunger. This is opposed to Heidegger's view that subjective individuation is supremely brought about in the context of my 'ownmost' confrontation with death. For Heidegger, as read by Lévinas, to grasp death as my ultimate possibility is to find the 'supreme lucidity' and 'supreme virility' of freedom towards death.[37] But, versus Heidegger, Lévinas sees death as, precisely, 'the limit of the subject's

---

[33] Lévinas, *De Dieu qui vient*, 118. The phrase "Geläut der Stille" is an allusion to Heidegger's claim that the origin of language relates to the poetic attention to the silence that precedes all utterance. Cf. my *Heidegger on Death. A Critical Theological Essay* (Farnham: Ashgate, 2013), 129–45.

[34] Lévinas, *De Dieu qui vient*, 119.

[35] Emmanuel Lévinas, *Le temps et l'autre* (Paris: Presses universitaires de France, 1983), 9–10.

[36] Lévinas, *Le temps et l'autre*, 19–20.   [37] Lévinas, *Le temps et l'autre*, 57.

virility and heroism'.[38] The sob of the dying—and Lévinas implies that we can none of us eliminate the possibility of such an irrepressible sob—marks an ineluctable return to infantile loss of responsibility.[39] At a certain moment in the process of dying 'we are not able to be able' and lose all mastery over our selves; therefore death can never—as Heidegger suggests—become a 'project', 'death is the impossibility of having a project'.[40] The will to pure subjectivity is therefore confronted with sheer alterity and sheer mystery (although not necessarily nothingness). But this discovery of the limits of what the solitary 'I' can achieve also opens the way to a new relation to others, when the mystery of death is experienced as our common human condition. Recognizing each other as sharing in the common fate of mortality constitutes a decisive reason—Lévinas says a 'call'—for becoming responsible to and caring for the neighbour. The bare existence of the other cries out voicelessly in the naked defencelessness of its humanity and summons me to its aid for no other reason than that it is the cry of the one who is both neighbour and one exposed constantly and utterly to death. This cry thus inaugurates a future that is very different from the Heideggerian future of death: 'the future is the other', Lévinas writes. When I join with others to work for a better common future a new relation to time becomes possible, namely, the time of history.[41]

But is this really a sufficient or actual basis for human community? Or can we envisage a historical community in which relations are built up as more than a series of rescue operations? Lévinas goes on to explore how this occurs through the relation to the feminine and through paternity, developing an alternative to what he sees as the Platonic community of the idea or the Heideggerian community of the will. But if this suggests a community defined by family ties and 'blood', it will also be a community that stands under the commandment to love the neighbour and to care not only for the widow and orphan but also for the alien in the land. As a reading of Judaism, it therefore points us to the community that welcomes Ruth more than to the exclusive community of Ezra and Nehemiah. And, as a community that allows justice to stand as the measure of its own life, it is a community under judgement and for which the day of the Lord brings with it the possibility of condemnation as well as of redemption. Precisely for this reason, prophetic religion calls this community to labour for eschatological peace within the nation and amongst the nations. This vision finds a partial expression in Marxism, at least in the kind of Marxism Lévinas saw in Ernst Bloch, whose utopian, almost religious philosophy of hope is affirmed by Lévinas—as far is it goes.[42] As far as it goes—but, according to Lévinas, even Bloch's extravagant hope lacks a dimension of '*hauteur*', the

---

[38] Lévinas, *Le temps et l'autre*, 59.   [39] Lévinas, *Le temps et l'autre*, 60.
[40] Lévinas, *Le temps et l'autre*, 62.   [41] Lévinas, *Le temps et l'autre*, 63–4.
[42] See, e.g. 'Sur la mort dans la Pensée de Ernst Bloch', in Lévinas, *De Dieu qui vient*, 62–76.

elevation of the divine that, in ways we have been seeing, is lifted above and beyond all human experience and knowledge, commanding from a time that always exceeds our time, prior to and beyond all history, immemorial and unhoped-for, as Chrétien has it. Beyond even death there is—or *may be*—'the Messianic time in which what is [merely] perpetual is transformed into the eternal'.[43] But, qua philosopher, Lévinas refrains from deciding whether this is indeed the advent of another order of time or what he calls 'an extreme vigilance on the part of the Messianic consciousness',[44] that is, a kind of regulative ideal which, as immemorial and unhoped-for, we can never presume to be a present reality.

THE VIRGIN IN HELL

In Memoriam Diane Oenning Thompson

In Chapter 2, we concluded the examination of Hegelian approaches to the question of time and eternity with reference to Proust's *In Search of Lost Time*. Here too it may prove helpful to supplement the philosophical arguments we have been surveying with the testimony of a great novel, namely, Dostoevsky's *The Brothers Karamazov*. This is a work in which issues of forgetfulness and memory, of the immemorial and the unhoped-for play a major and even decisive role. Indeed, it is possible to see the novel as a whole as a response to the question as to whether memory can make good its claims against the power of oblivion and, if so, just what kind of hope does memory enable us to hope for? If Proust is unsurpassed in testifying to the kind of transcendence of vanishing time that is possible within the limits of recollection, Dostoevsky (I suggest) raises the further question as to whether there might also be a way of transcendence leading beyond the timeless coincidence of temporally separate 'times' to a 'saving time' that is rooted in a sphere beyond recollection and beyond historical hope. Like Proust, Dostoevsky is concerned with the possibility that literature can both testify to and, in its way, even perform a work of resurrection. But if, as I suggested, Proust's resurrection is that of a Lazarus, Dostoevsky points us towards the resurrection that the raising of Lazarus only anticipated, that is, the resurrection of the one who performed that miracle, the Christ. If this is so, then Dostoevsky is not just aiming at liberation from time within and through time but at liberation into another time, a time freed from suffering and death, a time of infinite light and joy. This is not only a remembrance of things past—of 'time lost'—but, as the title of a study of Dostoevsky

---

[43] Emmanuel Lévinas, *Totalité et l'Infini: Essai sur l'exteriorité* (4th edn, Paris: Kluwer Academic, 1971), 318.

[44] Lévinas, *Totalité et l'Infini*, 318.

by P. Travis Kroeker and Bruce K. Ward nicely puts it, 'remembering the end'.[45] This, then, would be a form of remembrance in which the rupturing of time and eternity explored in the last chapter is both presupposed and transformed. Its time is the time of the apocalypse, it is Messianic time, the time to come that is in time.

Bakhtin taught us that since Dostoevsky's novels are built up of layers of dialogical and contrasting views and voices, we should not expect to find any simple or direct doctrine of remembrance in *The Brothers Karamazov*. However, if the novel suggests that faith in a heavenly other world may be integral to the realization of true humanity, this 'theological' answer is never offered as a doctrine separable from the dynamic movement of the novel as a whole. Another view is always possible. Consequently, the argument that follows is not that *The Brothers Karamazov* straightforwardly vindicates the possibility of 'remembering the end' but that Dostoevsky presents us with a certain experiential 'how', that is, a portrayal of how it might be and what it would mean for human life to realize or, alternatively, to forget its relation to the eternal.

Although *The Brothers Karamazov* picks out two particular human lives in which the relation to the eternal finds exemplary expression, the saintly elder Zosima and his pupil Alyosha Karamazov (whom Dostoevsky refers to as his 'hero'), it is a novel that is as much about communities as it is about individuals. Proust's novel too, as we saw, also mirrored the titanic upheavals in French society in his time. Yet Dostoevsky's novel is 'social' in another sense. It is not just that Dostoevsky's polyphonic method presents us with multiple personal histories, such as the stories of each of the brothers, of Zosima, and of other characters. It is also that both family relationships and the life and fate of Russia itself are explicitly thematized in such a way as to show the inextricable involvement of each individual with his or her encompassing micro- and macro-social environments. Consequently, the relation to the eternal that the novel comes to privilege is not a matter of individual spirituality. The forgetting of the eternal is a fateful event for communities and for Russia as a whole. In this perspective, then, the question of the eternal and of the possibility of becoming (once more) open to it also becomes a question of utopian politics, that is, a question as to whether the Kingdom of God is a real human possibility or a mere residue of antique other-worldliness. If Dostoevsky could be a sharp critic of secular utopianism, there are also grounds for seeing his work as expressing what Ernst Bloch would call the 'spirit of utopia' and the final triumph of the eternal value and truth of love in historical human existence.[46]

In her study of *The Brothers Karamazov and the Poetics of Memory* Diane Thompson argued that 'the interplay between cultural memory and the

---

[45] P. Travis Kroeker and Bruce K. Ward, *Remembering the End: Dostoevsky as Prophet to Modernity* (Boulder, CO: Westview, 2001).

[46] See Chapter 6.

individual memories of author and reader, narrator and characters has been so fully, subtly and variously developed that it can be seen as a dominant means of organising this novel's artistic system, structurally, aesthetically and semantically'[47]—and, we might add, metaphysically. Thompson herself makes just this last point when she continues by commenting that 'The whole novel constitutes a great eschatological dialogue in which, as Bakhtin put it, the questions of Dostoevsky's characters "sound before earth and heaven"'; it is, in other words, an exploration of a universe in which all human cultural life and meaning goes back to roots in 'divine memory'.[48] In the human situation as we find it and live it, however, not only have some individuals lost their sense for these divine roots but the basic 'divine memory' nourishing common life as a whole has been or is in the process of being erased, with what Dostoevsky sees as catastrophic results. The chaotic and violent 'family life' of the Karamazovs is not just the product of the vicious life of Fyodor Karamazov, father to the eponymous 'brothers', but a harbinger of what is to coming to pass in both Russia and the civilized world as a whole.[49] Yet the novel also opens the prospect of a rediscovery of what, in the closing pages is called 'eternal memory'.[50]

Thompson's study argues that although such 'eternal memory' is in the first instance enshrined and transmitted in the texts, traditions, and rituals of Russian society, it not only extends to but is ultimately governed by what she calls a 'Christocentric poetic memory system', that is, a memory of the God revealed in the Incarnation of Christ. For although the novel cannot itself decide the question as to whether such a 'cultural' memory might yield an actual revelation of human existence as grounded in the eternal memory of a divine time of origin and consummation it can—and Thompson argues that it does—offer a vision of what such a revelation might mean to those able to accept it.[51]

---

[47] Diane Oenning Thompson, *The Brothers Karamazov and the Poetics of Memory* (Cambridge: Cambridge University Press, 1991), 1. The reading that follows is heavily indebted to Thompson's work.

[48] Thompson, *Poetics of Memory*, 1.

[49] This may suggest that the problem is 'the West', but if Dosteovsky believes that Russia has resources with which to be healed of the ideological sickness originating in the West, he can also be read as wanting Russia's salvation to offer a way of salvation for the West itself. Against the nationalistic element in Dostoevsky's analysis, V. S. Soloviev argued that 'when Dostoevsky speaks of Russia, he was not able to limit his view to its national individuality. On the contrary, he judged the entire significance of the Russian people to subsist in the service of true Christianity, in which was neither Greek nor Jew. It is true that he regarded Russia as a people chosen by God, but this choice was not for the sake of vying with other peoples or of attaining rule or primacy over them, but for the sake of freely serving all peoples and of realizing, in brotherly union with them, the true universally human or ecumenical Church' (Vladimir Soloviev, 'Tri rechi v pamyat' Dostoevskogo' ['Three talks in memory of Dostoevsky'], in *Spor o Spravedlivosti* (Moscow: Eksmo-Press, 1999), 602).

[50] Fyodor Dostoevsky, *The Brothers Karamazov*, tr. Constance Garnett (London: Heinemann, 1912), 820.

[51] In Heideggerian terms, it offers a certain 'how' of religious existence.

The novel opens with 'The History of a Family', composed of a series of memories that set the stage for the main action of the story—although, as Thompson will also point out, this is a history punctuated by manifold acts of forgetting, not least Fyodor Karamazov's propensity for forgetting his own children.[52] Within this history, special emphasis is placed on a number of memories of special significance, of which one of Alyosha Karamazov's few remaining memories of his mother, who died when he was four years old, is amongst the most important.

> He remembered one still summer evening, an open window, the slanting rays of the setting sun (that he recalled most vividly of all); in a corner of the room the holy image, before it a lighted lamp, and on her knees before the image his mother, sobbing hysterically with cries and moans, snatching him up in both arms, squeezing him close till it hurt, and praying for him to the Mother of God, holding him out in both arms to the image as though to put him under the Mother's protection…and suddenly a nurse runs in and snatches him from her in terror. That was the picture![53]

This 'picture' alerts readers to the fact that Alyosha is in a special and even quite literal sense someone whose life is offered to God. But although the icon of the Mother of God is the most explicitly theological motif in this 'picture', we should also observe that what Alyosha recalled 'most vividly of all' were 'the slanting rays of the setting sun'. Both in Dostoevsky's previous novels and progressively in the course of *The Brothers Karamazov* the role of the sun and of light as symbols of the divine is of great importance probably reflecting the prominence of such 'Platonic' symbolism in St John's Gospel and Orthodox spirituality.[54]

However, this same passage also implicitly confronts us with the most intractable objection to such a possibility, namely, that we are temporal beings thrown towards death. For what the passage narrates is, precisely, Alyosha's memory of his dead mother, and it does so in such a way as to highlight the

---

[52] Thompson, *Poetics of Memory*, 161–9.

[53] Dostoevsky, *The Brothers Karamazov*, 13.

[54] See Plato, *The Republic*, 508a–509b and, e.g., John 1:4–5 and John 8:12 ('I am the light of the world')—though cf. also Hebrews 1:3 which speaks of Christ as 'the radiance of God's glory'. This theme was reinforced in Orthodoxy by the idea associated with the 14th century Greek theologian Gregory Palamas that light was not merely a creaturely analogy of divine being but the manifestation of a divine energy in the created order. See Vladimir Lossky, *In the Image and Likeness of God* (New York: St Vladimir's Seminary Press, 1974), Chapter 3 'The Theology of Light in the Thought of St Gregory Palamas', 45–70. Orthodoxy was not, of course, alone in associating the sun and the divine. We might, for example, think of its use in the painting of J. M. W. Turner. See John Gage, *J. M. W. Turner: 'A Wonderful Range of Mind'* (New Haven, CT: Yale University Press, 1987). On the importance of John's Gospel in Dostoevsky see, e.g. Irina Kirillova, 'Dostoevsky's Markings in the Gospel according to St John', in George Pattison and Diane Oenning Thompson (eds), *Dostoevsky and the Christian Tradition* (Cambridge: Cambridge University Press, 2001), 41–50.

grief of his early loss: it is, in fact, the memory of a light that has 'gone out'. As the narrator says in introducing the scene, 'Such memories may persist... even from two years old, but scarcely standing out through a whole life-time like spots of light out of darkness, like a corner torn from a huge picture, which has all faded and disappeared except that fragment'.[55] At the same time, we are told that Alyosha's mother suffered from symptoms that peasants regarded as indicating demonic possession, and the nurse's frightened reaction underlines that such moments of potential revelation may also be moments of 'fear and trembling' rather than 'sweetness and light'—what Rudolf Otto would describe as the *horror religiosus*. The divine is hidden from us not only by time and death but by the ills and harms consequent upon our fall away from the knowledge of God enjoyed in paradise.

Alyosha's mother's hysterical holiness is one of the first indications of the mood of trauma that pervades the novel. But this is not just a matter of accumulated individual traumas. Rather, it reveals a deeper trauma consequent upon human beings' fall away from God and the revelation of a world whose characters are mostly caught in a process of progressive amnesia regarding their divine origins or for whom the memory of that origin is obscure and enigmatic, a cause more of torment than joy. As Thompson shows, lapses of memory, false memories, and, in some case, total forgetfulness regarding large parts of their lives typify the psychological state of those she calls the novel's 'negative' characters. These may be seemingly trivial, as in Prince Miusov's ironically described 'memory' of 'almost taking part in the barricades' in Paris in 1848,[56] but cumulatively they amount to a systematic and metaphysical oblivion. 'All the negative characters in the novel', Thompson writes, 'forget their "Jerusalem," their children, family, neighbours, compatriots and Holy Russia'.[57] Ivan's devil is therefore appropriately depicted as rather absent-minded,[58] and Ivan himself 'works [oblivion] into an ideological system'.[59] By way of contrast, a redemptive possibility is revealed in the devastating grief of the pathetic drunkard Snegiryov at the death of his son Ilusha, when he cites the biblical verse 'If I forget thee, O Jerusalem' from the song of the exiles in Babylon.[60] Grotesque and inconsolable as he is, he has not forgotten that there may yet be something that gives life a more than transient and relative meaning, even if, for now, it is occluded by death.

Of course, Dostoevsky provides his readers with powerful reasons why we might forget God or consign him to oblivion. In the chapter entitled 'Rebellion', Ivan lists a series of terrible cruelties perpetrated upon children, challenging Alyosha to believe in a God who, though all-powerful, can apparently allow

[55] Dostoevsky, *The Brothers Karamazov*, 13.
[56] Thompson, *Poetics of Memory*, 103–4.
[57] Thompson, *Poetics of Memory*, 170.
[58] Thompson, *Poetics of Memory*, 203–4.
[59] Thompson, *Poetics of Memory*, 186.
[60] Thompson, *Poetics of Memory*, 169–70.

such things to happen in the world he has created. Ivan claims that he is only rejecting the world, not God—although the clear implication is that by rejecting the world created and ruled by God he is also, in reality, rejecting God.

But it is not only horrors that drive the thought of God from human minds. It is also the ugliness, squalor, and meanness of the lives of the inhabitants of this provincial town of which the name loosely translates as 'beast-pen'. In *The Idiot*, Dostoevsky penned the memorable line that 'Beauty will save the world', but here we are presented with an ugliness that threatens entirely to eclipse the pure and beautiful light of the divine sun that shone once on Alyosha's infancy.

Is this, then, simply a world that has abandoned or been abandoned by God? Are its inhabitants doomed never to see the sun? By no means—yet, when the divine light breaks in to the world it is not as a mighty sunburst but as an event that is, paradoxically, invisible to those who do not have eyes to see.

Thompson finds 'the symbolic centre' of the entire novel in the chapter entitled 'Cana of Galilee', which climaxes in Alyosha's vision of what St John depicts as Christ's first epiphanic sign.[61] The scene is set in a monastic cell in which the dead body of the Elder Zosima awaits burial. A monk is reading the prescribed scriptures over the body and Alyosha's disconnected thoughts about the text being read (the miracle at Cana), gradually pass into a fully-fledged dream-vision of the wedding feast, where, amongst the guests, he sees the radiant face of his beloved Elder, who calls to him to join them. 'It was his voice, the voice of Father Zossima. And it must be he, since he called him! The elder raised Alyosha by the hand and he rose from his knees'.[62] Urging Alyosha to 'begin his work', the Elder also points beyond himself to the one whose presence illuminates the whole scene. ' "Do you see our Sun, do you see Him?" ' he asks his disciple. But Alyosha is afraid to look. ' "Do not fear Him" ', continues the dream Zosima, ' "He is terrible in His greatness, awful in His sublimity, but infinitely merciful. He has made Himself like unto us from love and rejoices with us" '[63] We are not told whether or not Alyosha did look, only that 'Something glowed in [his] heart, something filled it till it ached, tears of rapture rose from his soul' and, waking, he rushes outside.[64] In a moment of ecstasy Alyosha falls to the ground, 'watering it with his tears', as Zosima had once counselled his listeners to do, experiencing a sense of connectedness to other worlds (again a motif of the Elder's teaching) and a desire for all-embracing forgiveness.[65]

Thompson aptly describes the Christ of Alyosha's vision as 'a vision of glory, of Christ Pantocrator, Sun of Justice, reigning in heaven and celebrated as His

---

[61] Thompson, *Poetics of Memory*, 293.
[62] Dostoevsky, The Brothers Karamazov, 377.
[63] Thompson, *Poetics of Memory*, 293–4.   [64] Thompson, *Poetics of Memory*, 294.
[65] For a fuller discussion of the theme of 'watering the earth' see my article ' "Water the Earth": Dostoevsky on Tears', *Litteraria Pragensia: Studies in Literature and Culture* 22/43 (July 2012): *Towards a Lachrymology: Tears in Literature and Cultural History*, 95–111.

divine Majesty', an image familiar to Dostoevsky's readers as the crowning image of every Orthodox iconostasis.[66] But it is not only a vision shaped by the formal iconography of the Church. It is also a vision that, in its own way, repeats and completes the first visionary moment of the novel, the memory of 'the slanting rays of the setting sun' that, in Thompson's words, can now 'be seen as a variation on and prefiguration of this image of the uncreated light'. And, she continues, 'From a child held up to the sacred image, Alyosha has now broken into the non-Euclidean space of the icon, he has entered the memory of the future and become a part of it, a part of the "whole"'.[67] The effect of this visionary memory is all-embracing and life-changing. The narrator comments that '[Alyosha] had fallen on the earth a weak boy, but he rose up a resolute champion, and he knew and felt it suddenly in his moment of ecstasy'.[68] Falling to the ground, watering it with his tears, and rising up 'a resolute champion' who is ready to 'begin his work', Alyosha thus fulfils the Johannine motto: 'unless a grain of wheat falls into the ground and dies, it remains but a single grain; but if it dies it bears much fruit'. And if we remember that, in the gospels, the good seed is said to represent the divine Word itself, we can see what is happening here as the coming to life of that Word in the incarnate form of a human life. At multiple levels, then, Alyosha's life remembers and in the double sense of the Hegelian *Er-innerung*, re-collects, repeats, and internalizes the evangelical word, now fused with his own earliest human memories of radiant light.[69]

And yet…isn't this all 'just fiction'? Even within the world of the novel, it is 'just' a vision, not seen by anyone else, and perhaps it is simply a result of the overwrought imagination of a grief-stricken and traumatized young man. Even if we allow that it is indeed a genuinely Christian experience, it does not break through the limits that circumscribe and condition all possible human experience. The fruit that is to be brought forth by the collaboration of sun, earth, tears, and seed is not a fruit that can be weighed and measured but a new 'how' of existence. The limit that divides this world from the next, from the unremembered and unhoped-for world of a lost paradise and a new creation, remains in place. But the question is not whether Alyosha 'really' had a vision of an other-worldly reality; it is whether his visionary memory is able to provide a new centre for re-ordering the chaotic, brutal, and ugly world of his 'beast-pen' of a town in which 'reptiles devour reptiles'.

We see the first fruits of Alyosha's 'work' in his adventures with a group of teenage boys. The public beating and humiliation of the previously-mentioned drunkard, Snegiryov, by Dmitri Karamazov has led to his son, the sickly Ilusha,

---

[66] Thompson, *Poetics of Memory*, 299.   [67] Thompson, *Poetics of Memory*, 299.

[68] Dostoevsky, *The Brothers Karamazov*, 379.

[69] As Dostoevsky's own word becomes a bearer of the divine Word through this process of remembering and incorporation.

being victimized by his classmates, led by the pretentiously nihilistic Kolya. Despite his brother's role in starting this affair, Alyosha becomes an agent of reconciliation, and Ilusha's last days are eased by his new friendship with Kolya and the others. At the funeral, Snegiryov cannot control his grief and his wild behaviour prevents this from becoming a sentimentalized Victorian death-scene. Yet for all the elements of scandal, the boys are given the possibility of learning something of decisive significance. Stopping by Ilusha's stone, Alyosha addresses them, urging on them the necessity and importance of good memories.

> Let us make a compact, here, at Ilusha's stone that we will never forget Ilusha and one another. And whatever happens to us later in life, if we don't meet for twenty years afterwards, let us always remember how we buried the poor boy at whom we once threw stones, do you remember, by the bridge?...You must know that there is nothing higher and stronger and more wholesome and good for life in the future than some good memory, especially a memory of childhood, of home. People talk to you a great deal about your education, but some good, sacred memory, preserved from childhood, is perhaps the best education. If a man carries many such memories with him into life, he is safe to the end of his days, and if one has only one good memory left in one's heart, even that may some time be the means of saving us.[70]

Although this 'good memory' may be interpreted in an exclusively secular or psychological horizon, it is not only the first-fruit of Alyosha's vision, it is also the figure of another level of memory, an eternal memory, that the boys may come to know and cherish through their remembrance of Ilusha.

> 'And may the dead boy's memory be eternal!' Alyosha added again, with feeling.
> 'An eternal memory!' the boys chimed in again.
> 'Karamazov,' cried Kolya, 'can it be true what's taught us in religion, that we shall all rise again from the dead and shall live and see each other again, all, Ilusha too?'
> 'Certainly we shall all rise again, certainly we shall see each other and shall tell each other with joy and gladness all that has happened!' Alyosha answered, half laughing, half enthusiastic.
> 'Ah, how splendid it will be!' broke from Kolya.
> 'Well, now we will finish talking and go to his funeral dinner. Don't be put out at our eating pancakes—it's a very old custom and there's something nice in that!' laughed Alyosha.[71]

---

[70] Dostoevsky, *The Brothers Karamazov*, 819.

[71] Dostoevsky, *The Brothers Karamazov*, 820–1. However, Garnett's 'For ever' does not quite convey the full metaphysical implications of the Russian *Vechnaya pamyat'*. Richard Pevear's and Larissa Volokhonsky's 'Memory eternal', though more correct, seems not what a group of boys would spontaneously shout (or not, at any rate, in English). Such are the trials of translators. That a more metaphysical reading is nevertheless justified follows not from this single passage but from the relationship between memory and the eternal in the novel as a whole.

Dostoevsky's observation that Alyosha was 'half laughing' and his talk of eating pancakes reminds us that the incognito of the limit is not broken. At the same time, the memory of their reconciliation with Ilusha points beyond itself to the 'eternal memory' revealed in and by the light of which his own childhood memory of the slanting rays of the setting sun was also a distant prefiguration. That memory, we recall, was the memory of his dead mother, a memory now revivified in the hope 'that we shall all rise from the dead and shall live and see each other again'. Holding to such memories is to remember the end, to live as connected to the irretrievable, immemorial time that is 'more ancient' than the time of human biography and is also anticipation of a 'time to come' whose times and seasons we may not know, the Messianic time that Berdyaev, one of Dostoevsky's most enthusiastic twentieth century readers, saw as the true theme of history.

Such memories would offer the ultimate resistance to oblivion and, in the course of Ivan's preamble to his 'poem' of the Grand Inquisitor he—and perhaps, through him, Dostoevsky himself—offers an extraordinary image by which to figure such resistance. It is an old Russian legend of 'The Wanderings of Our Lady through Hell' in which Mary is led by the Archangel Michael through all the regions of hell, including a fiery lake into whose depths some of the damned sink so far that they become known as those whom God forgets. Shocked at what she has seen, Mary returns to heaven and pleads with God for all the inhabitants of hell, even those forgotten by God. Her prayers are eventually heard and God grants to the damned an annual release from pain during the period from Good Friday through to the feast of the Trinity.[72]

As retold in the novel, this legend suggests that even in eternal time and even in the infinite space of divine memory there are void zones of oblivion, places—and times—forgotten even by God. But it also suggests that those condemned to such void zones may, after all, be remembered and even these unremembered ones may receive an unhoped-for release from suffering—suggesting, though by no means necessitating, the further possibility that divine judgement itself may be revisable in the direction of a greater, more all-encompassing remembrance of all suffering creatures, inclusive of the guilty and the damned. Even the Last Judgement is not 'the End'. A biographical indication of the importance Dostoevsky might have attached to this legend is that the icon he placed so as to overlook his work desk was precisely an icon of 'The Mother of God, the joy of all who sorrow'.[73] In the light of both legend

---

[72] Dostoevsky, *The Brothers Karamazov*, 253–4. An English translation of the Russian version of the legend, which is based upon an earlier apocryphal source, is found in S. A. Zenkovsky (ed. and tr.), *Medieval Russia's Epics, Chronicles, and Tales* (New York: Dutton, 1963), 122–9. In the legend itself Mary's prayers are limited to those of the damned who are Christian and Jews are expressly excluded. Interestingly, and despite charges that Dostoevsky himself was anti-Semitic, he here extends her intercession to all 'indiscriminately', thus universalizing the possibility of 'eternal memory' even for those forgotten by God.

[73] Illustrated on the cover of Pattison and Thompson, *Dostoevsky and the Christian Tradition*.

and icon, we may say that, for Dostoevsky, Messianic time is also Marian time and the Messiah himself, whose light streams into the darkened world from the bosom of his mother, is who he is only in relation to his mother—a comment that, in a certain perspective, is merely to state what is already said in the notion of Mary as God-bearer, 'Mother of God'.

We can, of course, read the legend of the Virgin's visit to Hell as imaginatively expressing the redemptive possibilities of Alyosha's own childhood memories of maternal love and, if we choose, we can then explain it away in terms of psychological compensation. But, in any case, further reflection on the temporality of the legend is instructive. In accordance with literary convention, it is told as an event that happened 'once upon a time', yet it is also an event of an essentially apocalyptic character and therefore occurring in a time that is not our time, a time beyond time, the time of the 'last things', and therefore, from our perspective, also 'time to come'. By projecting this childhood memory into such time to come, Dostoevsky novelistically anticipates the refiguring of time articulated by Ernst Bloch in the closing lines of his three-volume *Philosophy of Hope*. Bloch there spoke of utopia as a memory that has shone into the childhood of all but where no one has ever been—and it is just such an interplay of the archaic lost time of childhood or pre-humanity and eschatological redemption that is figured in Ivan's legend.[74] As remembering what is beyond time and history the Legend anticipates 'the end'. Paradoxically, then, Alyosha's memories of 'the slanting rays of the setting sun' can also become an image of the first, anticipated light of redemption's dawn. In this way Dostoevsky effectively reverses Hegel's dictum regarding the Owl of Minerva. If the goddess of wisdom must wait till dusk to spread her wings, it is only at the rising of the sun that the Dostoevskian hero learns to remember whence he has come so that he might then learn where he is to go.

All this is said, of course, in the manner of the novelist, free to tell whatever tales he likes. Dostoevsky assures us that Alyosha's visions of such an 'other' time can prove the starting-point for a life of active love in the world, but can we find a philosophical basis for hope that is more solid than the consolations of music or the stirrings of the night wind and, above all, for a hope sufficient to ground a real and living community in which none are forgotten and all are assisted through a common effort to fulfil their real human potential? We have just had occasion to cite Ernst Bloch, whose philosophy of hope was amongst the twentieth century's most sustained and consistent attempts to answer these questions with a resounding 'Yes'. It is therefore to Bloch and to his philosophy of hope that we now turn.

---

[74] Ernst Bloch, *The Principle of Hope*, tr. Neville Plaice, Stephen Plaice, and Paul Knight, 3 vols (Oxford: Blackwell, 1986), III. 1376.

6

The Call to Utopia

We have begun to consider whether time itself might yield intimations as to what is more than time and whether there is anything in time that might offer a counter-movement to time's otherwise universal solvent power. Might we be saved from time in time and even through time? Might the temporal phenomenon of remembrance, perhaps, redeem us from the temporal exile of forgetfulness? Following Jankélévitch and Chrétien, we tracked what I called, respectively, their aesthetic and mystical responses to this complex of questions. Where Jankélévitch seemed to end with a refined but ultimately aesthetic nostalgia, Chrétien raised the possibility that time itself might direct us towards the supra-temporal dimensions of the immemorial and the unhoped-for. Lévinas offered more, in the sense of emphasizing the need for any temporal salvation to be marked by the ethical care for the cry of the neighbour, although he remained decidedly apophatic with regard to the kind of time in which it could ever be appropriate to speak of redemption. An oblique attempt to address this question was made by means of reference to Dostoevsky's treatment of oblivion, memory, and hope in *The Brothers Karamazov*, culminating in his allusion to the legend of Mary's intercession for those 'forgotten by God' in the deepest depths of Hell. This story condensed a characteristically Dostoevskian kind of remembrance that has been called 'remembering the end', that is, a kind of remembrance in which the historical and metaphysical rupturing of the relationship between time and eternity is fully acknowledged but simultaneously countered by an appeal to the creative possibilities that the eschatological time of the legend might reveal in this-worldly, historical time. It is such remembrance that provides us with a hinge by which to turn from Hegelian recollection towards hope as a basic possibility of our human time-experience.

In this chapter, I therefore look more closely at a sequence of thinkers who especially look to the future and to the eschatological symbols of religion as an eminent way in which the eternal might become manifest. Each of these thinkers was deeply and directly affected by the cycle of twentieth-century catastrophes in which the rupturing of the relationship between time and eternity became a concrete historical experience. However, as the appeal to Dostoevsky might suggest, we are entering a domain in which the kind of knowledge that eschatological hope might support is inevitably open to question. Surely it would seem impossible for it to be knowledge of what is the case, since it is 'not yet'. But, then, is it knowledge only of what *may* be and, if so, just how much weight can that optative formulation support? Can 'it may be' point us to 'what shall be' and, if so, by what right? But if we do once become assured that 'it shall be so', then is this not a return to a kind of view *sub specie aeternitatis* in which the future is merely the form in which what is eternally, timelessly true appears to creatures of time such as we are? But does this then mean that we are back, once more, at a certain Hegelianism? Just how much or how little can the eschatological promises of religious faith deliver, if they are not to be exposed as mere wishful thinking on the one hand or a merely imaginative expression of ontological claims on the other? Or does the category of promise come with its own distinctive epistemological and ontological identity such that it cannot be thus reduced to one or other kind of psychological or meta-physical form? And if that is so, to what can promising appeal in order to validate its claims? These are the key questions to which this chapter will lead us, but first we step briefly back to look at how eschatology entered into the mainstream of twentieth-century religious thought.

Eschatology was, we now know, a salient feature of early Christianity, which inherited it from Jewish Messianism and prophecy, although it was largely only the nineteenth-century research on the life of Jesus that rediscovered the full scope of the early Church's eschatological thinking, beginning with Johannes Weiss's ground-breaking 1892 study of *The Preaching of Jesus concerning the Kingdom of God* and developing through Albert Schweitzer's apocalyptic life of Jesus.[1] As we noted in Chapter 4, this was influential in the beginnings of Barth's theology of crisis, but it has subsequently shaped modern theology in multiple ways. According to an influential view developed by the Church historian Franz Overbeck, this early eschatological impulse was significantly reduced when Christianity became the established religion of the Roman Empire and eschatological yearning gave way to defending the established imperial order. A complex thinker such as Augustine could not be unaware of the continuing injustice and lack of peace in the Christianized world, but in his *City of God* he provided a capacious theoretical framework within which to accommodate acceptance of the prevailing world-order with a continuing but weakened future eschatology. Yet, even if eschatological movements were thereafter regularly stigmatized as heretical, the Christian scriptures could not but stimulate repeated bursts of eschatological speculation and messianic claimants. Norman Cohn's *The Pursuit of the Millennium* is a classic study of

---

[1] See Albert Schweitzer, *The Quest of the Historical Jesus*, tr. William Montgomery (London: A. & C. Black, 1954). Weiss is discussed on 237ff.

such attempts to bring in the last days—beginning already with a messianic pretender recorded in the year 591.[2] Later movements such as the apocalyptic movements associated with the Protestant Reformation or English Civil War and early American 'Pilgrims' are well-known.[3]

But if the 'Age of Reason' seemed to put radical eschatology and other forms of 'enthusiasm' on the back foot, the decade following the French Revolution saw an extraordinary renaissance of eschatological speculation and a widespread expectation that Europe was on the brink of a new age. Such expectations were given more specific shape by the rediscovery of the medieval apocalyptic writer Joachim of Fiore, who had divided history into three ages, the age of the Father, of the Son, and of the Holy Spirit—the third age being precisely the age that Joachim believed was about to come in his own time and that Early Romantics such as Novalis again believed to be imminent in theirs. It is plausible that this becomes one of the decisive tropes of German Idealism, especially in Hegel and Schelling, who project the timeless Trinity of traditional theology onto the plane of historical time. Only, whereas Joachim's own speculations saw the third age as culminating in a kind of universal monasticism, the Idealists saw it as leading Christianity beyond all ecclesiastical forms into a new kind of freedom, whether in the life of the State (as in Hegel) or in some kind of open, de-institutionalized life of the free spirit.[4] For Hölderlin, it became connected with the expectation of a coming, Dionysian God, a motif renewed *mutatis mutandis* in Nietzsche's later Dionysian revolution.[5] And it is in connection with his meditations on Hölderlin that Heidegger too comes to posit the possibility of a 'last God' who is still 'to come'.[6] At the same time some popular religious cultures continued to produce apocalyptic expectations, as portrayed in Gerhart Hauptmann's novel *The Fool in Christ, Emanuel Quint.*[7]

---

[2] See Norman Cohn, *The Pursuit of the Millennium: Revolutionary Millenarians and Mystical Anarchists of the Middle Ages* (London: Secker and Warburg, 1957). As Cohn himself makes clear, his story is by no means complete, focusing largely on movements in France, Germany, and the Low Countries and therefore largely omitting related movements in the Latin South of Europe as well as in the Slavic East.

[3] Judaism had its own parallel movement amongst the followers of the Messianic pretender Sabbatai Sevi. See Gershom Scholem, *Sabbatai Sevi: The Mystical Messiah, 1626–1676*, tr. R. J. Zwi Werblowsky (London: Routledge and Kegan Paul, 1973).

[4] The story of Joachim's influence in nineteenth-century European culture is extensively chronicled in Warwick Gould and Marjorie Reeves, *Joachim of Fiore and the Myth of the Eternal Evangel in the Nineteenth and Twentieth Centuries* (rev. edn, Oxford: Clarendon Press, 2001).

[5] See Manfred Frank, *Der kommende Gott: Vorlesungen über die Neue Mythologie* (Frankfurt am Main: Suhrkamp, 1982).

[6] See for example Martin Heidegger, *Gesamtausgabe*, LXV: *Beiträge zur Philosophie (Vom Ereignis)* (Frankfurt am Main: Klostermann, 1994), Sections VI and VII, 395–417.

[7] G. Hauptmann, *Der Narr in Christo, Emanuel Quint* (Frankfurt am Main: Fischer Verlag, 1910). Karl Barth's admiration of the apocalypticism of Christoph Blumhardt the elder might also be mentioned here.

As Gould and Reeves observe,[8] and as histories of Russian thought confirm, apocalyptic speculation was also especially widespread in Eastern Europe and Russia.[9] In this context it not only fed into existing apocalyptic traditions that saw Russia as the Third Rome and the Russian Church as the third (Johannine) Church, destined to inherit and fulfil the spiritual destiny of the Petrine (Roman) and Pauline (Protestant) Churches of the West, but also gave a certain apocalyptic flavour to Russian Bolshevism. Via Arthur Moeller van den Bruck (editor of the German translation of Dostoevsky's political writings), this Russian Messianism returned to the West in the motif of a 'Third Reich', merging with a continuing stream of right-wing German Messianism.[10] At the same time connections to the Marxist vision of a future classless society seem unmistakable—except perhaps to a certain 'scientific' kind of Marxist. Here, however, eschatology has lost all connection with the eternal and has become a purely immanent, this-worldly 'end of history', although a Marxist thinker such as Walter Benjamin can still evoke something of the religious pathos of apocalyptic thought.[11] Yet even in—perhaps especially in—this Marxist transformation, eschatology may begin to offer a way towards a vision of what, in time, might save us from time. To see how this might be so, I turn now to the thought of Ernst Bloch, the twentieth-century philosopher who went furthest in making the future, and more specifically the utopian future, the defining point of reference for his entire intellectual work.

ERNST BLOCH: THE PHILOSOPHY OF HOPE

Bloch spoke of himself as representing the 'warm stream' of Marxism and whilst writing as a committed atheist he was prepared to offer a constructive interpretation of an extraordinary range of human cultural expressions as testifying to the 'spirit of utopia', including the symbols of Jewish-Christian eschatology. Where other versions of Marxism typically saw these as varieties of false consciousness, Bloch believed that they still had the power to nourish creative energies that could serve the advent of genuine, that is, materialistic,

---

[8] See Gould and Reeves, *Joachim of Fiore*, Chapter 11, 316–39.

[9] At the start of a chapter in *The Russian Idea* dedicated to 'The eschatological and prophetic character of Russian thought' Berdyaev writes that 'in accordance with their metaphysical nature and vocation the Russian people are a people of the End'. Nicholas Berdyaev, *The Russian Idea*, tr. R. M. French (London: Geoffrey Bles, 1947), 193.

[10] Much of this history is discussed in connection with Heidegger's own nationalism in Judith Wolfe, *Heidegger's Eschatology: Theological Horizons in Martin Heidegger's Early Work* (Oxford: Oxford University Press, 2013).

[11] See Walter Benjamin, 'Über den Begriff der Geschichte', in *Gesammelte Werke*, I.2 (Frankfurt am Main: Suhrkamp, 1974), 691–704; Eng. 'Theses on the Philosophy of History', in *Illuminations*, tr. H. Zohn (London: Pimlico, 1999).

socialism. Indeed, he even comes to the brink of affirming some kind of immortal existence beyond the life of what we now know as the body, saying, for example, that *'that we shall be saved, that there can be a kingdom of heaven*...this is not only conceivable, that is, formally possible; it is *downright necessary*...In the nature of the case, it is postulated *a priori*, and hence of a reality that is *utopian, essential'*.[12] And, a couple of pages later, commenting on a passage from the Zohar that speaks of the interdependence of work and prayer and of the outer and inner worlds, he adds that 'In such a functional relationship between unburdening and spirit—between Marxism and religion, joined in a will to the kingdom—flows the ultimate main system for all subsidiary currents. The soul, the Messiah, the Apocalypse which represents the act of awakening in totality—these impart the final impulses to action and cognition, and make up the a priori of all politics and all culture'.[13] Yet Bloch did not want to suggest some kind of dualism between spirit and matter: his point is precisely that matter itself, material life, is pregnant with possibilities that are far from exhausted by conventional materialism. Religion gives a certain anticipatory access to these, but they must be realized by real work in the real (i.e. material) world. Even with that caution, however, it is clear that in its breadth as well as in its conceptual innovations Bloch's work opens important paths of dialogue between socially progressive materialism and religion, that have still not been travelled to the end. Indeed, under the shadow of the so-called 'new atheism' these paths have become widely perceived as no longer passable, a perception that reflects the intellectual impoverishment of contemporary atheism.

Bloch's case for the primacy of hope draws on mythology, art, literature, religion, and philosophy, as well as on the explicitly utopian thinking seen in the medical, social, technological, architectural, and geographical utopias that he explores at length in the second of the three volumes of his *Philosophy of Hope*, including such diverse examples as Campanella, Aladdin, alchemy, and Prester John, alongside the more predictable cases of More, Bacon, Owen, Fourier, and Bakunin. However, that all of these came to be interpreted in a utopian perspective is possible only thanks to the basic existential and ontological structures that Bloch sets out in the first volume. This account is aimed at contesting an ontology in which all future possibilities of human development are constrained by 'what is' and 'what has been', that is, by the limits indicated by such Hegelian catchphrases as 'the real is the rational: the rational is what is real' and 'the owl of Minerva flies at dusk'. At the same time, Bloch offers a reading of Hegel himself that both draws him closer to the orbit of

---

[12] Ernst Bloch, *Man on his Own: Essays in the Philosophy of Religion*, tr. E. B. Ashton (New York: Herder and Herder, 1970), 70.

[13] Bloch, *Man on his Own*, 72.

Marx and discovers the utopistic potential in, especially, the *Phenomenology of Spirit*.[14]

Bloch's argument is in its early stages oriented by a critical reading of Freud and Jung. This is because he sees them as using the idea of the unconscious to limit the range of human beings' future possibilities to what is already given, especially when it is depicted by Jung as collective and expressed in archaic myths and symbols, thus closing off all significant novelty. Bloch concedes the reality of the unconscious, but he wants also to suggest that human beings have an unconscious future as well as an unconscious past. In our orientation towards the future there is much that we do not yet know and alongside the no-longer-conscious—the traumatic childhood material that, according to Freud, is the object of repression—there is also the not-yet-conscious. This is most clearly manifest in youth. As Bloch enthusiastically writes:

> Any young person who feels some hidden power within him knows what this means, the dawning, the expected, the voice of tomorrow. He feels called to something that is going on inside him, that is moving in his own freshness and overhauling what has previously become, the adult world. Bold youth imagines it has wings and all that is right awaits its swooping arrival, in fact can only be established, or at least set free, by youth... The green years are filled with forward dawning, they consist chiefly of not yet conscious states.[15]

In particular historical epochs this is not limited to individual aspirations but sets the tenor of the age. 'The dream then passes from vague, mainly private premonition to a more or less socially sharpened, socially mandated premonition'. One example is that of the Russian populists, of whom Bloch remarks that 'the conversations of young unmatriculated women and of male students utopianized on the dusty boulevards of Russian provincial towns' 'with sentimental or angry red dawn' in their minds (PH, i. 117).[16] The Renaissance too is said to have had an 'aurora quality' as 'the distant earth itself opened up and revealed new continents; the ceiling of the heavens cracked, leaving a clear view of infinity'. The range of ways in which this new feeling expressed itself was as diverse as the millenarian politics of Thomas Münzer and Bacon's

---

[14] In an essay on the *Phenomenology* he writes: 'It is utterly and incomparably youthful, fully and excessively aglow, thoroughly poetic, thoroughly scientific, a unique morning fermentation'. E. Bloch, 'Phänomenologie des Geistes', reproduced in the critical essays appended to G. W. F. Hegel, *Phänomenologie des Geistes*, ed. Gerhard Göhler (2nd edn, Frankfurt am Main: Ullstein, 1973), 847.

[15] Ernst Bloch, *The Principle of Hope*, tr. Neville Plaice, Stephen Plaice, and Paul Knight, 3 vols (Oxford: Blackwell, 1986), i. 117. (Further references are given in the text as 'PH'.)

[16] The populists were mostly young radicals of middle or upper class origin who went out to the people both to enlighten them through teaching and health education, but also to learn (or relearn their) solidarity with them. The populist impulse is manifest in Tolstoy's Levin (in *Anna Karenina*), eager to work alongside his serfs in the fields, although, as a landowner, Levin was in a rather different situation from the students or 'unmatriculated women' who took part in the populist adventure.

Novum Organon. Perhaps in such times no one knows exactly *what* is coming, but there is a sense of direction, a tendency, in which individuals and groups spontaneously intuit the impending realization of their own identity. Thus, whereas the Freudian and Jungian unconscious ties the subject back to the past (whether to its own childhood or to the collective unconscious of Jungian archetypes), the unconscious that is the not-yet-conscious reveals the future as the proper sphere in which individual and common fulfilment is to be brought about.

Consequently, where psychoanalysis directs us to nocturnal revelations of repressed materials, Bloch looks instead to daydreams, since, no matter how fantastic their content, they project a state of affairs that is envisaged as possibly coming to pass in the future: I might win the lottery, I might write a global best-seller, I might win the hand of a Helen...At least, the belief that 'I might' is integral to the dream—no matter how implausible this might seem in the cold light of day.[17] But precisely this 'cold light of day' is, in Bloch's eyes, a classic bourgeois ruse, designed to quash any hope that the world could become other than it currently is. Present-day 'practical' reality is made the yardstick of reality as such, shutting out all possibilities of future transformation. But the point of attending to the not-yet-conscious is precisely that we do not yet know reality for what it is or for all that it could be. And even if the particular content of the daydreamer's reverie is entirely fantastic, the act of dreaming is significant in itself by virtue of the way in which it contests the claims of the 'real world' of commonsense rationality to be the only 'real world' there is. And when the daydream does have a positive content, as in the Populists' fantasies of intellectuals and peasants sharing in a common experience of Russianness, then, as Bloch writes 'The waking dreams advance, provided they contain real future, collectively into this Not-yet-Conscious, into the unbecome-unfilled or utopian field' (PH, I. 113).

It is of course a familiar fact that, as spokesmen for 'reality' like to point out, the actual fulfilment of such fantasies is often disappointing. The unattainable and impossibly perfect maiden who is the object of the infatuated youth's fantasies turns out, in marriage, to be really rather ordinary.[18] Thus, suggests Bloch, the ploy of such late Romantic figures as Lenau and Kierkegaard to preserve the ideality of their beloved by *not* marrying her (PH, I. 180–3). In discussing this, he cites the legend of Egyptian Helen, used in Richard Strauss's eponymous opera. The story is that while returning from Troy, Menelaus comes ashore in Egypt, leaving on board the Helen whose face had launched a

[17] On the basis of personal experimentation, Bloch also articulates this difference in terms of the soporific effects of opium versus the arousing daydreams induced by hashish.

[18] It is perhaps a sad historical footnote that where Bloch can unabashedly hail the Russian Revolution as a genuine realization of the Populists' utopian dreams, subsequent experience suggest that—at least as regards its eventual descent into Stalinism—it is finally one more example of how realization fails to deliver all that had been expected of it.

thousand ships. On shore, however, he encounters the 'real' Helen, who never was abducted by Paris but, through divine intervention, had been hidden away in Egypt. This Helen is the model of true wifely faithfulness and, as Menelaus takes her, the dream-Helen on the ship vanishes. Returning home, all is once again as it should be. But, Bloch says, the memory of the dream-Helen and the ten years of fantasy, guilt, suffering, and hope invested in her remains 'dotted in outline' in the Egyptian (the 'real') Helen. Thanks to the dream, reality itself acquires a utopian dimension.

In fact, the moment of realization in which dreams come down to earth is never total. The act of realization itself contains aporias that philosophy has, until now, simply glossed over. The aristocratically leisured philosophers who inaugurated the history of Western philosophy were simply blind to the inherent difficulties of turning ideas into reality, since the real, actual work was invariably done by slaves and this oversight has passed on into subsequent tradition. What was made, represented in a work of art, or contemplated in the mind was seen entirely in terms of an 'idea' that was not envisaged as undergoing any *essential* change in translation from one medium to the other (i.e. from idea to reality). 'The realization comes out of the logical consistency of the matter itself', as Bloch puts it (PH, I. 191). Aristotle takes an important step forward in recognizing that realization, as the process by which potentiality (*dynamis*) becomes fully what it is (*energeia*) is a phenomenon in its own right. But Aristotle too sees basic continuity between the form or idea and its realized state: 'realization is solely self-realization of the form-idea or entelechy which is inherent in things; the entelechy is thus itself the energy (or the actus) towards its Realization' (PH, I. 191). Yet, at the same time, Aristotle does acknowledge that the conditions of materiality mean that this realization is never total or absolute but there is always an element of disruption or caprice in the actual outcome of the work. But Bloch nevertheless thinks that Aristotle did not sufficiently recognize this as a problem needing attention, whilst he himself sees it as being of supreme significance that '*in the realizing element itself there is something that has not yet realized itself.* The unrealized Realizing element brings its own most peculiar minus into the plus of the Realization as soon as the latter occurs' (PH, I. 193). This is the state of affairs that Goethe (a constant point of reference for Bloch) commented on when he said that 'nearness makes things difficult' and it is also why fulfilment 'brings a melancholy of fulfilment along with it' (PH, I. 193). But, for Bloch, the 'minus' is not really a negative and the 'melancholy' is by no means pathological. Rather, the element of unfulfilment in fulfilment is precisely a pointer to the not-yet character of the world, namely, that it really is not yet all it can be, and the minus and the melancholy are therefore indicative that there is still work to do and still a future to be attained, whilst simultaneously preventing consciousness from sinking down into contentment with the world as it is. For a philosophy of hope the point—as Bloch extensively comments with regard to Marx's eleventh thesis on Feuerbach—is not simply to understand the world but to change it.

This is possible—emphatically *possible*—because both within ourselves and in the external world there is much that is still unfinished: 'the Real is process; the latter is the widely ramified mediation between present, unfinished past, and above all: possible future' (PH, I. 196). But in what way is this going beyond Hegel who was already attentive to the process character of reality? Is this really offering more than the unfolding in time of what is knowable according to timeless categories of cognition? And how does Bloch make space for and make thinkable the possibility of a genuine 'Novum' in historical becoming?

In a first step towards substantiating his case, Bloch distinguishes between objective and real possibility. The former is basically what has not yet come about but which can safely be predicted on the basis of what is already knowable, that is, where a sufficient causal nexus is in place or is in the process of coming to be in place for the event to be precipitated. In the case of real possibility, however, the conditions in the object itself are not yet 'fully assembled; whether because they are still maturing, or above all because new conditions—though mediated with the existing ones—arise for the entry of the new Real' (PH, I. 196). In this regard the apocalyptic speculations of Joachim of Fiore or the English millenarians represent instances of conditions not being adequately assembled for the advent of the future towards which their thoughts were directed. New conditions must arise and must be adequately mediated with existing reality (i.e. must become concrete) for real social transformation of the kind symbolically anticipated in their utopian imaginings to take place.

But, again, we might ask how this is more than a materialist transformation of the Hegelian model. Where is the real novelty here? Bloch's answer is in terms of his distinction between 'What-is-according-to-possibility' and 'What-is-in-possibility'. The former is a state of affairs, past, present, or future, that is determined by the conditions of material existence—for example, human beings cannot fly unaided, due to the constraints of gravity and physiology. In this case, what might possibly come to pass is constrained by conditions external to it. In the case of What-is-in-possibility, however, that which is in a process of change actually generates the conditions for its own transformation. Bloch sees this as an idea favoured by those he elsewhere calls the 'Aristotelian Left', including especially Avicenna and Giordano Bruno.[19] Here, in the horizon of an essentially pantheistic religious world-view, matter is no longer seen as the passive material in and through which form comes to be realized or onto which it is to be stamped. Instead, it is matter itself that generates the forms that it brings to realization or, more precisely, that come to realization through the common work of matter and form. As Bloch puts it in

[19] See Ernst Bloch, *Avicenna und die Aristotelische Linke* (Berlin: Rütten and Loening, 1952).

more conventionally Marxist terminology, '*real possibility is nothing other than dialectical matter*' (PH, I. 206). That this can be anticipated in visionary art or apocalyptic expectations that look beyond the current configuration of matter constituting the objective world does not mean that a causal nexus is lacking. But such a nexus is not yet sufficiently in place to be observable according to the methods of scientific enquiry. Is there, nevertheless, space for a moment of unpredictability, of freedom, that would justify us in invoking the 'Novum', the 'new thing' of John's apocalypse?

The key, according to Bloch, is in human being itself since the human being is the eminent instance of 'what-is-in-possibility', that is, of self-realizing formal–material (dialectical) possibility. Although human beings are, as we have seen, constrained by objective material factors such that not everything is equally possible at any time and even socialist revolution must await its moment, we are ourselves agents in deciding what may be possible: 'man is the real possibility of everything which has become of him in his history and, above all, which can still become of him if his progress is not blocked. He is a possibility therefore which is not merely exhausted like an acorn in the enclosed realization of the oak-tree' (PH, I. 235). And, 'Man is that which still has much before it. He is repeatedly transformed in his work and by it. He repeatedly stands ahead on frontiers which are no longer such because he perceives them, he ventures beyond them... the Possible, as that which is not fully conditional, is that which is not settled' (PH, I. 246)—but which might, for example, be settled through work and revolution, as free acts that are also 'midwives of the future society with which the current one is pregnant... the world as mediated homeland, towards which nature is in possibility which has hardly even been entered upon' (PH, I. 247). The 'Archimedean point' of knowledge is therefore precisely coming to know not *what has been* (à la Hegel) but what is coming to pass, *what will be* (PH, I. 282ff).

But does Bloch's revised 'dialectical' Aristotelianism make the case for novelty in a decisive, ontological sense? What is to come may be beyond the scope of contemporary science and, for now, may be knowable only in the mirror of artistic and religious symbols.[20] But if this blows open the ideological limitations of a vulgar realism that restricts reality to 'what is', does it really speak of the new, of what neither the eye has seen, nor the ear heard, nor the heart of man conceived, that is, of what is not simply an object of hope but, even beyond all possible hope, the 'unhoped-for'? Bloch speaks of the eschatological

---

[20] In this regard Bloch's hermeneutics of hope is an extraordinarily productive tool, since, as his own practice demonstrates, it can be applied not only forwards but also backwards—but in a manner quite other than the retrospective view of the Owl of Minerva. For in revealing the utopian possibilities of past cultural productions it reveals them precisely as containing as-yet unfulfilled possibilities for our own future and therefore also as inspiring us to become alert to the utopian possibilities latent in our own present. Great art and powerful religious symbols are not only the record of what has been thought and done, they call us to what we may yet think and do.

impulse of Christianity as having 'stepped beyond the threshold of the previously known creature, of its anthropology and sociology' (PH, III. 1196) and we have seen that he has even questioned the necessary universality of death.

Although he several times acknowledges his appreciation of the general tendency of Bloch's thought, we have already seen that Emanuel Lévinas nevertheless argues that Bloch's extravagant hope lacks a dimension of '*hauteur*', the elevation of the divine that is lifted above and beyond all human experience and knowledge, commanding us from a time that always exceeds our time, prior to and beyond all history, immemorial and unhoped-for.[21] Beyond even death, Lévinas postulates that there is—or *may be*—a 'Messianic time in which what is [merely] perpetual is transformed into the eternal'.[22] If Blochian hope has a concrete social force that the aesthetic and mystical kinds of hope we found in Jankélévitch and Chrétien lack, Lévinas suggests that it is, in the end, unable to overcome the limits of its own material immanentism.[23]

Of course, the counter-argument is both predictable and forceful. As Kierkegaard argued concerning the Unknown towards which, he claimed, reason always advances, if it is unknown in absolutely every respect, then it would be unknown even if it we were to encounter it.[24] If there is no point of contact—Bloch would say concrete historical mediation—then the utopian imagination is utopian in what he too would agree was a bad sense, that is, it is purely vacuous. If, in Christian doctrine, the event of the resurrection can only occur by the power of God and is therefore not in itself a human possibility, it must still become a possibility for us in the sense that, in the words of the Creed, we are to live 'expecting' or 'looking for' 'the resurrection of the body and the life of the world to come'. And we can only do this if the world to come has some meaning in relation to the experienceable and knowable range of human life. But is it possible to combine the 'height' demanded by Lévinas with the concreteness of Blochian utopianism? One thinker who might be read as arguing for just such a combination is Paul Tillich, who both explicitly endorses Bloch's utopian imagination and also affirms the fundamental importance of 'the Eternal'. Can Tillich, then, model the step beyond

---

[21] See, e.g. Emmanuel Lévinas, 'Sur la mort dans la Pensée de Ernst Bloch', in *De Dieu qui vient à l'idée* (Paris: Vrin, 2004), 62–76.

[22] Emmanuel Lévinas, *Totalité et l'Infini: Essai sur l'extériorité* (4th edn, Paris: Kluwer Academic, 1971), 318. However, Lévinas refrains from deciding whether this is indeed the advent of another order of time or what he calls 'an extreme vigilance on the part of the Messianic consciousness' (318), i.e. a kind of regulative ideal which, as immemorial and unhoped-for, we can never presume to be a present reality. On key issues relating to Lévinas's Messianism, see Michael L. Morgan, *Discovering Levinas* (Cambridge: Cambridge University Press, 2007), Chapter 8, 208–27.

[23] In this regard, Lévinas's reserve recalls Kierkegaard's insistence on God's infinite, qualitative difference vis-à-vis human consciousness.

[24] SKS4, 249–50/PF, 44–5.

nostalgia and mysticism, and beyond material, temporal utopianism that we are seeking?

TILLICH: FROM UTOPIA TO ONTOLOGY

From the end of the First World War until his emigration to America in 1933, Tillich was an energetic advocate of religious socialism and identified with the Marxist interpretation of history.[25] At the same time, and like Bloch (whom he referred to in one article from this period as a friend), he was not persuaded that so-called 'scientific socialism' was an adequate response to the crisis of the age. According to scientific socialism (exemplifying what Bloch called the 'cold stream' of Marxism), the advent of a socialist society would come about through the working of laws of history that were as necessary, certain, and knowable as the laws of physics. In contrast to this, Tillich developed a philosophy of history that (in the spirit of Bloch's utopian imagination) was open to the influence of freedom in thought and action and that also sought to incorporate the spirit of biblical prophecy, so as to give social justice a moral as well as a 'scientific' ground. The congruence with Bloch is indicated in several writings from this period. In a 1931 article, for example, he wrote that utopia 'is the power that changes reality. It is the spring of all great historical move-ments; it is the tension which impels man beyond everything reassuring and safe to new uncertainty and unrest. Utopia is the power of renewal.'[26] He is also with Bloch in emphasizing the spirit of utopia rather than endorsing any actual utopian project. The spirit of utopia serves to break open the falsifica-tion of historical reality found in all forms of ideology and to make human beings free for alternative futures that are not conceivable within the horizons of the present. It reveals the 'unrest' at the heart of human being and our abil-ity to stretch ourselves out beyond the limits of the given.[27] What matters is not realization but anticipation or, more precisely, 'realization' in the sense of 'reali*zing*', as a present, active participle. Such anticipation is not an ineffective opiate but stirs us to actively contesting the prevailing power structures of the present. In another 1931 essay, 'The Protestant Principle and the Proletarian

[25] On the biographical aspects of Tillich's socialism see Ronald H. Stone, 'On the Boundary of Utopia and Politics', in Russell Re Manning (ed.), *The Cambridge Companion to Paul Tillich* (Cambridge: Cambridge University Press, 2009), 208–20.

[26] Paul Tillich, 'Mensch und Staat', in *Gesammelte Werke*, XIII: *Impressionen und Reflexionen: Ein Lebensbild in Aufsätzen, Reden und Stellungnahmen* (Stuttgart: Evangelisches Verlagswerk, 1972), 173.

[27] Paul Tillich, 'Ideologie und Utopie: Zum gleichnamigen Buch von Karl Mannheim' (1929), in *Gesammelte Werke*, XII: *Begegnungen: Paul Tillich über sich selbst und andere* (2nd edn, Stuttgart: Evangelisches Verlagswerk, 1980), 257.

Situation', he described the dialectical character of the relationship between anticipation and realization in the following terms:

> The attitude of anticipation develops into utopianism if it is allowed to lose its essentially dialectical character and is held as a precise and literal anticipation— an anticipation that at some time in the future is to be replaced by a tangible, objective possession. The thing ultimately referred to in all genuine anticipation remains transcendent; it transcends any concrete fulfilment of human destiny; it transcends the otherworldly utopias of religious fantasy as well as the this-worldly utopias of secular speculation. And yet this transcendence does not mean that distorted reality should be left unchanged; rather it looks forward to a continuous revolutionary shattering and transforming of the existing situation. Thus proletarian anticipation involves a real change in proletarian existence, a real shattering and overcoming of capitalism. But it does not and cannot involve the bringing-about of a situation that is exempt from the threat that always accompanies human existence.[28]

Or, as he succinctly puts it in his most substantial socialist work, *The Socialist Decision*, 'Anticipation involves action. Only active anticipation has reality'.[29]

There is, however, a tension in Tillich's thought that can scarcely escape notice and that is common to many currents of modern religious thought. On the one hand, he sees the 'ultimate concern' of human life as necessarily irreducible to any particular configuration of finite, temporal concerns. On the other, he does not want concern for ultimate things to weaken or disable a needful attention and commitment to changing the world. But Tillich is not going to argue for some kind of via media or 'balance'. Rather, like Bloch and many other interwar thinkers years (of both left and right), he has a far more catastrophic view, as can be inferred from his reference to anticipation 'shattering' the existing capitalist order. A genuine synthesis of the temporal and eternal must be more than a merely formal balancing-act. Instead, the eternal is only to be grasped in and through active participation in historical movements that rise and fall and that oscillate between creativity and chaos. Already in *The Religious Situation* (1925), Tillich's analysis of his present age is directed at what he calls 'unconditioned meaning' that, as such, 'transcends the process of mere becoming, the mere transition from past to future; it is to speak of that which supports the times but is not subject to them. If any present has meaning it has eternity'.[30] But if the present can reveal eternity at its own heart, this revelation does not issue in a purely theoretical 'interpretation' of the world

---

[28] Paul Tillich, *The Protestant Era*, tr. James Luther Adams (London: Nisbet and Co., 1951), 249. We might say that Tillich implicitly counters the reference in the Internationale to 'the last fight' with something in the spirit of Browning's 'I was always a fighter, so one fight more...'.

[29] Paul Tillich, *Die sozialistische Entscheidung* in *Gesammelte Werke*, II: *Christentum und soziale Gestaltung. Frühe Schriften zum Religiösen Sozialismus* (Stuttgart: Evangelisches Verlagswerk, 1962), 313.

[30] Paul Tillich, *The Religious Situation*, tr. H. Richard Niebuhr (New York: Meridian, 1956), 35.

since it is dependent on '*active* anticipation' and therefore involves commitment to concrete historical action (which, in the 1920s and 1930s, meant radical political engagement).

But action cannot be random. Against the kind of 'decisionism' current amongst radical political thinkers of both left and right, Tillich believed that right action depended also on the right moment—although it did then, at that point, demand free decision and commitment on the part of the agent (against the kind of passivity induced by the spectacle of the inflexible laws of history gradually working themselves out, irrespective of individual human decisions). Bloch, as we have seen, similarly held that utopian imaginings needed to be socially mediated if they were to be effective, in other words, the right constellation of material and historical forces had to coalesce if the corresponding action was to have constructive outcomes. Tillich didn't deny this, but he also argued for a theological dimension, appealing to the New Testament idea of the '*kairos*', the kind of time invoked at the beginning of Jesus' preaching ministry when he declares that 'The time (*kairos*) is fulfilled' (Mk. 1:15). This is the time that is the right time for action, as in such expressions as 'Seize the time!' or 'The hour has come!' By way of contrast, *chronos* or clock-time is time that is neutral and unchanging in its tempo regarding what is or isn't to be done in it. The clock gives no special privilege to any one time over any other, whereas in the ebb and flow of lived time we need not only to know what time it is but also whether it is the right time to initiate revolutionary action—or, to take a rather different example, whether it is the right time to 'pop the question'.[31] In the Christian view, however, there is one *kairos* that has a special eminence, namely, the 'time' of the Incarnation or Christ-event. This is 'the great *kairos*', the once-for-all and unique event that gives a centre to history and reveals the possibility of salvation from the destructive powers at work in history. But this is so far from negating the significance of other moments of destiny that it must rather be seen as enabling them, so that the great *kairos* 'is again and again re-experienced through relative "*kairoi*," in which the Kingdom of God manifests itself through a particular breakthrough'.[32] And precisely because *kairos*-time is qualitative rather than quantitative, 'Awareness of a kairos is a matter of vision. It is not an object of analysis and calculation such as could be given in psychological or sociological terms. It is not a matter of detached observation but of involved experience'.[33]

---

[31] It is a key to the climactic events of Victor Hugo's *Les Misérables* that the 1832 insurrection was fated to be no more than an insurrection rather than a revolution because for all the ardour of the insurgents, it was not (as in 1789 or 1848) a genuinely revolutionary moment. You cannot make history happen if history is not ready. See Victor Hugo, 'Le 5 juin 1832', in *Les Misérables II*, Book X (Paris: Gallimard, 1973/95, 395–421). This is also a text to which Derrida is attentive in *Specters of Marx*.

[32] Paul Tillich, *Systematic Theology* (One Volume Edition, Welwyn Garden City: James Nisbet, 1968), III. 395.

[33] Tillich, *Systematic Theology*, III. 395.

But can this appeal to the decisive time of the Christ-event give appropriate height to the utopian imagination?

There seem to be two issues here, one historical and one metaphysical. The first is that whilst the identification of Jesus as the Christ, the Messiah, is intended to incorporate the utopian prophetic symbol of the Messiah and his messianic reign (understood as the fulfilment of historical time), it does so by virtue of claims regarding a particular complex of past events. Christian faith, as opposed to Jewish Messianism, holds that Christ, the Messiah, has already come and that all the fulfilment that is possible within historical time has occurred already in history, namely, in his life. It is this logic that Tillich affirms when he writes that 'In terms of the eschatological symbolism it can also be said that Christ is the end of existence. He is the end of existence lived in estrangement, conflicts, and self-destruction... in him the New Being is present. His appearance is "realised eschatology" (Dodd)... in so far as no other principle of fulfilment can be expected. In him has appeared what fulfilment qualitatively means'.[34]

Of course, Tillich is not denying that there remains an outstanding future subsequent to the events of *c*. AD 1–33. As he puts it 'In the sense of "finish", history has not yet come to an end. It goes on and shows all the characteristics of existential estrangement... [But] In the sense of "aim", history has come to an intrinsic end qualitatively, namely, in the appearance of the New Being as a historical reality'.[35] In other words, Jesus as the Christ is understood as the end of history in the sense of *telos* rather than conclusion. In him history has arrived at a *ne plus ultra*, even though history itself continues and will continue to run on. It is in this sense that Jesus Christ is also the centre of history. But this would seem to limit the essential openness of utopian hope by tying it back to a past event and thus subordinating hope for a radical historical Novum to the recollection of the past, sapping the future of its decisive, utopian force.[36]

The objection might be countered by pointing out that Christianity does, nevertheless, retain the dimension of futurity with regard to the belief that Christ has yet to come again and to bring about a final and definitive judgement of historical life. Analogously to Bloch's example of the Egyptian Helen, Tillich continues to regard the 'not yet' of eschatological expectation as being just as integral to Christian experience as the 'already' of the Incarnation. There is, he says, an 'oscillation' between the 'already' (the Christ has come in time) and the 'not yet' (he will return to deliver a final judgement) that 'belongs inseparably to Christian existence'.[37] Christ is the end of history—*and*

---

[34] Tillich, *Systematic Theology*, II. 137.   [35] Tillich, *Systematic Theology*, II. 138.

[36] In this connection we might recall Bloch's etymological explanation of the term 'religion' in terms of re-ligio or tying the future 'back' to the past. See Ernst Bloch, *Atheismus in Christentum: Zu Religion des Exodus und des Reichs* (Frankfurt am Main: Suhrkamp, 1973), 6.

[37] Tillich, *Systematic Theology*, II. 138.

yet the end is still to come. It is no coincidence, then, that, in the highly organized structure of Tillich's system the doctrine of Christ, 'the centre of history', constitutes the second, central volume of the system, whilst the third and final volume culminates and—literally—ends with the question of 'History and the Kingdom of God'.

Tillich here portrays history as the sphere of life by which all other spheres are encompassed and embraced and he re-affirms that the symbol of the Kingdom of God is a specifically political symbol, since it is in the common and collective life of the political world that human fulfilment finds its ultimate sphere. Thus, Tillich reclaims the future-orientation of the utopian imagination: 'Like historical time, historical causality is future-directed; it creates the new ... Historical causality drives towards the new beyond every particular new ... Therefore man's historical consciousness has always looked ahead beyond any particular new to the absolutely new, symbolically expressed as "New Creation." '[38] But, as he immediately cautions, although the study of history and historical causality can lead up to this point 'it cannot give an answer to the "New-Itself" '.[39] Nor is this 'New-Itself' to be sought in technical innovations (in these perhaps least of all), in programmes of social progress, or in this-worldly utopias, since it has already appeared in Jesus as the Christ, specifically as the revelation of the 'New Being' and 'eternal God-Manhood'.

However, the characteristic term 'New Being' brings us to the second issue that Tillich's particular account of historical redemption raises, since it is a term that presupposes a distinctive metaphysical and ontological interpretation of existence in which God is ultimately seen less in terms of 'the God who is to come' but 'Being-Itself'. That God is Being-Itself, he would say, is the one non-symbolic statement that we can make about God. But if Being-Itself is to become manifest in existence, it can do so only under the finite, temporal conditions of existence, thereby introducing a split between essential Being (what Being always is in-itself) and existing Being, Being as it exists. This division is not immediately or necessarily conflictual—essential being does not have to be in conflict with existence—but, in fact, it is: existence as we know it is invariably 'fallen'. Once Being is divided within itself, the structures that belong together in the fullness of Being become opposed to each other and are transformed into 'structures of destruction'. Thus, each of freedom and destiny, dynamics and form, participation and individualization essentially belong together, but we experience them in existence as incessantly at war: I have to assert my freedom by opposing the 'destiny' mapped out by my class, parentage, or schooling—or I submit to my destiny at the cost of my freedom. Similarly with the other ontological polarities. Form can only come about through a process of becoming (dynamics) and dynamics disintegrates

---

[38] Tillich, *Systematic Theology*, III. 347.   [39] Tillich, *Systematic Theology*, III. 347.

into 'formless' chaos without form, but in existence we find ourselves having to choose between change and stability, or between revolution and the status quo. And such splits not only characterize individuals' external life-situations, they also occur within individuals themselves, ultimately manifesting themselves in extreme forms of mental disorder.

A spontaneous response to this situation is to long for the fullness and wholeness of Being that is denied us in existence. In place of the Being that has disintegrated into a cacophony of warring elements, we long for a 'New Being' in which the polarities are once more reconciled and brought into harmony and we again find the 'fullness of life' that we sense to be the real *telos* of human existence. And this, of course, is precisely the place at which the utopian imagination comes into its own, offering visions of the fulfilment that we believe to be our proper destiny. In these terms, then, utopian visions are never just visions relating the present historical time to a possible historical future but are always, at the same time, ontologically (and therefore, ultimately, theologically) relating the present longing for fulfilment to a trans-symbolic relation to Being-Itself—to the God who is our ultimate concern. But this means that it is not just that socialism's classless society 'answers to' or can be 'correlated' with the prophetic symbol of the Kingdom of God but that this prophetic symbol itself is to be interpreted in terms of the manifestation of Being-Itself. And this, as Tillich consistently argues, is as much the *ground* as it is the *goal* of existence.

Consequently, whilst the power of this New Being may be manifest in an open-ended range of *kairoi* or moments of revelation and decision, the claim that it is, precisely, a matter of New *Being* that is at issue reinforces the claim that it has, once and for all time, been definitively manifested in Jesus as the Christ. But this seem not only to mean that futurity yields, in the end, to 'height' (i.e. the height of Christ's sovereignty as 'Lord of time') but that 'height' itself is surrendered to an all-embracing system in which the *ontological* possibilities of existence are always already determined in advance of their occurrence.

Being-Itself is 'always already' complete and whole in-itself. It does not simply become what it is, but it eternally is that and it is therefore significant for this study that Tillich will consistently insist on interpreting the 'divinity' of Christ in terms of the 'eternal' character of the Being revealed in him. He asserts that 'God has no essence separated from existence, he is beyond essence and existence. He is what he is, eternally by himself',[40] and he calls for the replacement of the idea of Christ as having a 'divine nature' 'by the concepts "eternal God-man-unity" or "Eternal God-Manhood"',[41] commenting that '"Eternal" points to the general presupposition of the unique event of

---

[40] By 'by himself' I take Tillich to mean 'by means of himself' rather than 'on his own'.
[41] Tillich, *Systematic Theology*, II. 170

Jesus as the Christ. This event could not have taken place if there had not been an eternal unity of God and man within the divine life'.[42]

Does Tillich, then, finally remain within the orbit of German Idealism? Like German Idealism (as well as some other versions of philosophical theology), he demythologizes the view of eschatology as some kind of future historical end-time. On Tillich's terms the Last Judgement (or as he calls it, 'ultimate judgement') is not and cannot be a temporal event in history. What is of eternal value in human life is not decided by the possibility of some future extension into a temporal infinity, but the actual content of our historically lived lives in time.

> Time is the form of the created finite...and eternity is the inner aim, the telos of the created finite, permanently elevating the finite into itself. With a bold metaphor one could say that the temporal, in a continuous process, becomes 'eternal memory'. But eternal memory is living retention of the remembered thing. It is together past, present, and future in a transcendent unity of the three modes of time. More cannot be said—except in poetic imagery.[43]

Is this, then, a philosophy of recollection? A view of life *sub specie aeternitatis*? A positive answer seems to be suggested by Tillich's comments on eternal life, which he glosses as life 'in' God: 'in ultimate fulfilment God shall be everything in (or for) everything', a view Tillich calls 'eschatological pan-entheism'.[44] But this is precisely to say that it is not some possible future temporal state. Rather, it is a dimension of everything that is occurring now, in the present. The phrase 'in God', he says, 'points to the presence of everything that has being in the divine ground of being... [and] to the inability of anything finite to be without the supporting power of permanent divine creativity—even in the state of estrangement and despair... [it is] the state of essentialization of all creatures'.[45] Even conceding that, for Tillich, 'essentialization' is to be taken in the sense of 'eternal memory' and therefore as inclusive of the temporal action, passion, and struggle of existence, we do seem to be close to returning to the logic of *sub specie aeternitatis*. If Tillich's eternity has a dynamic character distinct from that of Spinoza (for example), this dynamism seems 'in the end', as on a certain reading of Hegel, to be subordinated to the unchanging form of eternity. If Tillich has helped us find a way to speak of the eternal in 'the now' that affirms and confirms our capacity for finding and enacting meaning in time, does he do so at the cost of sacrificing 'the new'? An alternative approach to this dilemma is that of Nicholas Berdyaev, a thinker contemporary with

---

[42] Tillich, *Systematic Theology*, II. 171.

[43] Tillich, *Systematic Theology*, III. 426. Conversely, what fails to find historical significance within history, that is, what fails to contribute towards the approach of the Kingdom of God, 'is not remembered at all. It is acknowledged for what it is, non-being' (Ibid.)—although, again, this is a starkly minimalist statement that can only be further glossed in poetic language.

[44] Tillich, *Systematic Theology*, III. 450.   [45] Tillich, *Systematic Theology*, III. 450.

Bloch and Tillich and one, like them, whose thought was shaped by the historical convulsion of the First World War and its revolutionary and catastrophic aftermath. It is to Berdyaev, then, that we turn next.

BERDYAEV: TRAGIC HOPE

Berdyaev's first book-length study devoted primarily to questions of time and eternity was his 1923 work *The Meaning of History*, a work thus nearly contemporaneous with the second edition of Barth's commentary on Romans—although at that point neither of them had knowledge of the other.[46] Berdyaev saw himself more as a religious philosopher than a dogmatic theologian and, in particular, as a critical inheritor of German Idealism, which he believed was an attempt to develop philosophy on the basis of a Christian concept of subjectivity and therefore more truly deserving the title of a Christian philosophy than the philosophy of the Middle Ages (which he regarded as philosophically compromised by the heritage of ancient philosophy). His reception of the German Idealists was, however, mediated both by Russian religious philosophy, notably the work of Vladimir Soloviev, and by Marxism (which, for a while, he embraced). At the same time, he was influenced by the messianic and apocalyptic elements in Russian popular religion and literature and, like Bloch and Tillich, sought to illuminate the crises of the years from 1914 onwards through the figures of eschatological thought. Arguably, as we shall see, the position at which he finally arrived was both more tragic and yet more profoundly affirming of the possibility of radical novelty than either Bloch's utopian imagination or Tillich's ontological eternity. This affirmation involved a move beyond both materialism (Bloch) and ontology (Tillich), and it is in this regard that Berdyaev can take us a step further in exploring ways of thinking the meaning of the eternal in the mode of hope.[47]

Berdyaev treats questions of time and eternity in his characteristically imaginative, aphoristic, and polemical style. In complete distinction from Tillich, this means that it is always going to be hard to distil any systematic teaching

---

[46] It is, however, striking that an early collection of essays on literary, political, and religious themes was published under the joint title *Sub Specie Aeternitatis*. See Nicholas Berdyaev, *Sub Specie Aeternitatis: Ópyty Filosofskie, Sotsialnye i Literaturnye 1900–1906* (St Petersburg: M. P. Pirozhkova, 1907).

[47] There is a history of ideas argument to be developed here that goes far beyond the scope of this present study. In brief, it would be an argument concerning the influence of the Russian emigration, including the religious philosophers, on the development of modern French philosophy, not least with regard to the radical emphasis on freedom, the suspicion of ontology, personalism, and, most pertinently, a certain eschatological tendency. In addition to Berdyaev, Shestov, Bulgakov, the two Losskys, Kojève, Koyré, and, in a later generation, Jankélévitch, we might also include Lévinas, a citizen of the Russian Empire prior to his emigration.

on these subjects. Nevertheless, the basic principles and intellectual perspectives with which he approached these themes are substantially consistent, although we may see an increasingly clearer statement of the negative element in the relations of time and eternity in the course of his life. This darker view is reflected in one of the two texts that will be the main focus of this brief discussion, *The Beginning and the End*, written in Paris in 1941.[48] But although Berdyaev himself often referred to his thought as essentially tragic I shall be arguing that the way in which he articulates the unattainability of the eternal in historical time also, if paradoxically, allows his work to be read as a philosophy of hope, and perhaps in a sense more radical than that of Bloch or Tillich.

Many of Berdyaev's key ideas relating to time and eternity are developed in *The Meaning of History*, a work he wrote in Berlin shortly after his expulsion from Russia in 1922. He begins with the observation that he and his contemporaries are living though a truly catastrophic era in which a prevailing world-order is in the process of breaking down. This leads to comments suggesting an essentially cyclical view of historical development of order, disruption, and restoration. For the present, however, what is decisive is that humanity is living through an age in which, as Marxism especially forcefully reveals, history has been denuded of mystery and 'de-animated'. In this situation a philosophy of history is needed that will be able to do justice to man 'in the concrete fulness of his spiritual being'.[49] This means that we must lift our eyes above the horizon of the present, as we do when we visit a landscape rich with historical memories like that of the Roman Campagna. In such a place 'we commune with another sort of life, with the mysteries of the past, with those of the after-world; we commune with the mysteries in which eternity is triumphant over corruption and death' (MH, 19).

Memory of this kind will prove a major theme in Berdyaev. Later in *The Meaning of History*, reflecting on the dialectic of creativity and conservation, he sketches a view of memory itself as creative and not merely a means of conserving the past. It is not the kind of memory invoked by Hegel's Owl of Minerva. Rather, it is memory that is active and creative and, as such, effects 'the union between the new world of the future and the old world of the past. The process occurs in eternity; and the union in eternal life fulfils itself in a sort of unique historical creative and dynamic movement' (MH, 39). But because memory thus serves to usher in the new world as well as to conserve the past and, in both functions, is integral to personal identity, it acquires an eschatological aspect. Berdyaev will write that the past is 'eternal reality in which each of us in the depths of his spiritual experience achieves a victory over the

---

[48] It is probably no accident that its more pessimistic tone reflects the situation of the Nazi occupation.

[49] Nicolas Berdyaev, *The Meaning of History*, tr. G. Reavey (London: Geoffrey Bles, 1936), 14. (Further references are given in the text as 'MH'.)

corruption and disintegration of history' (MH, 71–2). In its supreme expression this victory is nothing less than resurrection. As he writes elsewhere:

> Memory of the past is spiritual; it conquers historical time…It carries forward into eternal life not that which is dead in the past but what is alive, not that which is static in the past but what is dynamic. This spiritual memory reminds man, engulfed in his historical time, that in the past there have been great creative movements of the spirit and that they ought to inherit eternity. It reminds him also of the fact that in the past there lived concrete beings, living personalities, with whom we ought in existential time to have a link no less than with those who are living now. Society is always a society not only of the living but also of the dead; and this memory of the dead…is a creative dynamic memory. The last word belongs not to death but to resurrection. But resurrection is not a restoration of the past in its evil and untruth, but transfiguration.[50]

The memories stirred by wandering amongst the ruins of an ancient civilization, then, are not a matter of nostalgia and certainly not a matter of sentiment but of being aroused to the possibilities of creative existence to which the monuments and records of the past bear witness. It is through such arousal that we discover our own capacities for creative activity, so that we too are released into the resurrection world. It is in this sense that Berdyaev says that memory is the ontological basis of all history—not just as conserving the past but as bringing it into a creative relation to the present.[51]

In elevating us to a plane on which time is no longer experienced as sheer transience, creative memory initiates us into what Berdyaev calls the 'celestial history' that accompanies and precedes cosmic and historical time. This concept of celestial history is strongly reminiscent of much we have read in Boehme and Schelling, whom Berdyaev gratefully acknowledges as pioneers in this regard. Against a view of God as essentially immobile and immutable, Berdyaev tells us that Christianity's God is—or should have been—a dynamic God living a life of 'interior mystery, drama and tragedy', a truly Trinitarian God expressing 'the inner passionate divine thirst and longing for an other self' (MH, 48). Celestial history is the history of this God and, as such, can be seen by analogy with the Prologue to Goethe's *Faust* as the 'heavenly prologue' to historical existence; it is 'that predetermination which reveals and manifests itself in the terrestrial life, destiny and history of mankind' (MH, 40). But, in explicit accord with Schelling, this is a history that can only be spoken of mythologically. It can never become the matter of any objective science.

But what is the relationship between this celestial history and historical human existence? Starting from the latter (i.e. from our experience of time),

---

[50] Nicolas Berdyaev, *Slavery and Freedom*, tr. R. M. French (London: Geoffrey Bles, 1943), 111.

[51] Berdyaev, *The Meaning of History*, 73. Berdyaev seems close to Tillich's idea of eternal memory, but, for Berdyaev, this is a specifically human work: through the free act of remembering we ourselves become co-creators of the creative forward movement that is being remembered.

Berdyaev argues that time is itself metaphysically and ontologically significant and is so far from being a merely accidental feature of existence that we can say of it that it is rooted in eternity (MH, 63–4). However, if this might seem to suggest that Berdyaev is reiterating the summons of German Idealism to think time and all that happens in time *sub specie aeternitatis*, he qualifies his remarks by distinguishing two kinds of time. Even if there is a dimension of time that is rooted and grounded in eternity in such a way that our historical time can be understood as 'a period, an aeon in the life of eternity' (MH, 66), time as we experience it in this world is typically 'fallen' time. Such fallen time is time that is divided against itself, time in which each of the time dimensions of past, present, and future break away from the others and war against them. We reject the future for the sake of loyalty to the past or lose ourselves in the present to forget the past or yearn for a future utopia to escape the misery of the present.

Christianity therefore has a twofold relation to time. On the one hand it affirms that time is rooted in eternity (that is, in celestial history) but on the other it struggles against time (that is, fallen time) for the sake of eternity. This is not a struggle against time that will end with time itself being vanquished. It is a struggle for the victory of the eternal in time, within history itself. In this sense it is precisely a struggle against 'the final sundering of time from eternity' since this would be 'a victory of time over the eternal' and 'would signify the triumph of death over life' (MH, 68). At the time of *The Meaning of History*, Heidegger had not yet published *Being and Time*, but Berdyaev would later come to see Heidegger's philosophy as the ultimate expression of such a victory of time over eternity and being towards death as precisely articulating what a view of history that has become entirely forgetful of the eternal might mean.[52]

Berdyaev concludes a synoptic narrative of the successive periods of European history and their respective understandings of time and history with a clear statement as to the failure of all views of historical progress. The various nineteenth-century doctrines of progress (Berdyaev names Comte, Hegel, Spencer, and Marx) all fail because they are limited to the plane of fallen time and on this plane the question posed by the historical experience of time is insoluble. This failure is especially marked by the way in which all of them see death as an impassable limit to human existence. 'The idea of progress bases its expectation on death itself. Its promise is not of resurrection in eternal life, but of the incessant extermination of past by future, of preceding by succeeding generations' (MH, 190). Doctrines of progress are therefore 'a sin before eternity' (MH, 196). Properly understood, 'History is in truth the path to another world. It is in this sense that its content is religious. But the

---

Nicolas Berdyaev, *The Beginning and the End*, tr. R. M. French (London: Geoffrey Bles, 1952), 212. (Further references are given in the text as 'BE'.)

perfect state is impossible within history itself; it can only be realized outside its framework' (MH, 197). 'Man's historical experience has been one of steady failure and there are no grounds for supposing that it will ever be anything else' (MH, 198). And this failure is, in the end, theological: it is 'the failure to realize the Kingdom of God'. 'History is pre-eminently destiny, tragic destiny' (MH, 206). Insoluble at its own level, the question of history must turn to eternity: 'our function at every period, at every moment of our historical destiny, is to determine our relation to the problem of life and history in the terms and according to the criteria of eternity' (MH, 196). And, for Berdyaev, this means the apocalyptic reintegration of human or terrestrial history with celestial history, an event that can only be known in, with, and under the figures of myth.

If there is an eternal solution to the question of history and if that solution qua 'celestial history' is also *in some way* temporal, time as we know it in the course of human history must nevertheless be radically transformed so as to be taken up into that celestial history. And this is not simply a matter of mystically communing in the 'now' with the eternal but of an eschatological, future-directed reorientation of existence. Crucially—and as distinct from Tillich—this turn towards the eternal qua future eschatological possibility is not underwritten by a doctrine of Being. The free decision of a free human personality in which and only in which the turn to eternity is enacted cannot be interpreted within the horizons of any ontology. Berdyaev thus states that his is a philosophy 'which recognizes the supremacy of freedom over being'.[53] 'Freedom is without foundation; it is not determined by being nor born of it. There is no compact, uninterrupted being. There are breaks, fractions, abysses, paradoxes; there are transcensions. There exist, therefore, only freedom and personality'.[54] And, he adds, 'Personality is outside of all being. It stands in opposition to being...its principle is dissimilarity'.[55] Each personal existence is 'something new' in nature that never has been before and never will be again: 'Personality is the exception, not the rule. The secret of the existence of personality lies in its absolute irreplaceability, its happening but once, its uniqueness, its incomparableness'.[56] The time of celestial history, then, is accessible only under the conditions of such personal freedom and singularity. But if the turn towards celestial history is a genuine possibility of personal life then eschatological hope is no longer limited by the constraints of either materialistic (Bloch) or ontological (Tillich) conditions. Universal despair does not destroy the possibility of singular hope.

These points emerge especially forcefully in *The Beginning and the End*, a work said by Berdyaev to summarize his metaphysical position as a whole. Here he

---

[53] Berdyaev, *Slavery and Freedom*, 75–6.   [54] Berdyaev, *Slavery and Freedom*, 76.
[55] Berdyaev, *Slavery and Freedom*, 80.
[56] Berdyaev, *Slavery and Freedom*, 23; cf. Nicolas Berdyaev, *Solitude and Society*, tr. George Reavey (London: Geoffrey Bles, 1938), 68.

remarks that 'Truth is saving, but it saves for another world, for the eternal world which begins in temporal life, but begins with suffering, with grief and frequently with what seems like hopelessness' (BE, 49). It is not hopelessness, however, because 'supreme Truth is eschatological' (BE, 49). Berdyaev therefore sets himself the task of developing the idea of eschatology philosophically, that is, to develop a philosophy and not just a theology on the basis of eschatological thinking.

In this work, as in many of its predecessors, the world as we know it in historical time is a world that has fallen under the spell of objectification. Like the German Idealists, Berdyaev sees this objectification—in knowledge, ethical and historical action, technology, art, and religious life—as ultimately rooted in the creative self-expression of the self. However, when the self becomes externalized in its own products, it comes to experience the resulting situation as one of self-estrangement.[57] Human spirit is trapped by the world it has created for itself and become the servant of its own idols. This 'object world' 'is infinite, but eternity is lost to sight in it' (BE, 79).

In philosophy, Berdyaev sees both Thomism and Hegelianism as trapped in objectification. Hegel may have wanted 'to put life into numbed and ossified being' but he never attained to 'real concreteness' (BE, 94). And this will always happen as long as philosophy looks to being and not to 'that to which and to whom being belongs, that is, the existent, that which exists' (BE, 95). Consequently, Berdyaev proposes an alternative 'eschatological metaphysics' that 'denies the stabilization of being and foresees the end of being, because it regards it as objectification' (BE, 99). Against ontological theologies, then, Berdyaev prefers apophatic traditions, which he sees as convergent with his own eschatological metaphysics. 'Nothingness', he suggest, with reference to Boehme, 'is deeper down and more original than some-thing' (BE, 100) because nothingness allows for the possibility of freedom, which—with a flourish that recalls a theme in Michalski's interpretation of Nietzsche—is also associated with the spiritual symbol of fire (BE, 112f.). In fact, Berdyaev radicalizes Boehme's idea of the *Ungrund*, since where Boehme portrays this as an element within the life of God, Berdyaev suggests that it is *external* to God in such a way that God has to humble himself under this 'nothing'. In other words, there is a kind of kenosis inherent in the very moment of divine self-genesis, underwriting the absence of any determining ontological structure in God. Not only human freedom but divine freedom is itself ungrounded. The fire flares up and nourishes itself solely on empty air, on nothingness.

Turning again to history, Berdyaev sees the unavoidable presence of evil. 'The principal content of history continues to be war' (BE, 205), he writes,

---

[57] The idea is basically Hegelian and recurs for example in Tillich's idea of the transformation of the structures of Being into structures of destruction as well as in Marxist theories of alienation.

and, he insists, we cannot and should not say that God is the creator of this fallen world. God does not act in plagues, wars, or acts of individual violence. 'God does not govern this world...God is not "the world", and the revelation of God in the world is an eschatological revelation' (BE, 152). Freedom, love, creativeness, and the value or personality truly exist and can 'invade' and 'act creatively' within the objectified, fallen world of historical time, yet Berdyaev, like Barth, gives a negative emphasis to Goethe's line that 'Everything transient is but a symbol', stressing the 'but' (or 'only') (BE, 154).[58] 'Time', as we mostly know it, 'is not the image of eternity...time is eternity which has collapsed in ruin' (BE, 207). It is Berdyaev's refrain that, nevertheless, 'we do not live in an aeon which is absolutely shut in on itself. It is possible for the world to enter into the eschatological era, into the time of the Paraclete, and then the face of the world and the character of history will be essentially changed' (BE, 167).

This suggests the possibility of real newness, newness that would not be a product of the past (as in the evolution of 'new' forms of life), but 'real newness' that 'always arrives, as it were, from another world' (BE, 167). It is a product of a paradoxical time no longer constrained by causality (BE, 169). In this respect his account seems, against Tillich, to position the 'New-Itself' as prior to 'Being-Itself'. The future manifest in creative activity is eschatological, 'an upward flight towards a different world' (BE, 183), a future 'which has withdrawn beyond the bounds of this empirical world' (BE, 178).

According to Berdyaev the impulse of messianic thinking is 'the basic theme of history' (BE, 200), since 'there is no Kingdom of God as yet, it has not come. "Thy Kingdom come"!' (BE, 203). Consequently, the messianic spirit refuses to be reconciled with any version of realized eschatology. But although Christ in the wilderness resisted the temptation to rule over the kingdoms of this world, the Church has repeatedly succumbed to just this theocratic temptation. Maybe it is inevitable that any historical manifestation of spirit will always fall prey to objectification and, as such, pass out of the realm of spirit into that of fallen time. But perhaps, as Berdyaev suggests, 'historical Christianity is coming to an end, and that a rebirth is to be looked for only from a religion of the Holy Spirit which will bring Christianity itself to birth again, since it is the fulfilment of Christianity'.[59] Such a Christianity of the future (and here Berdyaev nods towards the influence of Joachim of Fiore), will be 'eschatological', a 'religion of the Spirit, a Trinitarian religion which is a fulfilment of promises, hopes, and expectations'.[60] It is 'the end' not in the sense of 'a final completeness and consummation' but as 'eternal newness, eternal creative ecstasy, the

---

[58] Berdyaev himself is generally negative towards Barth, whose work he sees as involving a 'rejection of man' (BE, 234).

[59] Nicolas Berdyaev, *The Divine and the Human*, tr. R. M. French (London: Geoffrey Bles, 1939), 1.

[60] Berdyaev, *The Divine and the Human*, 1.

dissolving in being, in divine freedom' (BE, 169–70). Such a Christianity, we may say, is not an account of how the natural or supernatural world 'is' but a summons to hope, even in situations of entire hopelessness. But this doesn't mean we should jettison the past for the sake of the future. As Berdyaev says with reference to Nietzsche's Zarathustra, 'It is only eternity which is good and to be loved' (BE, 170).

Under the twin signs of eternal memory and eschatological hope, the suffering experience of time is not a basis from which to project an earthly or celestial future of fulfilment but, nevertheless, it cannot exclude the possibility that, beyond all historical probability or possibility, such time may be transcended and time itself, in the mode of messianic longing that the Kingdom might come, may 'move out into eternity' (BE, 230). But, once more, 'It is only eschatologically, only in the Kingdom of God and not in the earthly realm, that God can be all in all' (BE, 251). Acts of genuine human creativity, of love and art, will be taken up into that Kingdom, but how such transfigurations might occur and what they might mean is beyond our present knowledge. Messianism of Berdyaev's kind is inherently and determinedly apophatic. Neither knowledge nor will can leap over the void that still, for now, divides us from eschatological time and the eschaton is itself the possibility of the impossible. Berdyaev's work therefore not only marks a certain 'end' of classical German philosophy, it also anticipates the more recent phenomenon of 'a certain' messianic thinking (Derrida) that has dispensed with the figure of a historical Messiah and becomes, perhaps, eschatology without an eschaton. But does Berdyaev's position come down to more than the resolve to remain hopeful and expectant even though nothing in the realms of biological, historical, or metaphysical being can provide us with reasons for doing so? Isn't his kind of hope entirely arbitrary and unmotivated—apart from the sheer wish that things could be otherwise than they are? And even if we can distinguish Berdyaev's Messianism from the self-assertion of Nietzsche's will-to-power, isn't this just a matter of taste or pathos, since both might seem, in the end, to offer mere variations on a common vision of the tragic freedom of human existence? But if this is so, then, once more, the 'height' of hope demanded by Lévinas would seem to have been lost.

A THEOLOGY OF PROMISE?

But what other grounds could there be for any hope that is realistically cognizant of the terrors and uncertainties of history? Theology's customary answer to this question is that we have the promise of God and the testimony of history. But how does that promise offer a better ground than the kind of exigency of hope offered by a philosophy of hope, tragic or otherwise? I shall address this question by briefly commenting on Jürgen Moltmann's theology of hope, which was strongly influenced by and in dialogue with Ernst Bloch.[61] Certainly, it is Moltmann who did most to restore the category of hope to the centre of Christian discourse in the late twentieth century. But where Bloch developed the case for hope from within the horizons of a strictly materialist world-view (such that, as he put it, 'What is decisive is transcending without Transcendence'[62]), Moltmann argued on the basis of what he took to be the defining biblical category of promise.

Like many twentieth-century Protestant theologians, Moltmann assumed that there was a basic difference between the categories of Greek and Hebrew thought.[63] The former is characterized as 'static' or 'metaphysical' and tied to a 'cyclical' view of time, whilst the latter is seen as dynamic and historical. In terms of the revelation of God, this plays out in terms of the Greek tendency to speak of 'epiphanies' of God, manifestations of the divine that sanctify the special places and times at which revelation occurs and which are connected with ritual repetitions of the founding myths of the community. Of course, the Bible too tells of various manifestations of God, such as the revelation to Moses at the Burning Bush (Exodus 3). But, on Moltmann's reading, there is a fundamental difference in that 'Israel was but little concerned to understand the essential meaning of the "appearances" of Yahweh in terms of such hallowing of places and times, but for Israel the "appearing" of God is immediately linked up with the uttering of a word of divine promise'.[64] The fact of this revelation is not to be honoured by turning the site of the bush into a holy place of pilgrimage but by attending to and believing the promise of liberation that God gives to Moses. 'The point of the appearance to particular men in particular situations lies in the promise. The promise, however, points away from the appearances into which it is uttered, into the as yet unrealized future which it announces.'[65] Where the Greek epiphany brings the worshipper and the time of worship into congruence with the presence of the god and of the divine eternity, the Hebrew revelation is 'on the contrary' 'that the hearers of the promise become incongruous with the reality around them, as they strike out in hope towards the promised new future'.[66]

---

[61] See, e.g. Bloch, Introduction to *Man on his Own*, 19–29.

[62] This is the first of two mottoes on the title page of Bloch, *Atheismus in Christentum*. The two parts of the second motto, 'Only an atheist can be a good Christian; only a Christian can be a good atheist' reflect the respective positions of Bloch and Moltmann—although Bloch himself is here prepared to present them in paradoxical combination.

[63] This is also true of Tillich, although his work can be seen, precisely, as an attempt to synthesize the two. See especially Paul Tillich, *Biblical Religion and the Search for Ultimate Reality* (Welwyn Garden City: James Nisbet, 1955). This correlates closely with the typology of religions developed and popularized in many works by Mircea Eliade.

[64] Jürgen Moltmann, *Theology of Hope: On the Ground and Implications of a Christian Eschatology*, tr. James W. Leitch (London: SCM Press, 1967), 99.

[65] Moltmann, *Theology of Hope*, 100.   [66] Moltmann, *Theology of Hope*, 101.

Moltmann continues by sketching seven features of 'promise' that mark out a religion of promise from a religion of epiphany: promise denotes 'the coming of a reality that does not yet exist'; it 'binds man to the future and gives him a sense for history'; it signifies that history is not cyclical but 'has a definite trend towards the promised and outstanding fulfilment'; it 'stands in contradiction to the reality open to experience now and heretofore' (i.e. it marks the advent of a 'new thing'); it 'creates an interval of tension between the uttering and the redeeming of the promise'; since only God can deliver what God has promised, faith in the divine promise is inherently critical of any attempts to schematize the course of world-history in any way that would make this self-fulfilling or comprehensible without reference to God's promises; and, finally, in its Christian appropriation and transformation the promise is extended beyond Israel to include all peoples and, at the same time, to include the overcoming of death.[67] Steering away from the Greek epiphany and therefore from any timeless concept of divine eternity, Moltmann believes that faith in the divine promise places the process of this-worldly and material self-transcendence towards a better human future in its proper context, namely, the transcendence of the truly divine other.

Although undeveloped in *Theology of Hope* itself, it is clear that Moltmann does not want to understand the dynamics of promise and hope as simply possibilities accruing to a subject who is in all other regards already complete. It is not as if, first, we have the biological entity *homo sapiens*, then the historical individual shaped by the rise and fall of civilizations and cultures, and then (and only then) the emergence of hope in the promise as a response to a specific civilizational crisis such as the enslavement of Israel. If promise is to be a fully theological category, it has to be understood as revealing who God truly is from the beginning. God didn't make the world and then add on time, history, and a promised eschatological fulfilment as an afterthought, like the discovery of hope at the bottom of Pandora's box. Creation itself is from the beginning oriented towards what will be given in the promise. This is what Barth means when he states that 'the covenant of grace, and therefore history, is the aim of creation', leading him to reject all views that make the relationship between God and creation a timeless constant. 'For this timeless relation has nothing whatever to do with God's decree of grace in which God from all eternity has condescended to His creature in His Son in order to exalt it in His Son'.[68] And, he adds, 'Creation as history fashions in every sense and dimension the pre-form (as yet concealed) of the work and Word of God which is to consist in the accomplishment of a series of histories of the

---

[67] Moltmann, *Theology of Hope*, 103–6.
[68] Karl Barth, *Church Dogmatics*, ed. G. W. Bromiley and T. F. Torrance (London: T. & T. Clark, 2004), III.1. 60.

revelation, representation and communication of God's grace to man'.[69] Yet, as we have seen, although Barth develops layers of complexity with which to expound the relationship between time and eternity, this remains essentially contrastive, leaving Moltmann closer to the Marxist Bloch than to his theological predecessor.[70]

Moltmann's Gifford lectures on the theme of God in creation make more explicit what is touched on in *Theology of Hope*: that (with Barth) creation is *from* the beginning and *in* the beginning oriented towards the making and consummation of the divine covenant and that therefore (and going beyond what is expressly said in Barth) the life of creation is itself the history of the promise that becomes explicit in the prophets. Taking another idea from Barth (in turn alluding to Goethe), namely, that the world and worldly experience may be interpretable as parables of the Kingdom, Moltmann suggests that this analogy (and more than an analogy) shows how, within the structures of creation—prior to what we normally demarcate as 'history'—there may be parables revealing 'the hidden presence of the future of the coming Kingdom' and, if we then understand the parable 'as the hidden presence of a qualitatively new, redeeming future in the everyday experiences of this world, then the parable becomes the promise'.[71] And this, he emphasizes, is also true of what we call 'nature', often seen by theological commentators as trapped within cyclical, atemporal structures: 'all systems of life...are all—each in its own way—open to the future'.[72]

But what kind of knowledge is given in the promise? The question is sharpened by Moltmann's explicit rejection of the conventional definition of truth as the *adequatio* of thought and object, suggesting that promise and hope are categories that generate an *inadequatio* between the human being who, inspired by the promise, reaches out beyond the environing present.[73] With further echoes of Bloch (but also, more remotely, of Heidegger), it follows that whatever kind of knowledge it is, it will not be a knowledge of the

---

[69] Barth, *Church Dogmatics*, iii.1. 67.

[70] However, Barth speaks in terms that Moltmann could have endorsed when he uses an image from the creation narrative to underline the view that promise does not merely supervene on the given categories of created life but goes all the way down, right back to the beginning, to the point at which it can be said that it *is* the beginning. This is the image of the tree of life, of which he says that 'It is a sign which speaks for itself...Its presence means that man lives in and with and by its promise without any need for the latter to find expression. He does not need to eat of its fruits...It is a confirmation of the fact that he may really live here where God has given him rest' (*Church Dogmatics*, iii.1. 256–7). To be created, that is, to understand one's life as given by God, is to exist in and through the promise of life, and the utopian imagination of the prophetic religion inherited and transformed by Christianity is that this promise points beyond itself to a fuller and richer abundance of life, advancing 'from glory to glory' (2 Corinthians 3:18).

[71] Jürgen Moltmann, *God in Creation: An Ecological Doctrine of Creation (The Gifford Lectures 1984–1985)*, tr. Margaret Kohl (London: SCM Press, 1985), 62.

[72] Moltmann, *God in Creation*, 63.

[73] Moltmann, *Theology of Hope*, 118.

world as it is (i.e. of what is currently a part or even the sum of the facts that may be known about the world). In the light of Moltmann's remarks about nature, this may even apply to aspects of natural science.[74] But more urgently, and not least with regard to his own political interests, Moltmann's point relates especially forcefully to claims regarding knowledge of history, where, as we have seen, he rejects the idea that world-history can be mapped out or 'schematized'. History is an open field of the free possibilities opened up by the promise, a view that guards against both scientific socialism and liberal democratic doctrines of progress. However, it also, importantly, guards against Christian claims to have privileged access to the future unfolding of history. Faith in the promise does not give insight into how things are going to develop or turn out in history. Faith in the promise is just that: faith in the promise, no more, no less. But, then, what exactly is the 'knowledge' offered by such faith?

The obvious answer is, perhaps, 'knowledge of God'. But, then, who is the God revealed in the promise, and what kind of knowledge could that be? Some views see such a revelation as primarily and distinctively the revelation of God's *person*, that is, as revealing *who* God is for us. Quoting the Old Testament theologian Walter Zimmerli, Moltmann says that this means that 'This self-disclosure of Yahweh is a "word of revelation in which the 'I' discloses itself in its 'I'-character"'.[75] On this basis, the fulfilment of the promise is simply what is given in the revelation itself: knowledge of God. But, for Moltmann, even if this does not make the mistake of separating God off into a timeless metaphysical dimension, it does downgrade the reality *and the utopian possibilities* of all that human beings live through in history. That is, it absorbs future possibilities into a horizon determined by a fulfilled present-ness. Although the God known in historical revelation is revealed as the same God who has acted in the past and who continues to act in the present so as to open a path towards the promised future, this does not mean that we can only know God in the rear-view mirror of philosophy's grey on grey. Instead, God becomes primarily knowable in and through our concrete commitment to the life and suffering of the present, in which the promise to which the past bears witness comes alive as a promise of hope. 'Knowledge of God will then anticipate the promised future of God in constant remembrance of the past emergence of God's election, his covenant, his promises and his faithfulness... [T]o know about God is always at the same time to know ourselves called in history to God.'[76] But this also means that knowledge of God is inseparable from and can only be reached through a fundamental creaturely human orientation towards

---

[74] That even the most elemental structures of the natural world are incomplete without regard to their teleological orientation towards a fulfilment still to come is a central claim of the visionary cosmology of Teilhard de Chardin.

[75] Moltmann, *Theology of Hope*, 113.   [76] Moltmann, *Theology of Hope*, 118.

the future, as that deepens into messianic, eschatological, and 'eternal time'. Although it is unclear what sense Moltmann allows to eternity, other than that he wishes to dissociate it from the timeless eternity of metaphysics, it is clear that the only access to it is precisely through turning towards what Bloch called utopia and he himself speaks of as eschatology.[77]

But what distinguishes Moltmann's theological thematization of promise from Bloch's secular utopianism? And in what way can a knowledge of God grounded neither in facts of history nor metaphysical principles but solely in the word of promise really count as 'knowledge'? And if the claim is that it does give a kind of 'knowledge' doesn't it then lose the force of its own futurity and revert to a kind of ontology, an implicit statement as to how things are in the world? Giorgio Agamben, for example (although not specifically referring to Moltmann), suggests that when God swears an oath to Moses in the power of his own name, this becomes the primary historical form of onto-theology. Why? Because it affirms the fundamental identity of word (logos) and Being and makes this identity basic to human beings' entire God- and world- (and therefore also self-) relationship.[78] In that case, 'promise' would be interpretable in Tillichian terms as a symbolic representation of a more fundamental and, could we say, immutable ontological condition.

I suggest two preliminary responses to Agamben's objection. The first is to draw attention to the 'gift' character of promise, as explored by Gregory Walter. Here the paradigmatic case is the promise of the mysterious visitors to Abraham at Mamre that he and Sarah will have a son. On Walter's interpretation, this promise is offered as a gift given in response to Abraham's hospitality. Yet, as Sarah's laughter reveals, it is a gift that both falls short of the requirements of standard economic exchange whilst also exceeding these same requirements infinitely—it is, after all, the promise of what seems impossible.[79] And, in a further reflection on the Pentecostal giving of the Spirit, Walter suggests that the time opened up by such a promise is precisely a 'weak' time characterizable in terms of possibility. Why? Because the gifts of the Spirit are not, as it were, tools fitted to particular tasks but reveal the openness of the apostles' new possibilities. Both the promise to Abraham and Sarah and the gift of the Spirit thus open a future that is not determined by past possibilities but is open in all directions to the new.[80]

However, this might seem to rescue the promise only by stripping it entirely of content and, so far from bringing us into a hopeful relation to the future,

---

[77] See Moltmann, *God in Creation*, Chapter V 'The Time of Creation', 104–39.

[78] Giorgio Agamben, *Il sacramento del linguaggio: Archeologia del giuramento* [The sacrament of language. The archaeology of oath-swearing] (Bari: Laterza, 2008), 67–73.

[79] This example is also used by Philo, cited by Chrétien. See Chapter 5 'Eternally Unhoped-for'.

[80] Gregory Walter, *Being Promised: Theology, Gift, and Practice* (Grand Rapids, MI: Eerdmans, 2013). Walter, perhaps ironically in the context of my present argument, also draws on Agamben in emphasizing the primacy of possibility over actuality (56–8).

placing us before a future that was devoid of all possible horizons. However, a second response to Agamben is to apply the Kantian criterion of universalizability to what is said in the promise. Kant is often accused of having been the source of the peculiar sense of moral duty expressed by some Nazi war criminals, as Adolf Eichmann (who had been responsible for transportations to Auschwitz) himself claimed, but a rigorous application of Kantian principles would suggest that any promise worth making or receiving would need to be directed towards a universalizable (i.e. cosmopolitan) human good and that on these grounds Eichmann's repeated asseveration that 'An oath is an oath' constitutes a merely specious line of defence.[81] On Kantian terms, the content of any promise would have to be both for the good of the one to whom the promise was made and for the good of all, thus pointing to the universal horizon towards which the messianic promises themselves point (in Kantian terminology, a 'Kingdom of Ends'). And this, in fact, accords with the divine promise to Abraham, since this is not only a promise relating to what we could think of as the patriarch's 'selfish' desire for progeny but points to the fulfilment of the original promise that all nations are to be blessed through him (Gen 12:3). That the promise is made in the name of God or is received as God's own promise is to affirm that we may give ourselves to it and to what it requires of us without reserve, since it will and can only be for the good of all to whom God's word is spoken—and, in the Christian dispensation, that must be, quite simply, 'all'.

'A CERTAIN MESSIANICITY'

Our argument has moved into 'a certain' proximity to the later writings of Jacques Derrida, who also endorses 'a certain Messianicity' that he identifies as a part of the continuing heritage of Marx, as in the parallels between Marx's expectation of a coming classless society and biblical symbols of the Kingdom of God drawn by Bloch and Tillich. At the same time, this Messianicity is also shaped by what we might think of as more Kantian demands concerning justice and responsibility. And in *Specters of Marx*, in which he examines the continuing 'spectral' presence of Marx in contemporary political discourse, Derrida also looks back to Shakespeare, with Hamlet providing an important and continuous point of reference. In this connection he asks whether 'before

---

[81] Eichmann's argument was that that by swearing an oath of loyalty to Hitler he had foresworn the possibility of acting otherwise than as he was ordered to act. But, on Kantian terms, an oath that is in this way abstracted from universalizable good becomes a case of heteronomous subjection. This, I suggest, is precisely what we see happening in Eichmann's repeated insistence that 'An oath is an oath'. See Hannah Arendt, *Eichmann in Jerusalem: A Report on the Banality of Evil* (New York: Penguin, 1964), 135ff.

Nietzsche, before Heidegger, before Benjamin—can one not yearn for a justice that one day, a day belonging no longer to history, a quasi-messianic day, would finally be removed from the fatality of vengeance?'[82] Later, in linking Messianism to both Marxism and deconstruction, he drops the 'quasi-', stating that 'the Marxist ontology grounding the project of Marxist science and critique also itself carries with it and must carry with it, necessarily, despite so many modern or post-modern denials a messianic eschatology'.[83] Still more strikingly, deconstruction, a 'nihilistic' concept according to its critics, now shows itself to be inseparable from the same Messianism. As Derrida says, 'what remains irreducible to any deconstruction, what remains as undeconstructible as the possibility itself of deconstruction is, perhaps, a certain experience of the emancipatory promise; it is perhaps even the formality of a certain Messianism, an idea of justice—which we distinguish from law or right and even from human rights—and an idea of democracy—which we distinguish from its current concepts'.[84] He even notes that since the epicentre of global instability has become identified with what he calls the 'the war for the "appropriation of Jerusalem"', this provides a concrete, if 'elliptical', connection to the Messianisms of the Abrahamic religions.[85] In fact, it is just this link that calls for a 'certain' Marxist eschatology in order to liberate us from the violence of some other messianic ideologies.

In the lecture 'How to Avoid Speaking: Denials', in which he clarified the difference between deconstruction and the negative theology with which it had often been associated, Derrida had already linked the idea of Jerusalem with the idea of 'promise' via the ritual greeting 'next year in Jerusalem'.[86] Like Moltmann (of whom I suppose him to have been entirely unaware), Derrida explicitly sees the element of promise as integral to his 'certain Messianism'. It too must not become identified with any given state of affairs or any historically achieved goals, since 'Otherwise it rests on the good conscience of having

---

[82] Jacques Derrida, *Specters of Marx: The State of the Debt, the Work of Mourning, and the New International*, tr. Peggy Kamuf (London: Routledge, 1994), 21. The use of *Hamlet* is suggested to Derrida by the fact that *The Communist Manifesto* opens with the statement that 'A spectre is haunting all of Europe' and he sees in this an analogy to the role of the ghost at the beginning of Shakespeare's play, summoning Hamlet to put right the time that is 'out of joint'.

[83] Derrida, *Specters of Marx*, 59.

[84] Derrida, *Specters of Marx*, 59. The idea of a 'democracy to come (à-venir)' is a thread running through many of Derrida's later writings. Specifically, here, he is at pains to distinguish this from the idea of democracy propagated by defenders of the American global hegemony such as Francis Fukuyama. As he points out, the supposed 'end of history' proclaimed by Fukuyama and identified with the universal acceptance of liberal democracy is highly implausible, given the range of both domestic and foreign policy challenges to Western claims to instantiate the democratic ideal—some, at the time of writing, being in their own terms somewhat successful.

[85] Derrida, *Specters of Marx*, 58.

[86] J. Derrida, 'How to Avoid Speaking: Denials', tr. Ken Frieden, in Harold Coward and Toby Foshay (eds), *Derrida and Negative Theology* (Albany, NY: State University of New York Press, 1992).

done one's duty, it loses the chance of the future, of the promise or the appeal, of the desire also (that is its "own" possibility), of this desert-like Messianism (without content and without identifiable messiah)'.[87] Pushing even further the logic of such an 'atheological heritage of the messianic', Derrida speaks of an asceticism that 'strips the messianic hope of all biblical forms, and even all determinable figures of the wait or expectation; it thus denudes itself in view of responding to that which must be absolute hospitality, the "yes" to the *arrivant(e)*, the "come" to the future that cannot be anticipated'.[88] And, he adds,

> Open, waiting for the event as justice, this hospitality is absolute only if it keeps watch over its own universality. The messianic, including its revolutionary forms (and the messianic is always revolutionary, it has to be), would be urgency, immi-nence but, irreducible paradox, a waiting without horizon of expectation. One may always take the quasi-atheistic dryness of the messianic to be the condition of the religions of the Book, a desert that was not even theirs (but the earth is always borrowed, on loan from God, it is never possessed by the occupier ...)[89]

As John D. Caputo comments, this is an openness towards and a readi-ness to welcome 'I-know-not-what'.[90] But to what does such readiness actually commit us?

We have already seen why and how the logic of expectation—as opposed to the execution of a five-year plan according to the 'scientific' laws of socio-economic development—must contain a dimension of unknowability. If theology speaks of its 'knowledge' of what is given in the promise, then it is using 'knowledge' in a distinct and unusual sense. It is knowledge of things hoped for, not of things that *are*. Yet we cannot avoid the critical comment that this seems to undermine the grounds for Derrida's own rejection of contempo-rary capitalism. The hungry need to be fed, the homeless housed, the refugees welcomed, the sick tended, the victims of torture released and their torturers be compelled to answer for their crimes by a justice that is distinct from any spirit of revenge. And even if any single act of feeding, housing, welcoming, caring, release, or prosecution cannot—as, of course, it cannot—be identified with the arrival of the Kingdom, that the Kingdom will be known by just such marks is, in Jewish and Christian traditions, surely undeniable. If this is not so, then we seem to be back once more in the domain of empty wishes discon-nected from the exigencies of real possibilities.[91]

---

[87] Derrida, *Specters of Marx*, 29.   [88] Derrida, *Specters of Marx*, 168.

[89] Derrida, *Specters of Marx*, 168. Connections between the nomadic 'desert' religion of pro-phetism and its conflict with the fertility cults of settled, agricultural communities have long been a commonplace of Christian and Jewish commentary—see, e.g. Martin Buber, *The Prophetic Faith*, tr. Carlyle Witton-Davies (London: Macmillan, 1949), 70–80.

[90] John D. Caputo, *The Tears and Prayers of Jacques Derrida: Religion without Religion* (Bloomington, IN: Indiana University Press, 1997), 135.

[91] This seems to be implicit in Agamben's objection that 'Deconstruction is thwarted Messianism, a suspension of the Messianic'. See Giorgio Agamben, *The Time that Remains: A*

Let us grant the point. Vacuous wishing would not be a helpful outcome of this 'certain' Messianism. But must we immediately equate an element of indeterminacy with vacuity? Isn't the insistence on the dimension of unknowability in a genuine messianic expectation necessary in order to build in safeguards against religious triumphalism? The hungry have been fed, but wounds remain unhealed, the scars, perhaps, of traumatic memory. There will always be more work to do and it would seem to be unrealistic to suppose that, in any foreseeable historical timespan, we will individually or collectively be able to say that we have done enough or that the last fight has been fought. Even when all seems well, and perhaps most of all when all seems most well, we will still have to remain vigilant and to keep open the question as to whether there remains some unidentified injustice, oppression, or hurt that needs our attention. Responsibility is never finished with—and, as Derrida points out, the claims of justice and responsibility will often place us in impossible dilemmas: 'I cannot respond to the call, the request, the obligation, or even the love of another without sacrificing the other other, the other others'—which, again, pushes responsibility beyond the sphere in which it could be guided by any programme of ethics.[92] We will always be having to be on the alert, and always having to decide—with no time off for good conduct!

In *Politics of Friendship*, Derrida quotes a Messianic legend told (invented?) by Blanchot, in which the Messiah arrives at the gates of Rome 'among the beggars and lepers', as 'incognito' as even Kierkegaard might have wished. Nevertheless he is recognized, and asked 'When will you come?'[93] In the same spirit, even when we see the works of the Kingdom (cf. Matthew 11:2-6), we might—*must*—still ask: what *more* is to be done, and how are we to hasten the coming of the Kingdom *even as we experience the light of its coming*?

This is not only a question of specifying the ethical content of Derridian Messianicity. It is also a question of time itself—of the possibility of time and of present temporal experience being able to inform our view of human flourishing. If Tillichian ontology ends by tying hope back to a presupposed ontology, doesn't Derridian Messianism deny any possibility of time opening any kind of path towards a future worth hoping for? As we shall later find Michael Theunissen arguing, the experience of time as exercising a tyrannical power of human life is not just a matter of how time throws us towards an ineluctable annihilation, it is also the experience of how time fragments and congeals our lives, of how we feel ourselves to be robbed of our past, without a future, or unable to be in our own present—or, alternatively, trapped in one or other

Commentary on the Letter to the Romans, tr. Patricia Dailey (Stanford, CA: Stanford University Press, 2005), 102-3.

[92] See Jacques Derrida, *The Gift of Death*, tr. David Wills (Chicago: University of Chicago Press, 1995), 68. Throughout this discussion Derrida is at his closest to Lévinas.

[93] Jacques Derrida, *Politics of Friendship*, tr. George Collins (London: Verso, 1997), 46. For discussion see Caputo, *Tears and Prayers*, 78-81.

temporal dimension, prisoners of our past or present circumstances, or believing our happiness to be bound up with unrealizable ambitions for the future.[94] A salvific experience of time, by way of contrast, would be one in which the present became rich with valued memories that sustained it in its quest for future fulfilment. Such time would be an enriched and complex time, rather than the no-time of an impossible utopia. In similar terms, Giorgio Agamben interprets the messianic time of Paul's theology as a time *within* the secular era, and, as such, having a uni-dual structure such that the messianic kairos is enfolded *within* a chronological sequence.[95] In these ways Theunissen and Agamben give voice to an element of 'arrival' or 'realization' ('eschatology in the process of realization') that Derrida assiduously resists. But does this mean that we are back to the structure of recollection implied in Tillichian ontology? No, because eschatology-in-the-process-of-realization of this kind hinges on internalizing a future that is apophatically guarded against becoming an object of cognition, even the kind of cognition that can be described as knowledge of Being-Itself. While such a future can inform life in the present, it cannot define it or predict its existential possibilities. Even in heaven, God remains beyond all comprehension.[96] The most we can say, it seems, is that 'It may be so'. But doesn't religion require more? Doesn't it need to be able to say 'It shall be so'— 'All shall be well'? So how might we ground the promise so as to empower it to speak such assurance?

Derrida offers an important hint as to how we might address this question by his emphasis on the theme of the 'call', a theme closely related to that of promise. In the 1980 lecture 'On an apocalyptic tone recently adopted in philosophy', he had already condensed the exigency of messianic attentiveness into the summons to 'Come!' that he also identified explicitly with the usage of the book of Revelation that becomes a major inter-text in the latter part of the lecture. Eschatological, messianic thinking is thinking that is always vigilant, always on the watch, always listening out for the call to 'Come!' Such a call calls us beyond any ontology, calls us only on condition of there being an other who, in calling, unsettles any achieved identity and security. It is an 'apocalypse without apocalypse' and, equally, 'the apocalypse of the apocalypse, *Come* is apocalyptic'.[97]

---

[94] See Chapter 7.

[95] Agamben, *The Time that Remains*, 70. The complex of questions opened up by the themes we are now addressing is also dealt with (and from related angles) by several of the essays in Neal DeRoo and John Panteleimon Manoussakis (eds), *Phenomenology and Eschatology: Not Yet in the Now* (Farnham: Ashgate, 2009). See also my review of this collection in *Theologische Literaturzeitung* 135/9 (2010), 1007–8.

[96] Which also indicates why John Hick's notion of an 'eschatological verification' won't work.

[97] Jaques Derrida, 'D'un ton apocalyptique adopté naguère en philosophie', in Philippe Lacoue-Labarthe and Jean-Luc Nancy (eds), *Les fins de l'homme : À partir du travail de Jacques Derrida* (Paris: Hermann, 2013 [1981]), 477.

But, again, called to—what? Derrida would say this cannot be answered, at least, not universally or once and for all. For you or I it may, however, have a very concrete answer: to give succour to this person or group who, even now, are calling out for aid. And it is no less concrete if, in making that decision I have perhaps to sacrifice the time I used to give to the sick animals shelter in order to devote it to working with refugees. That is a decision entirely within the domain of my responsibility, and who could object to giving time to sick animals—but this lack of objective or even 'ethical' determination does not mean that the call, the summons to 'Come!' is merely some event in my own mind.

And, remarkably, the lack of definable content in the apocalyptic summons to 'Come!', the entirely open and unlimited nature of this certain Messianism leads Derrida, as it does Bloch, Tillich and, especially, Berdyaev, to extend the summons to and beyond the limits of death. This is not to say that he ever supposes there to be any kind of after-life or conscious survival of death, but that the past—or, more concretely, those who have died in the past—are still within the scope of the call. Once more, it is the logic of what Derrida calls 'the non-contemporaneity with itself of the living present' that requires responsibility to open out towards both past and future, since it is 'a responsibility and…respect for justice concerning those who *are not there*, of those who are no longer or who are not yet *present and living*'. That is to say, justice requires responsibility 'before the ghosts of those who are not yet born or who are already dead, be they victims of wars, political or other kinds of violence, nationalist, racist, colonialist, sexist, or other kinds of extermination, victims of the oppressions of capitalist imperialism or any of the forms of totalitarianism'.[98] Derrida will not propose anything as ambitious as Berdyaev's creative memory, transforming time into eternity and living in a living communion with those we call the dead, but he will propose, as integral to the inexhaustibility of responsibility, that we commit to a counter-movement against the forgetfulness that otherwise sweeps all our past and our past history of injustice into oblivion. 'In forgetfulness lies exile; in remembrance is the secret of redemption.' Testimony must be borne. That most basic of communist values, solidarity, may not be withheld even from the dead.

In the chapters that now follow, each of the thinkers we shall engage will help further deepen the question as to what it might mean to speak of a universal hope as grounded in the phenomenon of a 'call', but I now conclude with a powerful literary statement of the intertwining of call, promise, and universal solidarity. It is taken from Hermann Broch's essay on 'The Disintegration of Values', interspersed into the pages of his novel *The Sleepwalkers* and which, I think, perfectly epitomizes the point. Describing the situation in which reason and autonomy have been installed as society's ultimate values, Broch sees

---

[98] Derrida, *Specters of Marx*, xix.

this as leading to 'the unaccented vacuum of a ruthless absoluteness, in which the abstract Spirit of God is enthroned … reigning in sorrow amid the terror of dreamless, unbroken silence that constitutes the pure [i.e. abstract, form-less] Logos'.[99] Yet, Broch adds, no matter how far we go in the direction of the 'muteness of the abstract', there remains 'the voice that binds our loneliness to all other lonelinesses, and it is not the voice of dread and doom; it falters in the silence of the Logos and yet is borne by it, raised over the clamour of the non-existent. It is the voice of man and of the tribes of men, the voice of comfort and hope and immediate love: "Do thyself no harm! for we are all here!"'.[100]

The three thinkers we shall consider in the following chapters are Michael Theunissen, Søren Kierkegaard, and Franz Rosenzweig. Each of these may be read as offering something like a phenomenological account of how our human time-experience points to what we might call 'saving' possibilities, that is, possibilities for living time otherwise than as mere thrownness towards death. Each of them also helps us to go further in exploring how the themes of hope, promise, and call might inform and shape these possibilities. Such fur-ther exploration is perhaps especially urgent in the case of the 'call'. For if the hope of which religion speaks is indeed founded on a call, what might make us think that this was in truth a *call* in which we are claimed by a power 'not ourselves'? None of the thinkers to whom we now turn will attempt to offer anything like a proof that there is indeed a Supreme Being 'out there' who calls us in one way or other, but, in accordance with the loosely phenomenologi-cal and soteriological approach taken by this study, they do offer testimony as to what it could mean to understand ourselves as 'called' and what difference such a self-understanding might make to the manner of our lives.

I acknowledge at once that they are not the only thinkers who might have been selected for further study at this point, and there are also many liter-ary works that would serve the same purpose. We have already drawn on Proust, Dostoevsky, and Broch.[101] Alternatives can be argued over endlessly, but I think that the three figures I have chosen mark out a set of interrelated approaches that, in their similarities and in their differences, offer, if not com-prehensiveness, a fullness of possibilities and directions. Each is a critical heir of Hegel and both Rosenzweig and Theunissen are also heirs of Kierkegaard,

---

[99] Hermann Broch, *The Sleepwalkers*, tr. Willa and Edwin Muir (London: Quartet, 1986), 639–40.

[100] Broch, *The Sleepwalkers*, 648.

[101] In my book *'The Heart Could Never Speak': Existentialism and Faith in a Poem of Edwin Muir* (Eugene, OR: Cascade Books, 2013), which can be read as an important supplement to the present work, I offer an account of the Orcadian poet Edwin Muir as another paradigmatic thinker in this respect. Amongst the philosophers, it has been suggested to me that William Desmond might also be an appropriate conversation-partner, and amongst the poets both Wordsworth and T. S. Eliot (both of whom, like Muir, are discussed in my *God and Being*)—all of which are, of course, excellent suggestions, and the list could easily be extended.

Theunissen having devoted several monographs as well as important interpretative essays to the Dane. And, since Theunissen's work culminates in a monumental study of Pindar, they might also be taken as respectively representing a Pagan, Christian, and Jewish approach. Of course, each set of writings constitutes a distinctive and individual voice that cannot be subsumed into some general classification without significant loss. Theunissen does not speak for either Philosophy or Paganism, nor Kierkegaard for Christianity, nor Rosenzweig for Judaism. Yet there are in each case both natural and effective affinities to such larger currents of thought, action, and vision. Pindar's narrations of the divine genealogies of his sporting heroes, Kierkegaard's disjunction of time and eternity, and Rosenzweig's focusing of the question of the Eternal on the life of 'the eternal people' strongly reflect their respective religious outlooks. At the same time (or so I believe), each of them also illuminates basic human possibilities that are not limited to these religious positions.

Non-Western traditions would, of course, yield further exemplification (and perhaps qualification) of what will have been arrived at. 'What will have been arrived at'—not a consensus, to be sure, but a significantly convergent set of testimonies to how time may be experienced and lived as more than time, drawn towards and pouring out beyond itself towards an immeasurable fullness of time to be enjoyed as what John's gospel variously calls 'abundance' and 'eternal life'.

7

Theunissen: Pindar, Poet of Hope

A NEGATIVE THEOLOGY OF TIME

Michael Theunissen describes himself as coming from the Hegelian-Marxist tradition and his work has been developed in dialogue with existential philosophy, psychiatry and theology, with special emphasis on Kierkegaard, whom he several times pits against Heidegger (typically in Kierkegaard's favour).[1] Time is a major theme in Theunissen's thought and his most ambitious work is a study of what he calls the 'turning of time' that he finds in ancient Greek poetry, especially Pindar.[2] It is here that many of the often diverse threads of his work are brought together in a coherent and detailed account of the hope he finds in the Pindaric odes. As a postscript on 'Heidegger, Hölderlin and the Greeks' makes clear, this is also a critical response to Heidegger's narrative of the 'fall' of Western metaphysics after the Presocratics. An earlier collection of papers, *Negative Theology of Time*, indicates the range and thrust of Theunissen's engagement with questions of time and highlights some of the key themes recurring in the Pindar study that will be the main focus of this chapter. It is therefore with this collection that I shall now begin.

It is clear from the outset that, for Theunissen, the question of time is not just one philosophical problem amongst others but raises basic issues about the possibility and nature of philosophy today. As the opening paper in *Negative Theology of Time* makes clear,[3] Theunissen understands the task of philosophy

---

[1] See Michael Theunissen, 'The Upbuilding in the Thought of Death: Traditional Elements, Innovative Ideas, and Unexhausted Possibilities', tr. George Pattison, in Robert L. Perkins (ed.) *International Kierkegaard Commentary*, IX: *Prefaces and Writing Sampler* and X: *Three Discourses on Imagined Occasions* (combined volumes) (Macon, GA: Mercer University Press, 2006), 321–58.

[2] The idea of a 'turning' in time is also important for Theunissen's sometime colleague at Berlin's Freie Universität, Jacob Taubes. For a brief discussion of this term, see David Ratmoko, 'Preface' to J. Taubes, *Occidental Eschatology*, tr. David Ratmoko (Stanford, CA: Stanford University Press, 2009), xv and, in Taubes's text, 10, 54.

[3] Michael Theunissen, 'Möglichkeiten des Philosophierens Heute' [Possibilities for philosophizing today], in *Negative Theologie der Zeit* (Frankfurt am Main: Suhrkamp, 1991), 13–36. (Further references are given in the text as 'NTZ'.)

today in a historical perspective that is shaped by Hegel's failure to provide a unified logical structure by which to explain—arguably, as we have seen, *sub specie aeternitatis*—the whole history and development of Western thought in its manifold forms. The fact of this failure does not, however, absolve philosophy from reckoning with its own past, although what exactly this means must be recast post-Hegel. One thing it does mean, according to Theunissen, is that philosophy must take seriously the task of research, a move for which the paradigm case is Marx's turn from philosophy to economics. The relationships between phenomenology and psychology or philosophy of language and linguistics provide Theunissen with other examples in which this can be seen as necessary. But, versus various kinds of reductionism, Theunissen does not recommend that philosophy simply dissolves itself in favour of one or other specialized science. The turn to research must therefore be complemented by a supplementary questioning as to the truth of the deliverances of the specialist sciences. This gives philosophy a different and a more modest task than attempting to establish or critique the foundations of scientific enquiry, as (variously) in Hegel, Husserl, or Heidegger. Such a post-metaphysical metaphysics is 'last' rather than (as in Aristotle) 'first' philosophy. In terms of the kind of truth that such 'last philosophy' seeks, it is helped by remembering its ancient kinship with both art and religion since this is truth about the human condition, about what it means to be human and, ultimately, what human flourishing might mean. In the fragile space between empirical reality and fantasy, and beyond both fact and value, philosophy seeks to give an account of human flourishing as involving not just morality but also love and grace. But—for reasons that will become ever clearer as we proceed—this means that the problem of time becomes central to philosophy, since, as the title of the second essay suggests, the question as to the possibility of true human happiness is also the question: 'Can we be happy in time?'[4] But what is it to live in time or, still more basically, what is time?

Theunissen declines the option of defining time. We cannot know what time is, he says. Yet, in the light of the preceding comments about the situation of philosophy today, we can make a beginning by seeing how philosophy has understood it in the past. If we do this, he argues, we see that the basic questions of metaphysics have from the start involved time. Whether in its Parmenidean beginning or its Aristotelian development, metaphysics does not offer a route to what is beyond being, but it does propose understanding time in terms of what is beyond time: eternity. The distinction of time and eternity thus gets established as the horizon within which the subsequent history of classical and pre-modern metaphysics moves. Time itself is real but the things in it, subject to coming into and passing out of being, are not. These include

---

[4] Michael Theunissen, 'Können wir in der Zeit glücklich sein?' [Can we be happy in time?] (NTZ, 37–86).

human beings, who are conceived as 'in' time and thus subject to its dominion. This changes in the modern period when it is no longer a question of human beings living 'in' time but of time being 'in' us. This change is accompanied by a multiplication of different kinds of time, so that we come to distinguish between, for example, astronomical time, historical time, and subjective time. Time is now universalized in such a way that eternity disappears and there is nothing that is not in some way temporal. And, finally, whereas the older metaphysical view saw life in time as falling away from the fullness and perfection of being, time is now seen as something to be affirmed and valued positively. Theunissen notes a minority report amongst Jewish thinkers such as Adorno and Benjamin for whom, as he puts it, transience (*Vergängnis*) becomes fate (*Verhängnis*). This fate encompasses eternity itself, which now comes to be seen as unmoving time or time in which things reproduce themselves without genuine historical progression. Such a move does not simply reject metaphysics as nonsense (à la Ayer), but it does see it as involving a misreading of human time-experience.

Theunissen's response to the tradition and its modern transformation is to propose a hypothesis against which what philosophy has said about time can be tested. This hypothesis is that 'time rules over us human beings, just as it does over things' and it does so in an alienating and not a liberating way. Indeed it is the eminent way in which the world as a whole rules over us and in us (NTZ, 40). But if this is so, then the modern thesis as to the subjective origin of time is as false as the thesis concerning the plurality of times: it is time (singular) that rules over us. However, the thesis as to the universality of time fares somewhat better, and Theunissen concludes that time just is 'the encompassing', that is, the medium in which everything is and everything happens. This, of course, makes defining time exceptionally difficult since it is always presupposed in any definition we may give. Time is pre-subjective: we live time before we develop or take up a subjective relation to it, a distinction Theunissen makes by contrasting lived time (*gelebte Zeit*) with experienced time (*erlebte Zeit*) (NTZ, 43). But this also underlines how it is that we can suffer from time and experience it as an alien power that rules over us. In its extreme form this alienation leads to time itself becoming the object of our discontent, namely, in tedium or, as the German so expressively puts it, *Langeweile*. All of which, of course, militates against the fourth aspect of the modern relation to time, namely, affirmation. The eternal recurrence of the same is not liberation but boredom.

When we come to Theunissen's Pindar-interpretation we shall see several examples of how the dominion of time is experienced as suffering. In *Negative Theology of Time*, he examines the negative time-experiences of sufferers from melancholia and schizophrenia as extreme illustrations of possibilities latent in normal time-experience.[5] They reveal the ever-present possibility that time

---

[5] See Theunissen, 'Können wir in die Zeit glücklich sein?', and 'Melancholisches Leiden unter der Herrschaft der Zeit' [Melancholic suffering under the dominion of time], (NTZ, 218–81). In

might seem to stop moving or that the relationship between past, present, and future might fall apart in such a way that the future is no longer experienced or conceived as potentially fulfilling our present wants and may even take on a threatening aspect. Such reflections illustrate Theunissen's negative method very precisely since, the suggestion is, such time-experiences negatively mirror what a healthy or fulfilled time-experience would be, namely, one in which past, present, and future were mutually enriching. In such an experience the present would be meaningful both in relation to valued past experiences and as directed towards anticipated good outcomes. This kind of time is not the merely linear movement of past to future but involves the creative action of the future in the present.

In relation to time, then, it seems that the possibility that our lives will fail is always present although, despite our being subject to the dominion of time, it doesn't follow that that we *must* fail. But, clearly, success too will, on Theunissen's premises, involve time, even if, as he puts it 'Human life succeeds, if it succeeds, not thanks to time, but despite it' (NTZ, 56). He then goes on to suggest three forms that such success may take: dominion over time, as when time is instrumentalized in relation to one or other subjective purpose; freedom from time, as envisaged in Aristotelian *theoria* or Schopenhauer's aesthetic intuition; and reconciliation with time. All three are forms of freedom, (1) and (2) being, respectively, freedom to do something with time and freedom from time. Yet these ((1) and (2)) are only ever relatively free in relation to time. I can make or take time to meet a person I like or play a round of golf, but meanwhile the clock ticks away; similarly, the sage and the aesthete must both come down from their 'timeless' moments and return to the flux of daily life—such moments of release are, precisely, 'temporary'. Nevertheless, they do indicate that even if we are powerless to tear ourselves loose from time, the grip of time can, sometimes, be loosened.

But what would reconciliation with time be and how might that shift our relation to time? As distinct from classical accounts of timeless contemplation or aesthetic rapture that turn away from time, Theunissen proposes the possibility of a time-experience in which 'In its depths, the present appears as being other than time: this is what the tradition called "eternity"' (NTZ, 60). As his references to Goethe and to Nietzsche show, this is the kind of deepening of time that we have encountered in some interpretations of Nietzsche's eternity.[6] However, Theunissen himself demurs from calling this 'eternity'

these essays we can see how Theunissen makes good his methodological principle that philosophy has to attend to the deliverances of the concrete sciences by engaging with psychiatric and psychoanalytic approaches to mental illness. From a different angle it is also taken up in his study of Kierkegaard's *Sickness unto Death*: Michael Theunissen, *Der Begriff Verzweifelung* (Frankfurt am Main: Suhrkamp, 1993).

[6] See the section 'Eternal Recurrence' in Chapter 3.

because if we are no longer operating with a classical metaphysical concept of eternity then the mere word leaves unexplained how or in what way this eternity within the moment can be the 'other' of time. What is clear is that, for Theunissen, seeking access to an eternity that is the other of time 'as its own depth dimension' cannot require us to deny time; if he appeals to eternity as an 'other' of time, it is an other that is only ever revealed as such in time and therefore it too must be a kind of time or capable of occurring in time. This therefore suggests both a transcendence of time and also a certain mimesis of time. Why 'mimesis'? Because it wins freedom from time by turning towards and not away from time.

That this 'other' of time is revealed *in* time is possible because it is precisely the whole of time, time as such. Consequently, what we win in coming into a creative relation to it is not just a momentary experience of freedom but something that can be described equally well in terms of any of the dimensions of time. This is not an 'eternity' bound exclusively to any one of past, present, or future. As Theunissen puts it with reference to the idea of utopia, 'The other time that is the hidden other of time itself within time is eternity as future, future as eternity', but, as such, it is also the basis of genuine historical action in the present (NTZ, 65).[7] Therapeutically, such a time-experience reverses the 'negative' experience of the melancholic or schizophrenic for whom past, present, and future have become emotionally and practically disconnected. But it is also the apostle Paul's 'future *aiōn* [that], although eternal, is not just this but is also, and primarily, historical' (NTZ, 65).[8]

Aiōn becomes a key term in Theunissen's interpretation of Parmenides' Fragment 8.5-6a[9] and, as we shall see, will prove central to his critique of traditional metaphysical approaches to the question of time. This critique is largely developed in terms of the distinction between *aiōn* and *chronos*, the former having the sense of creative or fulfilled time, the latter denoting time as unalterable duration. In a conventional translation the fragment reads: 'It was not in the past, nor shall it be, since it is now, all at once, one, continuous' (*homou pan, hen, suneches*).[10] How is this to be understood? Theunissen distinguishes two possibilities with which we are already familiar: timeless eternity (for which he reserves the term 'eternity' in the full sense) and temporal duration.

---

[7] This clearly corresponds closely with the structure both of Bloch's spirit of utopia and certain applications of the biblical category of promise. See Chapter 6.

[8] Here the closest parallel is clearly to Agamben and his notion of Messianic time as the 'uni-dual' coincidence of two heterogeneous time, *chronos* and *kairos*. Again, see the section 'A Certain Messianicity' in Chapter 6.

[9] 'Die Zeitvergessenheit der Metaphysik. Zum Streit um Parmenides. Fr. 8.5–6a' [Metaphysics' forgetting of time. On the debate concerning Parmenides' Fr. 8.5–6a], (NTZ, 89–130). In the following discussion my primary aim is to outline Theunissen's own position and not to enter into discussion as to whether it is philologically well-grounded.

[10] From G. S. Kirk and J. E. Raven, *The Presocratic Philosophers: A Critical History with a Selection of Texts* (Cambridge: Cambridge University Press, 1971), 273.

On the first view, which comprises metaphysical, mathematical, and logical timelessness, the statement must, according to a logic we have already encountered in Chapter 1, be seen as relating less to the 'it' (i.e. being) and more as a rule determining how not to speak of it, since ascribing tenses to it in any way is already to misrepresent it. Nor is Parmenides denying the reality of time but, more precisely, substantial change in time—that being either came or comes into being or passes or will pass away. However, Theunissen thinks that it is historically implausible that a fully developed concept of timeless eternity was available to Parmenides or that he used it. But if it is a matter of duration, what concept of duration might Parmenides have had?

For Theunissen, in close dialogue with critical discussions amongst classical philosophers, it is important to see that the two parts of the extracted verse are not saying the same thing and that the second part, 'since it is now, all at once, one, continuous', adds something significantly new to the first, 'It was not in the past, nor shall it be'. But this second part also provides the basis for what Theunissen sees as an alternative view of eternity (which he calls E_2) that sees Parmenides as teaching 'the absolute identity of the eternal present', 'pure presentness', or 'the pure now of epiphany in which everything is at once and in one' (NTZ, 104).[11] When this is then identified with the *aiōn* of Plato's *Timaeus* and with the Latin *aeternitas* (as distinguished from *sempiternitas*) it comes to be understood as *aidion ousia*, everlasting or eternal being, although Theunissen distinguishes this *aidion ousia* from his own interpretation of *aiōn* since the former lacks the quality of 'life' that is crucial to the latter. As opposed to the simple denial of time, E_2 sees being as abiding in unity with itself in such a way that its present never becomes a past or a transition to a future. Like the timeless view it denies time, yet it also sees time as, somehow, preserved within the unity of the present.

If such a view is to be consistent, it must refrain from commenting on how this unity in the now will continue to be maintained in the future since, if it does so, it will slip into a version of duration. This, Theunissen thinks, is what we see in Augustine's concept of a *semper praesentis aeternitatis*, the ever-present eternity or *nunc stans* of the scholastic metaphysics of time. If or when this involves duration it can be classified as a third (E_3) concept of eternity whereas, in its more rigorous form, it proposes a 'non-durational eternity'. This, as Theunissen notes, is often identified with the *aiōn* of Plato's *Timaeus*, which, as remaining-in-unity-with-itself and therefore also remaining-in-itself is taken by Plotinus (for example) to be definitive of eternity. Crucial for both (E_2 and E_3) is that this involves 'life'. But, he adds, 'Parmenides' being (*Seiende*)…is neither endowed with positive attributes nor divine nor living'; consequently, 'Parmenides can only have in view endurance in time and not

---

[11] The citations refer respectively to the positions of Calogero, Gigon, and Picht.

some kind of time-transcending abiding, since the latter would have to be the life of the highest divine spirit that abides in itself and this is impossible for him' (NTZ, 107). Indeed, Parmenides' critical relation to the cosmological views of his predecessors shows that his intention in substituting 'being' for 'God' and/or 'world' is precisely to exclude the divine and the living. 'The reality that Parmenides displaces by means of being is, however, life, the life of God and of the world. Accordingly, the negations by means of which he wants to purge being so that it is only being [and nothing more] aim in the end at a No to life. An *akinētos* (8.38), an unmoved entity, is what being, in its life-lessness, is' (NTZ, 108). Even more dramatically, Theunissen states that what Parmenides' negations are specifically against is coming-into-being or growth and that his thought therefore manifests a fundamental 'will to annihilate life' (NTZ, 109).

A key element in the Platonists' attempts to read Parmenides as anticipating their own views is the phrase *homou pan* in the sequence *homou pan, hen, suneches* ('all at once, one, continuous'). Noting some of the philological issues (such as the placing of the comma and textual variants), Theunissen suggests that, contrary to the Platonist reading, this should not be read temporally but spatially—not as 'all at once' but 'all together in space'. What it all means, therefore, is that 'Being is the whole, the unitary meaning of which consists in its existing uninterruptedly in time', that is, duration (NTZ, 111). As such it reveals the ontology of the Eleatic school to be deeply anti-utopistic: 'There's nothing more to look for', as Theunissen sums up the thrust of his interpretation (NTZ, 114).

In terms of Theunissen's hypothesis concerning the 'dominion of time' and of the human condition as 'suffering from time', Parmenides can therefore be taken as unwittingly exemplifying just this subjection to time. For, in reality, there is no such thing as what merely is or exists. The 'perfection' of such a being would be no more than the 'immobility of rigidity' and a metaphysics that takes this as its leading idea will be a metaphysics of nihilism since what it calls 'being' is, in fact, no being at all: 'Parmenides not only denies the reality of flowing time, he also and especially denies the time that belongs to flowing reality, reality itself, apprehended as temporal, living, unfinished, and as a whole' (NTZ, 116).[12] This also, he says, is relevant to Parmenides' silence concerning the knowing subject, namely, himself. Such metaphysics, Theunissen says, is actually a 'sickness'; rather than denying death for the sake of life, Parmenides 'denies life in the name of death. This "No" issues from "suffering from life"' (NTZ, 117).

---

[12] By way of contrast, Theunissen will say later that modern philosophy is a fruit of *Lebensphilosophie* and that it accepts life as self-transcending and, more precisely, life itself and not just 'spirit' as self-transcending (cf. NTZ, 213).

This establishes a theme that will run through many of Theunissen's later essays in *Negative Theology of Time* and that will also prove relevant to his approach to the ancient Greek poets, who reflect the time-experience that Parmenides sought to negate. One important thread is provided by the special emphasis Theunissen gives to the idea of *aiōn*. As we have seen, he argues that the Platonic *aiōn*, with its connotations of life ('a unity of time and eternity that is grounded in life' (NTZ, 301)[13]), is not applicable to Parmenides' view of time, which privileges *chronos* at the expense of *aiōn*. Where *chronos* is limited to the horizon of worldly time, which it construes in essentially linear terms, aionic time points beyond the world and construes time as dimensional, co-implicating the three dimensions of past, present, and future. As such, it is therefore also time as we live it: it is not just the linear measurement of our life but our 'lifetime' or 'the time of our lives', that is, what we do with our lives and how we, as it were, translate the objective time that exists prior to our becoming aware of it into subjectively lived and meaningful time (NTZ, 303–4).[14] Thus, biological time is transformed into biographical time and rather than just being carried forward by a causal chain established by a preceding sequence of events we learn to live in creative relation to our future. This is not, it seems, a mere reiteration of the nineteenth century distinction between nature and spirit, since Theunissen denies that 'nature' is to be identified with either the eternal laws of Hegel's philosophy of nature or with the cyclical time of eternal recurrence. On the contrary, every day brings something new in the life of nature as well as in historical time. It is only when nature itself has been reified (i.e., only in a certain conception of nature) that it can appear as eternally recurring. More pertinent instances of eternal recurrence are the negative experiences of time as empty repetition generated by capitalism, schizophrenia and melancholia. The task, then, is not to deny the reality of chronological time but to contest its exclusive dominion over our lives.

In this regard, Theunissen affirms Plato's intention to point to an eternity that can be characterized in terms of life and freedom, but what he delivered was, in the end, only necessary being. In any case, we have no alternative but to start with chronological time and find in it an 'other' of time, another time, time as saving—the eternity that is the trace of aionic time left in modernity. Proust offers one example of how this might be done, since his great work of recollection is occasioned precisely by an involuntary association of ideas and, as it continues, discloses a world in which subjectively lived time is intertwined with the world of things subject to chronological time. Music suggests another aesthetic mode in which something similar might be achieved (NTZ, 313). Yet such aesthetic 'redemptions' seem fated to fall back once more into

---

[13] As we shall see, he also relates this to the New Testament idea of 'eternal life'.

[14] Cf. the earlier remarks regarding the distinction between *gelebte* and *erlebte* (*lived* and *experienced*) time, to which Theunissen also returns in this passage.

forgetfulness. If we are to speak of a time that might, in time, rightly bear the appellation 'eternal' it would be a time that, coming from the future, also had power over the future.

Such saving time is not a time that philosophy can develop out of its own resources, however, and one instance that especially engages Theunissen's reflections is that of Christian faith, the subject of the last essay in *Negative Theology of Time*, '*Ho aitōn lambanei*. Jesus' Prayer of Faith and the Temporality of Being Christian' (NTZ, 321–77).

Jesus, as Theunissen acknowledges, represents a problem for philosophy. The claims made for him and that he seems to have made for himself relate not only to qualities such as goodness, which might be open to philosophical inspection, but also to his claims to sovereignty: that he doesn't merely discourse about truth but claims *to be* the truth. Nevertheless, his claim that this truth is a truth that sets human beings free does open a path towards dialogue with philosophy. In modernity, of course, freedom is largely conceived in terms of self-realization: freedom means being free to fulfil the projects I set myself (i.e. being maximally autonomous). However, the gospels' idea of freedom is not based on autonomy but on faith. How then might we decide between these? Theology might appeal to the teaching of the Church. But philosophy, if it has not committed itself in advance to the cause of autonomy (and much contemporary philosophy seems to have done just that), will be limited to experience, which Theunissen understands in the specifically Hegelian sense of the experiences 'that human beings make with themselves', that is, the testing out of alternative ways of understanding ourselves and our lives in the world (NTZ, 324–5).

As we might expect, a decisive issue is the relation to the future. Jesus' preaching of the Kingdom of God speaks of the Kingdom as 'near' or 'at hand' and, indeed, already here—precisely by virtue of the way in which it is 'at hand'. As Theunissen glosses Jesus' message, 'in the not-yet of fulfilment the promise itself is already fulfilment' (NTZ, 326). It is this proleptic character of gospel-existence that makes it possible for the believer to be patient and free from anxiety in relation to time. Again, when Jesus says 'He who seeks, finds', Theunissen notes that 'finds' is spoken in the present tense, suggesting that the future itself is already in some way contemporary with the seeking. 'The one who asks receives already in the asking' (NTZ, 332).[15] Asking is not only necessary but sufficient for fulfilment and, as such, provides us with a way of understanding faith as 'the self-reflection and the unnaturally direct connection between asking and its acceptance', as in another of Jesus' sayings: 'Your faith has made you whole' (NTZ, 335–6). It is not that the future disappears, but it is now seen in the perspective of faith, that is, of an attitude

---

[15] Theunissen understands texts that suggest a separate future fulfilment (such as 'Ask and you *will* receive') as needing to be interpreted in the light of 'He who seeks, finds'.

in which it is no longer a cause of anxiety. The one who has faith is therefore freed from preoccupation with himself and his future. And, with Kierkegaard but against Heidegger, Theunissen argues that this must be more than a kind of self-relationship.[16] The liberation of the self from its self so as to be itself involves more than the self. In faith the self does not make itself free but chooses itself as having been set free.[17]

But can this also have a positive content? Is it only a matter of 'freedom from' or can it also be 'freedom for'? Theunissen's answer is that it is also positive, since when we are set free from the future we are set free for world-changing action, albeit not in the manner of pure autonomous action. In terms of a distinction he has drawn previously, we are free in a positive sense because our relation to the future has changed from protensive to prospective. That is to say, the future is no longer simply what we relate to as the sphere of our protensive future action or 'what is to be done or what is to come'. Although this is an integral element in any conception of the future, it becomes the basis for various forms of stress and pathology if it is isolated and made into the sole measure of our relation to the future, as when I am stressed out at the thought of all that I've still got to do today or what I'm going to say to x when next I meet him after our violent disagreement. By way of contrast, the future is prospective when it offers a view as to how I can move forward in my life. As such, the attitude of faith therefore has a certain 'lightness' of touch, since those who have faith are no longer burdened by a future that seems to weigh down upon them.

But what is the positive content that such an attitude is able to embrace? Theunissen finds the answer to this in Paul's characterization of Christian life as a life of faith, hope, and love. In terms of temporality he further glosses this in terms of faith in the Father as the original heavenly power, hope in the Son as the one who is to come, and love in the Spirit, who is the presence, now, of what is to be.[18] But as the present power of love both the Spirit and the love that it communicates are not detached from the future and the totality

---

[16] In NTZ, 345 Theunissen says that everything valid in Heidegger's concept of original temporality is owing to Kierkegaard.

[17] This may be compared with the account of faith offered by Bultmann. See especially the last of his Gifford lectures in which he describes the view of the self that is implicit in the act of Christian preaching in the following terms: 'The man who understands his historicity radically, that is, the man who radically understands himself as someone future, or in other words, who understands his genuine self as an ever-future one, has to know that his genuine self can only be offered to him as a gift by the future' (Rudolf Bultmann, *History and Eschatology: The Gifford Lectures, 1955* (Edinburgh: University of Edinburgh Press, 1957), 150). And, Bultmann adds, 'To be historical means to live from the future' (152).

[18] Theunissen acknowledges that, in these terms, and if the Father is identified as the Creator who was 'in the beginning', then faith becomes correlated with a relation to the past, which is problematic, since the overall movement of the argument is to point to the co-implication of the time dimensions in the humanly lived life of the Spirit, which is also the unifying power of the Godhead.

of faith-hope-love is anticipation. With specific reference to love, this can be seen in how the requirement of love is the requirement to attend to and to wait upon the neighbour's articulation of their need and of the help they seek, a structure of relationship in which present and future are intensely reciprocal. The temporal form of love of neighbour is that of a future present, which is also and as such anticipation of the future in which the believer hopes to see God face to face. For now, of course, *that* future exists only in the mode of anticipation that is the concrete love of neighbour. With Buber Theunissen affirms that the relation to the neighbour is not the 'mediation' of divine love but is the way in which the eternal You is present now. The immediacy of neighbour-love is the surety of the eschatological fulfilment of love of God. At the same time, love is not simply a mode of being present. Love *is* presence or shows what presence is, but, we should note, it does so precisely by showing it to be an essentially 'futural present'.[19]

Theunissen again echoes Buber, although this time without explicit reference, in making this the occasion for a critical comment regarding Kierkegaard. Kierkegaard's account of being or becoming a self, he says, is one of constant striving. The Kierkegaardian self can never let up on its efforts to avoid sinking into despair, whereas love has an element of receptivity or devotedness that gives a ground, 'not ourselves', making self-affirmation possible. Love, that is, has a quality of *Gelassenheit*, a key term in relation to Heidegger's later thought that is notoriously difficult to translate but has a range of meanings including equanimity and letting-be. Love doesn't have to work all the time: it can just accept. But although the self-relationship that is the engine of Kierkegaard's account of being a self is not sufficient to ground an account of love, love is able to encompass and ground the self-relationship described by Kierkegaard. To love the other involves a self-relationship such that we can choose or allow ourselves to get involved with the other.

Love is also explained by Theunissen in terms of having time for the other or giving our time to the other, as in the New Testament example of the Good Samaritan or as implied in Jesus' command to go the extra mile. But this further indicates a certain transcendence of time. 'If I *have* time, I *am* not solely *in* time but raise myself above it. I possess the power and the freedom to take my time and to win something abiding from what is passing away. But the eternity that thus comes to appearance in love is still connected to the condition of *having* time and cannot be separated from its dialectical unity with time' (NTZ, 361)—a reminder that what Theunissen is proposing is precisely a view of eternity that is neither timelessly detached from time nor sempiternal duration through time but an experience of time itself as saving, as liberating

---

[19] See also the discussion of Buber and other dialogical philosophers (including Rosenzweig) in Michael Theunissen, *The Other: Studies in the Social Ontology of Husserl, Heidegger, Sartre, and Buber*, tr. Christopher Macann (Cambridge, MA: MIT Press, 1984), 307–15.

us from what is otherwise the oppressive dominion of time. Love is a way of being in time that offers just this experience of saving time, time itself as salvific. Love's way of knowing the present is to know it as reflecting and grounding hopeful expectation, whilst the past is no longer something from which it has to strive to break free but a history to which it can return with endless gratitude.[20]

This brief survey of *Negative Theology of Time* suggests something of the range and ambition of Theunissen's philosophy of time as involving a reckoning with the history of philosophy and as marked by an interest—a decisive interest, I suggest—in theology's testimony to experiences of time that can rightly be called salvific. His most sustained discussion of this latter theme, however, is in his monumental study of Pindar. As we now move to consider this extraordinary work, it will perhaps be helpful to make a couple of comments about its relation to *Negative Theology of Time*. Firstly, we have seen how Theunissen indicts the metaphysical tradition since Parmenides of taking a view of time that is essentially life-denying. In doing so, he sees philosophy, in its beginnings and since, as deliberately turning away from the contributions of poets and cosmologists who, instead, prioritized life. In returning 'behind' philosophy to the poetic world of Pindar and his predecessors Theunissen is attempting to reconnect philosophy to the mythical world that it has persistently seen as its other. But he is also, in a sense, out-manoeuvring Heidegger, whose critique of the history of Western metaphysics results in an attempt to rethink the beginning of philosophy made by the Presocratics. For Theunissen, however, the 'fall' of philosophy has already taken place in precisely those thinkers to whom Heidegger looks for this new beginning. Heidegger too, of course, turns to the poetic, above all to Hölderlin (whose own work also stands in a significant relation to that of Pindar). However, as an epilogue to Theunissen's work will make clear, he sees Heidegger's reading of Hölderlin as vitiated by a view of the Greeks that privileges the beginnings of what would become philosophy. Secondly, by thus positioning his own thought by reference to Pindar, Theunissen makes the poets rather than the philosophers the locus of his *preparatio evangelii*. Although this is by no means unprecedented (Hölderlin himself might be read as doing something similar), it results, as we shall see, in an original way of approaching a theological interpretation of time and of the relationship between time and eternity. And this, in turn, also opens a path towards a further rethinking of the relationship between philosophical and theological ways of addressing time, which is to say that it also means rethinking the relationship between philosophy and theology as such.

---

[20] Although Theunissen does comment on the relation to the past in something like these terms (see NTZ, 363), this is not given equal weight with his discussion of the relationship between present and future.

PINDAR: POET OF HOPE

The sub-title of Theunissen's Pindar study is 'The Human Condition and the Turning of Time'. Theunissen explains this last phrase in terms of a distinction between the change or turn from one era or epoch to another (*Zeitenwende*), as we might talk of the change from the medieval to the modern world, and a change or turning within time itself (i.e. a change in the character of human time-experience).[21] As we have seen, he regards human life as exposed to a range of negative experiences of time that are historically and intellectually reinforced by Western philosophy's privileging of *chronos* over *aiōn*, that is, its denial of real, flowing, living time in favour of a 'time' that is really no time at all but an empty unchanging timeless eternity. Amongst alternative interpretations of temporality he gives special weight to the teaching of Jesus, but, in quantitative terms, it is the poetic writings of Pindar (522–443 BC) that occupy the chief place. These exemplify a 'transcendence' of time of which he writes that 'Transcendence, however, does not mean breaking out of time altogether into a supposed timelessness, as in Parmenidean-Platonic metaphysics, but a transformation of time's dominion into another [time]. In this way the change is a turning, a turnaround of time in itself'.[22]

In broad terms, Theunissen's programme might seem reminiscent of Heidegger's 'destruction' of the tradition of Western metaphysics in order to return to a point of origin from which philosophy might find a new or second beginning. But where Heidegger looks back to the Presocratic philosophers to find this point of origin, Theunissen needs to go back further still, since, in his view, Parmenides actually inaugurates the time-sickness that has plagued Western philosophy ever since. And, unlike Heidegger, he does not take Parmenides as exemplary for a supposed 'Greek' way of thinking, since, in fact, philosophers were by no means the only kind of Greeks and, in addition to and preceding the philosophers, we also have the testimony of the poets.[23] And it is in the poets, in this case Pindar, that he sees a kind of time-experience and time-interpretation that points away from the lifeless time of the philosophers to the living and vital time that will also come to expression in Jesus' proclamation of the Kingdom of God. Thus, it is Pindar, rather than the philosophical tradition, who becomes a kind of preparation for the gospel. And this suggests a further difference from Heidegger (for whom neither the New

---

[21] See note 2.

[22] Michael Theunissen, *Pindar: Menschenlos und Wende der Zeit* (München: C. H. Beck, 2000), 1. (Further references are given in the text as 'P'.)

[23] In his essay 'The Ancient Concept of Progress' E. R. Dodds long ago pointed to the tendency to identify the 'Greek' view of time with the views of the philosophers, suggesting that scientific and historical writers in particular, and some of the poets, had a more positive view of time. See E. R. Dodds, *The Ancient Concept of Progress and other Essays on Greek Literature and Belief* (Oxford: Clarendon Press, 1973).

Testament texts nor subsequent theological reflection on them reach the level of fundamental metaphysical thinking), since Pindar's distinctive contribution will prove to be precisely religious and, in a very definite sense of the term, theological. For the possibility of the turning of time that he finds in Pindar is the possibility of movement from life lived under the dominion of time to life lived under the rule of God (P, 3).[24]

Yet many—not least those philosophers and some theologians in the Anglo-Saxon tradition who are inclined to eschew large-scale historical narratives—might feel that all of this sounds rather too grandiose. Indeed, if they read beyond the headlines they may well feel that their initial suspicions are justified since it could seem that the outcome of Theunissen's monumental study is, in truth, somewhat modest. However, this modesty of outcome might also be seen as entirely in accordance both with Theunissen's fundamental idea of what it is to practise philosophy today and with the actual content of the position he ascribes to Pindar himself. Regarding the first of these points, we have already commented on how Theunissen sees contemporary philosophy as living in the aftermath of attempts (epitomized by Hegel) to develop a philosophy that would serve as the first or basic or universal science of sciences, grounding and guaranteeing the ontological, epistemological, and moral legitimacy of all forms of human knowledge. Instead, he counsels that philosophy must wait upon the deliverances of the special sciences, providing an interpretative commentary on them and open to revision in the light of their development. In the present case, this means that the philosophical interpretation of Pindar must engage with the work of classics scholars and be attentive to the philologists' accounts of what is at issue in any given word, phrase, or reference.[25] Regarding the second point (the content of Pindar's thought), modesty is arguably obligatory since (on Theunissen's reading) this content is itself centrally associated with the idea of 'measure' and the limitation of human powers to what concerns their proximate material and temporal reality. With that proviso—that the outcome will be necessarily and intrinsically 'modest'—Theunissen will also (and no less emphatically) say that 'Pindar's thought is, in its foundations and as a whole, hopeful thinking about hope (*Hoffnungsdenken*)' (P, 346).

---

[24] This is not to be understood as exchanging one heteronomy for another, however, since the point is precisely that, for Theunissen, the 'rule of God' is a rule that gives and sustains us in life and, as such, the 'fulfilment' of time.

[25] Throughout *Pindar*, Theunissen is meticulous in following-up on the classicist literature relevant to the poem under consideration. It is in large measure the thoroughness of this 'footnote' scholarship that contributes to the great bulk of the work—and, at many points, to its extraordinary interest. However, both for reasons of space and because the present author is not competent to enter into these kinds of debates, this aspect of the book will not get due attention in the following presentation.

But if the philosopher must be attentive to the deliverances of the philologist, what—on Theunissen's assumptions—will be distinctive about a philosophical interpretation?

This is a question to which he does not, perhaps, give a direct or extended answer. His practice, however, suggests that what he has in mind is something like the following. Where the classicist will be occupied by philological, contextual, and formal considerations, the philosopher will be looking in the first instance to the noetic content of the poem. If this is so, then the philosopher will presuppose (as, perhaps, other interpreters need not) that the poetic work will, in each case, be articulating a distinct and coherent thought, saying something that can be restated in non-poetic and propositional form. This does not mean—as it might on Hegelian or positivist principles—that this non-poetic restatement is in any way 'higher' or more adequate than the poetic statement: in fact, Theunissen seems open to the possibility that the poets may indeed have chosen the better way. But it does mean that the philosopher may have a distinctive role in clarifying both what the poet is saying and what the poet might say to us and to the questions—such as the question of time and eternity—that the histories of philosophy and religion have bequeathed to us. Although the poet speaks to local and time-bound situations and as the individual that he is, Theunissen at several points indicates that there is a universalizing current in the work (cf. P, 61) that makes what is being said of universal human significance. In these terms, the poet is also speaking in such a way as to both merit and demand the attention of philosophy.

CHRONOS

We have seen how Theunissen regards the chronos-dominated thinking of the philosophical tradition as enshrining a basic hostility to time, but if *Pindar* too moves to a position in which *kairos* is privileged over *chronos*, the treatment of *chronos* is in this case more nuanced than in the essay on Parmenides. Although he is in principle committed to his interpretation being guided by the text and by its philological commentary, and therefore as needing to accommodate the range of usage actually found in Pindar, Theunissen does not see this as undermining the fundamental distinction between *chronos* and *aiōn* (*kairos*). The latter relates to a 'lyrical utopia of a turning of time in which life that is dominated by an alien time is freed to be itself and wins power from itself' (P. 31), whilst the former, *chronos*, is time that exercises an alien dominion over human life. Yet (and unlike in Parmenides' world-view), although *chronos* is described as 'sovereign' over human life, this is not a sovereignty that is exercised in complete independence of human actions or attitudes. Time is not just a feature of an indifferent universe but is itself interdependent with the positions that human beings take towards it, a feature to which epic poetry already witnesses (cf., e.g. P, 42).

The sovereignty of time is consequently revealed in the different ways that it is experienced and construed by human beings. Prominent amongst these, and with a characteristically negative meaning, is 'the day', an experience summed up in the eighth Pythian Ode: 'Creatures of a day? What is someone? What is no one? A dream of a shadow'.[26] As these lines suggest, the ephemeral character of human existence is revealed to us as meaning not only short-lived but also unstable and fluctuating, being-towards-death, and ignorant of the future (P, 46–9). Illustrating how Pindar inherits this understanding from his predecessors, Theunissen cites Archilochus' graphic image of the power of the day as a wave that, rising up over a ship doomed to shipwreck, already has the boat in its power, even if it has not yet fallen and destroyed it (P, 174–6). As such, it is a power against which humans have no protection or recourse. An unexpected eclipse functions similarly, revealing to Archilochus the arbitrariness and unpredictability of the gods (P, 197). For another poet, Alkaios, it is the inconstancy of the wind that most figures the rule of time (P, 205).

It is against such background assumptions that Pindar is writing, but his teaching, Theunissen will claim, is quite different. He too, as in the lines quoted, knows that human life is inconstant and, to put it in Heideggerian terms, thrown towards death (cf. P, 53–4, 65–7). But for him the gods are not just the agents of the arbitrary, impenetrable, and annihilating power of 'the day' and, following the depiction of human life as 'a shadow', he nevertheless proceeds to speak of the 'Zeus-given brightness' that may come as 'a shining light' to 'rest upon men, and a gentle life' (*Pythian* 8.96–7). In this regard, Theunissen notes that just as the poem here ends by pointing to the possibility of divine intervention, so it began as a prayer to 'Kindly Peace'. At the same time—and we shall come to say more of how—'peace' is not God-given in such a way as to rule out the need for human action. To receive peace, we must also struggle for it—just as the athlete Pindar is celebrating has struggled for his victory.

Another approach to the negative sovereignty of time is through the spatialized image of time 'hanging', as in *Olympian* 7.24–6: 'But about the minds of humans hang numberless errors, and it is impossible to discover what now and also in the end is best to happen to a man'. Again, this is a theme found already in Pindar's predecessors, as when Simonides speaks of death as hanging over all (P, 123) or in the dispute between Solon and Mimnermos as to whether old age is to be dreaded as a condition of unmitigated decline, with Mimnermos taking the view that old age 'hangs over' human life. On this bleak view (and

---

[26] Unless otherwise started, translations of Pindar's Odes are taken from the two-volume Loeb Classics edition, *Pindar*, I: *Olympian Odes; Pythian Odes* and II: *Nemean Odes; Isthmian Odes; Fragments*, ed. and tr. William H. Race (Cambridge, MA: Harvard University Press, 1997).

unlike in the Hebrew tradition) there is no consolation to be looked for in the love of family or in the prospective flourishing of future generations. Nor is there any point in keeping alive the praise of famous men or recording the history that binds one generation to another. Instead, life in time is seen as a matter of individual fate, de-historicized and naturalized (P, 145). When time hangs over human life in this way it signals the errors, ignorance, and guilt to which we are prone, a state that is not a matter of some failing in our individual subjective response but, as Theunissen puts it, 'an essential trait in the conception of reality itself' (P, 85).

Pindar, again, both inherits but also contests this interpretation, and *Olympian* 7 itself shows how such a state can be turned around. The Ode rehearses three myths relating to the foundation of Rhodes. The first two are instances of human error resulting from confusion of mind and forgetfulness, yet each time the gods turn the potentially disastrous consequences into the occasion for blessing, effecting a transition from a state of *Unheil* to *Heil*. The third error, the gods' failure to give Helios, the sun, a land of his own, ends (again felicitously) with Helios himself ordaining Rhodes to be his own possession. The mind that contemplates these events is delivered from its entanglement in the errors of time through enlightenment as to the *telos* of events and the failures recounted in the poem are framed at beginning and end by the grace (*charis*) of the gods.

The issue seems nicely balanced, however. The uncertainty accompanying the seeming treachery of time cannot be entirely removed. We are delivered from the negativity of time only to the extent of being able to see what is 'before our feet': as we turn from the past towards the future, it is especially the future that is, in some way, already present that 'saves'—a pattern that Theunissen sees being offered by Pindar as universally human. Similarly, we cannot expect to control the suffering that comes to us in time, but we can look for healing from it and freedom in relation to it, a freedom that is itself indicative of the action of a god (P, 120). Pindaric hope is thus, implicitly, 'an expression for a utopian consciousness' (P, 122), even if we have no immediate prospect of a fully utopian time. 'Although time hangs over us afterwards as before, it no longer imprisons us...The myth encrypts its utopian hopes in [the prospect of] an alternative future' (P, 122), that is, one that has been freed from the errors and suffering of the past.

In Pindar, then, we see the possibility that the hold of 'the day' over human life can be broken. This, Theunissen says, is 'transcendence in the original processual sense of the word' (P, 217). This transcendence is initially experienced in the two modes that he calls 'irruption' and 'surmounting'. The former signifies moments when other possibilities (not limited by the hegemony of 'the day') break into human time, like the 'Zeus-given brightness' or 'shining light' of *Pythian* 8. In the first instance such irruptions have the character of suddenness ('the time of radiance is the moment' (P, 219)). Yet, as *Pythian* 8 also suggests, the divine radiance may continue in the gift of 'a gentle life' (*meilichos aiōn*), a 'sweet time' that is not immune from change or undisturbed by what happens in time and that is not for ever but that, nevertheless, provides the abiding solace of a tranquil, gentle, and joyous *charis* that, as it were, wipes away the tears of ephemeral troubles. At the same time, the way in which the poem narrates this transformation through the story of Herakles reminds us, again, that such a state cannot be looked for without the cooperation of human beings themselves. The advent of grace is also a story of human freedom, a pattern that is reflected in the narrative reach of the poem that, originating in the world of myth, ends, via the legends of Herakles, in the historical time measured by the cycle of games that he inaugurated (P, 225).

The role of human freedom also indicates how the transcendence of the day may be figured as a 'surmounting' of time. We have seen that the human domain is what lies before our feet, both spatially and temporally, and, as Theunissen comments, 'what is mortal befits mortals' could be a motto for Pindar's entire work (P, 225). If there are moments in which we do resemble the gods (as in the moment of athletic victory), this, strictly speaking, only amounts to achieving the maximum possibilities of life in time and is not a surmounting of time. It is, after all, something we can do—although only with the help of the gods. Whilst hubris is never in place, there is therefore no simple opposition between humans and gods but a pattern of unity and difference, and Pindar's anthropology is a 'theo-anthropology' (P, 234). We are not and never will be gods, yet we and the gods draw breath from the same mother (P, 227). Even the intention to restrain ourselves from hubristic excess reveals 'a moment of surmounting transcendence', since we could not limit ourselves to what is properly human if we could not attain those limits. This situation is poetically conveyed in the image of the pillars of Herakles, a symbol of the limits of human striving but also of how far human beings both can and should venture (P, 235).

Three elements are emerging in this account. With reference to *Olympian* 1.97–100,[27] for example, Theunissen points out that the victory enjoyed by the victor and acclaimed by the poet is itself the outcome of the athlete having borne the strain of the fight. But the joy of remembering this moment of greatness is only a limited joy, the joy of recollection. It is perfected when it becomes a matter of daily renewal, in which, as Theunissen emphasizes, Pindar's *aiei* does not mean 'always' (that is, it is not an inalterable state) but is a matter of 'one day at a time'. This is a happiness that not only remembers but that actively 'flowers' in time (P, 261). But if a constant state of abiding happiness is not possible for human beings, such daily renewal in happiness is possible, although—and this is the third element—only when it is received 'day by day,

---

[27] 'And for the rest of his life the victor | enjoys a honey-sweet calm, | so much as games can provide it. But the good that comes each day | is greatest for every mortal.'

ever anew...from the hand of the god and in equal measure constantly aware of doing so' (P, 265).

What, then, can we hope for and how does hope take shape in human life? Surveying the pre-Pindaric poets, Theunissen discerns a range of attitudes to hope. Hope can be something that grips us rather than an attitude we take up (P, 314–15). It is often illusory (P, 321–2), light, vain, and generating anxiety (P, 327), empty and unfulfillable (P, 332). Yet if the overwhelming tendency is towards a negative view of hope, it is also possible to see some texts, such as Bacchylides' Ode for the dying Hieron, that do express a hope that points beyond the ephemeral. Hope's ambiguity is perhaps best epitomized in the ascription to Apollo of the exhortation to think both that tomorrow we may not see the sun *and* that we may live another fifty years (P, 336).[28] Bacchylides also offers the reflection that 'whoso deals properly with the gods, salves his heart with a hope that is yet more divine' (P, 337), whilst Theognis comments that 'hope dwells with men as the sole good divinity' (P, 339). Yet where these split hope into, respectively, other-worldly and this-worldly expectations, Pindar consistently sees both the positive aspects of hope and combines other- and this-worldly elements. Pindar's hope bespeaks a genuine transcendence of ephemeral time and, as such, is a genuine surmounting of time rather than a merely speculative flight over its inescapable limits.[29] Hope, therefore, must fulfil certain conditions in order to be hope and not just wishful thinking. Chiefly, it must be guided by a prospect of the good it holds in view and must also be grounded in experiences of achievement. And, again, as we have seen, any view of the future requires the assistance of the divine (P, 343) whilst, again, real hope will be inseparable from effort on our own part (P, 348).

The realism of Pindar's hope is indicated by the strictly limited view we are able to have of the future, a view conveyed in the term *promatheia*, a looking-ahead that gives a limited illumination of the darkness ahead whilst at the same time serving as a 'corrective to forgetting'. Both looking ahead and remaining mindful of the past we must always keep to the just measure of human possibilities.[30] In this way, hope is something very different from measureless striving, as when we pursue wrong ambitions, illustrated by Asclepius' misdirected attempts to bring the dead back to life (P, 366).[31] Against such examples, well-founded hope is grounded in reality, as in the case of a well-prepared athlete's hope for victory. But even this well-grounded hope cannot become a stable or constant element in our lives unless we are aided by a power 'not ourselves'. But if such help comes to us, then 'In place of

---

[28] Cf. SKS5, 464/TDIO, 96.

[29] We might think again of Kierkegaard's complaint that Hegelian philosophy speculatively flies itself off to the moon.

[30] The account of *promatheia* runs through P, 252–65.

[31] The reference is to *Pythian* 3.

momentary self-elevation [such hope] represents the elevation of those who have been elevated above the ephemeral thanks to divine assistance' (P, 382). Such assistance can be looked for in prayer and received in constant gratitude, and it can be bestowed on both individuals and communities as well as relating both to this life and the life to come.[32] As Pindar states in *Isthmian* 8, 'A man must cherish good hope' (P, 391). This, to Theunissen, suggests both that we may hope *and that we ought* to do so and that the 'may' is underwritten by the 'ought' since, in the final accounting, our hope is a response not only to the negative aspect of life in time but also to what the gods have done for us (P, 392).

But how does divine action manifest itself in relation to human life? One kind of manifestation is what Theunissen calls 'the sudden', as when the irruption of new possibilities and the chance to surmount negativities comes in a moment of 'fateful time'. If 'the day' previously indicated the limitation of human powers, the hero or athlete now experiences the 'day of destiny' in which he is to achieve success as a privilege. It is the opportunity to live a turning in time and to transcend the fate that hangs about everyday experiences of ephemeral life. As such, and even if in the first instance this relates only to possibilities of exceptional achievement, it testifies to a more general possibility for the transcendence of time. It is, in a phrase Theunissen takes from Hegel, 'the lightning flash that reveals a new world' (P, 403). In Plato's expression, it is 'the sudden, *to exaiphnēs*, something 'that intersects time in the midst of time, vertically, from above' (P, 403).[33] And when the gods break *in* to time, humans can break *out* (P, 433). Poetically, this is articulated in images of the 'swiftness' of the gods, as Aphrodite comes quickly to Sappho (P, 414). Such rapidity constitutes the very being of the gods, for whom there is no gap between willing something and accomplishing it. As Pindar states in *Pythian* 9.67–8, 'Swift is the accomplishment once gods are in haste, and short are the ways' (P. 415). This is not the 'no time' of Plato (still less the timelessness of Christian theology), yet it is a kind of poetic approximation to that time, an acceleration rather than a negation of time, a transcendence that is also a turnaround in time itself (P, 418). Programmatically, it can be said that whereas prior to Pindar such a turnaround in time was normally construed negatively (as a change of fortune from good to bad), Pindar accentuates the positive possibilities it offers. Once more, we see that it involves the interplay of passive and active elements. Or, as Theunissen puts it in grammatical terms: we translate the presence of the divine irruption into human life into an a priori perfect—what the gods have always already done—as a way of expressing the 'suddenness' with which time reveals its internal alterity (P, 418). This can never be understood as some kind

---

[32] We shall return to Pindar's eschatology below.
[33] Theunissen does not refer explicitly to Barth, but the reference to a phrase that has often been taken as summarizing Barth's view of divine revelation is surely not accidental.

of temporal law: time doesn't have to change and doesn't have to present us with unexpected possibilities. Danae, adrift at sea with her infant child, has no rational expectation that Zeus might be transformed from rapist to gracious god—and yet she addresses him in a prayer full of hope (P, 425–6).

It has been mentioned that Theunissen does not see *chronos*, as found in Pindar, in as consistently negative terms as in his discussion of Parmenides (see P, 499ff.). *Chronos* is not only time but also, as time, sovereign over the lives of human beings and perhaps even of the gods, as in Pindar's *Fragment* 33, which speaks of 'Chronos, the ruler, who surpasses all the blessed ones', even Zeus (P, 485). Yet already before Pindar, Solon is making a connection between time and truth, speaking of 'the tribunal (*dikē*) of time' (P, 492), to which Theunissen comments that 'It was already the case for [Solon] that world-history was the last judgement' (P, 493). Yet, in Solon (it will prove otherwise in Pindar) the interconnection of time, truth, and justice has an almost mechanical aspect, which has led some scholars to see in it the necessity of a natural process or even a secularization. Theunissen, however, argues that even here, to the extent that the dimension of truth is emphasized, there is not only a kind of automatism but also a kind of synergy. It is not time alone that reveals truth but the god, acting in time (P, 501). And for Pindar, time will not only disclose truth but also realize and accomplish it (P, 653), testing, preserving, and maintaining it (P, 656–7).

Moving beyond mere process to the dimension of the personal, time blends into the figure of 'Lord Destiny' (*Nemean* 4.42), the power of fate (*Geschick*) that prescribes each individual's gifts and powers and circumscribes the individual in relation to others. But time as 'Lord Destiny' is also immediately linked to the future, the 'coming time' of *Nemean* 4.43, which, to Theunissen, indicates a kind of time that is no longer bounded by the fluctuating and inconstant time of 'the day' but has acquired a structural and linear orientation towards the future. Where, previously, we heard of the divine power that 'turns' time as striking suddenly, time is now said to move towards the future 'creepingly'—the verb *herpein* being used also of the movement of a serpent. When time such as this is accompanied by a god, then, the emphasis is less on how the god comes but on how he goes with those he protects, moving forward with them. But, in its fullest development, time is not solely linear but multi-dimensional, a mutual coinherence of past, present, and future, as the victory of the present repeats the mythical and historical triumphs of the athlete's ancestors and gives a promise of a good time still to come—a structure reflected in the complex narrative development of the poem as the pre-history is narrated, so to speak, backwards (P, 638). With reference to *Olympian* 10, this provides occasion to note the role of Herakles as representing both the mythical world of pre-history and the inauguration of history and the historical chronology defined by the Olympic Games themselves (P, 644). Yet even in thus establishing the historical time of a social order that secures freedom from the preceding age of arbitrary violence, Herakles' acknowledged dependence on Zeus also witnesses that all time is under the rule of the latter. The Ode thus narrates a constellation of three elements—violence, order, and grace (P, 640)—that, in the victory of Hagesidamos, is taken up into the time of a fulfilled present connecting to both past and future (P, 675). Again, this is not an automatic functioning of time, but what can happen to and with time in its interrelationship with human agency.

But the involvement of human agency signifies that the condition of violence from which the advent of social order promises deliverance is not just the violence of a state of nature. It is the wilful violence of human beings themselves, as exemplified in Herakles (again), that is responsible for the condition from which we need to be saved; and the transformation of that condition therefore also involves the need for cleansing from guilt. Subsequently, this is discussed with reference to Pindar's presentation in *Olympian* 2 of the hubristic passion of Semele and the crazed action of her sister Ino, as well as to the Oedipus cycle (P, 698–733) and the divine actions that amend the errors and outcomes of these stories. In relation to a history that is a history of guilt, it is not enough to be able to forget—the hurt must also be reconciled and made good. But this can only be the work or the result of the work of a god.[34] As such it is also a work that introduces a new dimension that may be called 'love', as in the displacement of Zeus's sexual desire for Semele with the beneficent love shown her and her son by Athena (P, 729). For Semele the outcome is that she is raised to live amongst the immortals.

In *Olympian* 2 Time itself is named the 'All-Father' and seems to be the agent of this change, bringing about 'another time' that is the 'other' of negative time (P, 731). Precisely by securing the continuum of generations through the house of Laios, time makes possible the final liberation of that house from its successive catastrophes in the figure of a descendant, Thersandros, who has escaped the concatenation of guilt and error (P, 730–6). But the victory (*tuchē*) through which Thersandros and, subsequently, his descendant Theron achieve this liberation is, again, both a human act and a divine gift. But how far does the salvific time envisaged in these Odes extend? Is our possible happiness for this life only, or does it extend into another life? And, if the just in fact fail to find happiness in this life, what, if any recourse might they expect?

Such questions find their natural focus in the eschatological passage of *Olympian* 2 where Pindar speaks extensively of the afterlife and the judgement that awaits us there. Theunissen emphasizes from the outset that, whatever else we make of this, it is speaking of a future life and therefore of a kind of time, as opposed to liberation from temporality as such (P, 741). The poem speaks of how the good are 'forever having sunshine in equal nights and in equal

---

[34] Again we might think of the analogy with Bultmann's account of the interrelationship of temporality, guilt, and forgiveness discussed in n. 32.

days' (*Olympian* 2.61–2), on which Theunissen comments that this 'mystical equinox' (P, 759) marks a qualitative transformation of the negative features of ephemeral life such that even the nights are now illuminated by the sun, a poetic figure that does not suggest freedom from time as such but from the alternation of happiness and unhappiness. Rather than the bliss ascribed to Christian saints in Paradise this is more a matter of equanimity and content-ment (P, 761–2). And it certainly can't be taken as indicative of translation into a timeless state of existence.

Although the passage could easily be read as a reversion to myth, Theunissen maintains that it is in fact profoundly anti-mythical and even the expression of a 'proto-philosophical faith' (P, 747). Why? Because the emphasis is not on what we might call the mythical furniture of the narrative but the attainment of insight into what Fichte would call a 'moral world-order' (P, 747—the refer-ence to Fichte is mine, not Theunissen's). It is about accountability and, as such, adumbrates a universal possibility concerning liberation from the guilt that is woven into historical existence. In a much-discussed passage that speaks of 'those with the courage to have lived three times in either realm, while keeping their souls free from all unjust deeds' travelling 'the road of Zeus to the tower of Kronos' (*Olympian* 2.68–71a), Theunissen sees the obscure image of the tower of Kronos as representing 'an ideal and primordial situation' and 'a sym-bol for that towards which *chronos* surmounts itself and into which it is there-with transformed' (P, 772). On the one hand, the way to the tower is linear and therefore historical, requiring effort on the part of humans, whilst on the other it is said to be 'of Zeus'. In this last respect, to be on this way is therefore paradoxically to have already arrived at the goal, since it means moving from a time that is 'merely' historical and, as such, characterized by chronic injustice to a time that is under the benign governance of Zeus and therefore leading to the future fulfilment of a Kingdom of justice. But, as the examples of Peleus, Kadmos, and Achilles show, ultimate acceptance into Elysium is secured not by our own merits but by the intercession of another—and, for Theunissen, it is especially significant that, notably in the case of Achilles, this is the interces-sion of his mother, who, as the 'other' of the Father figure of Zeus is also the manifestation of love and loving favour (P, 777). Righteousness requires the supplement of grace (*charis*), as merit requires the supplement of hyperbolic joy (P, 778ff.). For Theunissen, then, the whole poem moves towards the need for *charis* and to a model of the good life that not only incorporates the moral and political virtues of kindness to strangers and good deeds to one's allies but also a comportment that brings joy to others (P, 783).[35]

---

[35] Inevitably, a theologically-trained reader will see this discussion as broaching classical theo-logical debates about merit and grace. Theunissen does not explicitly mention these, although it is clear that he is aware of them. If this suggests the charge that he is importing a Christian theological agenda into his reading of Pindar, he would presumably respond by commenting that the interesting question here is how precisely these 'Christian' themes are in fact grounded in a

Still, however, we might be uncertain as to just what is involved in making the transition from negative time to saving time and, to help clarify this, Theunissen turns to the role of *kairos* in Pindar.

KAIROS

Kairos is a term that has played a significant role in the history of twentieth-century theology, becoming prominent in early dialectical theology and being further developed in works such as Oscar Cullman's *Christ and Time* and, somewhat differently, in the theology of Paul Tillich.[36] We have already seen how Tillich takes the term from Jesus' preaching that 'The time (*ho kairos*) is fulfilled, and the Kingdom of God has come near; repent, and believe in the good news' (Mk 1:15) and extends it to a more general idea of time as 'the right time, the time in which something can be done' in contrast to the 'measured time or clock time' designated by *chronos*.[37] An eminent instance of such a 'right time'—the eminent instance, we might say—is the historical moment of the Incarnation itself, 'the moment which was selected to become the centre of history'.[38] Nevertheless (or rather, precisely on this basis of this exceptional moment), this 'great *kairos*' 'is again and again re-experienced through relative "*kairoi*," in which the Kingdom of God manifests itself through a particular breakthrough'.[39] And, each time, 'Awareness of a kairos is a matter of vision... It is not a matter of detached observation but of involved experience'.[40]

All of this could be said also of the *kairos* that Theunissen finds in Pindar.[41] The leading image of his interpretation is that of 'hitting the mark' or 'aptness'

---

universally human set of experiences. If well-grounded, such a claim is, of course, controversial for those who see the dialectic of merit and grace as distinctively and uniquely Christian. It is, to borrow the title of a collection of essays by Simone Weil, a matter of 'Intimations of Christianity amongst the Ancient Greeks'.

[36] See the section 'Tillich: From Utopia to Ontology' in Chapter 6.

[37] Paul Tillich, *Systematic Theology* (One Volume Edition, Welwyn Garden City: James Nisbet, 1968), III. 394. Cf. also Paul Tillich, *The Interpretation of History*, tr. N. A. Rasetzki and Elsa L. Talmey (New York: Charles Scriber's Sons, 1936), 123–75 and the sermon 'The Right Time' in *The New Being* (London: SCM Press, 1956); repr. in *The Boundaries of our Being* (London: Fontana, 1973), 270–7.

[38] Tillich, *Systematic Theology*, III. 394.

[39] Tillich, *Systematic Theology*, III. 395. This suggests (as I believe it is intended to) that the possibility of a new relation to time revealed in the Christ-event is a possibility that both prospectively and retrospectively reaches beyond the explicit confession of Jesus as the incarnate Son of God.

[40] Tillich, *Systematic Theology*, III. 395.

[41] Again, as with the discussion of sin and merit, this could be seen as indicative of Theunissen reading Christian concepts back into his Greek sources (see n. 51). However, here too the case could be made that the analogy is best seen in reverse, namely, that the possibility of the distinctively Christian concept of *kairos* is itself a modification of a broader human time-experience.

(*Treffen*),[42] which, he says, allows for all the other meanings usually associated with it: opportunity, chance, and fulfilled time (he too refers to the New Testament's 'the time is fulfilled'). This also indicates what we might call the virtuous aspect of *kairos*: the *kairos* does not come to those not willing to engage in earnest striving (P, 787–9). Starting with the term's etymological roots in such practices as weaving and archery, Theunissen sees these as ensuring its grounding in basic material life.[43] Time can only be or become *kairos* for those who live embodied lives in the world. Hölderlin's translation of *kairos* as *Glück*, happiness, also felicitously suggests how it indicates more than the success of a consciously intended act (as in hitting the mark in archery), but may also comprise gratuitous and unintended moments of happiness, gifted, perhaps, by the action of a god (P, 791). Although not originally a time-word, its early usage, focused on the notion of 'the point' at which weaver or archer should aim, expands to embrace 'the right point' and thus the right time, measure, and, most comprehensively, justice (P, 800). As such, and mindful of its constant relatedness to material life, it is 'a relative absolute': relative qua material and contingent on opportunity and occasion, absolute qua its intrinsic rightness; and, inseparably from its 'poietic intercourse with things, [and] remaining true to the world' it may also be an 'epiphany of the divine' (P, 804–5).

Yet the connection with justice and measure, found already in Hesiod, is primarily a negative idea, emphasizing the limits of human powers and negating the impulse to excess (P, 810–11). Before Pindar the prevailing usage stays close to Hesiod in this regard. Pindar, however, whilst not neglecting the dimension of measure, comes to inflect *kairos* with a more dynamic meaning in terms of 'the divine favour of the hour and the art of being able to strike...[that] qualifies *kairos* for being a history that mediates divine and human' (P, 828). This is also an approach that combines objective and subjective elements, indicating both that the time has come when we must act and that it is, as it were, 'our' time to act (P, 821). With regard to its relation to *chronos*, the general pattern is as we have seen described by Tillich. However, it is not a relationship of simple opposition. As Theunissen states: *kairos* is 'that in time for which it is time' whilst *chronos* is 'the encompassing time in which it is time for something' (P, 823). Or, since the linearity of *chronos* time may, in a sense, contain the 'point' of kairos time, the point could also be seen as 'owning' the line in the sense

---

[42] Cf. Kierkegaard's 1849 discourse on the obedience of the lilies and the birds (SKS11, 31/ WA, 26): 'The lily and the bird are unconditionally obedient to God, and they are masters in that art. As befits those who are masters in their art, they know how to hit the mark unconditionally, something that most human beings, alas, either miss or in some way bungle...What wonderful sureness, to be able to hit the mark like that and to live one's life so unconditionally!' Translation from Søren Kierkegaard, *Spiritual Writings: Gift, Creation, Love: Selections from the Upbuilding Discourses*, ed. and tr. G. Pattison (New York: Harper, 2010), 202.

[43] In reference to weaving, it means the opening that needs to be threaded and that only appears as such for a brief moment.

that it constitutes the start or apex of the line, thus determining its course as a whole (P, 823). But this also means that the 'line' is not really thinkable *more geometrico* and *chronos* itself must have the potential to become *kairos* (P, 824).

When this occurs, we once more encounter the fundamental paradox of the relative-absolute that is *kairos*, that it is both fleeting and normative. It is fleeting: 'As soon as the high time has come, it is always the highest time to use it' (P, 824, commenting on *Fragment* 168 'there is not much *chronos* in *kairos*'). But this is no longer just because of the ephemeral and transient nature of human reality: it also reflects the suddenness of the divine irruption into time and the 'speed' of the gods' action in relation to humans (P, 827). The transcendence of *chronos* is therefore not a transcendence in the direction of timelessness but the possibility of a history that mediates divine and human and, as such, 'a category of the theology of history' (P, 828).[44] Indeed, commenting on the Kyrene myth related in *Pythian* 9, Theunissen states that 'At the extreme horizon of [this] myth, the later, Christologically deployed concept emerged' and explicitly alludes to Mark 1:15 (P, 872)—although, as he also remarks, this is, in the Greek poem, essentially marginal to its main interest.

If a life lived constantly in the light of *kairos* time would be a life that unfailingly hit the mark, it is clear that this would be exceptionally difficult to achieve. As signifying what is truly real and truly the case, *kairos* is, as it were, embedded in the opaque and shifting patterns of the ephemeral world (P, 879–82). Under such conditions, a life lived in the light of *kairos* is a life that combines courage with measure and understanding: measure limits courage from becoming foolhardiness, whilst courage raises measure from acquiescence and complacency—it is a matter of remembering both the misery and the glory of being human (P, 885) and, as at *Nemean* 7.58–60, it requires the gift of Moira, fate. In a vivid image from *Isthmian* 2.21–2, the person who combines all these elements is like the charioteer who both whips the horse and protects his chariot, acting both courageously and wisely (P, 888). But—as would be the case in Plato's charioteer imagery—the point here is not only to celebrate a successful chariot-racer but also to signal the ultimate aim of self-governance in every endeavour. This, however, is not simply the autonomy of modern humanism but is both *auto*nomy and auto*nomy*, that is, both subjectively seizing the time and objectively internalizing the measure provided by the real limitations on human action, limitations that relate both to our material and temporal situation and to our dependence on divine favour for lasting happiness (P, 899). And, again, the final issue is not simply a matter of heroic self-control, but a sublimation of passionate desire into love (P, 901). What is at issue is a '*Zeitgerechtigkeit*' or being righteous with regard to time, a characteristic that is both theoretical and, more importantly perhaps, practical

---

[44] Theunissen acknowledges that little of this is stated explicitly in Pindar, but claims that it nevertheless brings out what is implicitly or weakly articulated by the Greek poet.

(P, 909). As involving or establishing the measure of a temporal movement, such 'righteousness' is itself temporal and, if it is a relation to an absolute, it is to an absolute of the second degree, as Theunissen calls it (P, 908). The person who practises this is no longer a 'slave' of time but nevertheless remains in the *service* of *kairos* (P, 910–11). And, remembering that the real 'sting' of our entanglement in time is not the fact of mortality but our guilt for the misuse or neglect of the time we have, the ultimate gift of such service is the possibility of forgiveness (P, 917–18).

As was said at the outset, this may seem a modest outcome for Theunissen's monumental endeavours, although, as I suggested before, it is a modesty that fits a view of life construed in terms of measured courage, serving time in time and looking to find tranquil contentment in time in the practice of forgiveness. And if this seems little enough, we may say that, on Theunissen's account, it is a prospect that was unavailable to the pre-Pindaric poets, for whom human beings are inextricably entangled in the impenetrable uncertainty of ephemeral existence—and we may also say that it is a view of life that has on a grand scale escaped the human beings of the last century, a century of wars, ideologies, and totalitarianisms that has matched the reckless material self-indulgence of the few with the utter impoverishment of many, many more. In face of all the harm done this last hundred years, there remain many for whom forgiveness is not even an option to be considered. Does this, then, justify Theunissen's view that seems, like Heidegger's philosophy, to culminate in the claim that 'only a god can save us'? And what can that mean if such a god, on his account, can only encounter and transform us within the parameters of time itself? Isn't this, in the end, a demythologized divinity who never escapes the 'godless' parameters of time as the universal horizon of human existence?

Before returning to these questions, I should like first to address a theme that runs through *Pindar*, namely, how all that we have been considering is taken up into and exemplified in poetic practice itself. This returns us—from another angle—to the questions of call and promise to which we were led in the last chapter by our reflections on the possibility of hope in the face of all known historical ill.

THE POET

Comments on Pindar's concept of poetic practice are found at many points in Theunissen's study and in Chapter 2 of Part 3 there is an extended discussion of the relationship between poetic art and *kairos* that, as the chapter title states, sees *kairos* itself as a principle of poetic production—an insight that Theunissen sees as an innovation on Pindar's part (P, 817). Like those whose achievements he celebrates, the poet too must hit the mark with his words, speaking the truth (P, 834), and, at one point, Pindar likens his work to that of a spear-thrower (P, 841). To succeed he must *see* (in a sense, Theunissen says, that anticipates philosophical ideas of intuition[45]) what he is aiming at and it is this that gives due measure to his words. Citing *Pythian* 2.34, Theunissen states that the task is 'to see the measure of everything' (P, 851). As one whose task is to speak truth, the poet's role is assimilated to that of time itself, underlining the point that time does not reveal truth merely automatically but also calls for the help of human agency (P, 676). Yet the poet does not have capacities above those of other human beings; if his vision has an eagle-like power that resembles that of the gods and that therefore enables him to become a divine instrument, his knowledge of the future is limited to the proximate future of his own poetic undertaking, as a singular instance of limiting ourselves to what lies before our feet (P, 538).

The point, then, is not to indulge in hyperbolic praise (that, as several of the odes suggest, soon becomes wearisome), but to get to the root of the matter. In the Ode for Xenophon, winner of the stadion and pentathlon, Corinth is praised not only because it is the athlete's own city but because it is seen as the home of order, justice, and peace and, in all these respects, a bulwark against hubris. Even when Pindar then seems to break his own rules against excess (as when, in *Olympian* 9, he casts aspersions on mere boasting before going on to list the athlete's triumphs), this can be seen as governed by the law of measure—in this case, the triumphs of the athlete are justly celebrated if, as happens in the poem, the deeds of the gods are extolled beforehand and the achievements of men put in their rightful (i.e. secondary) place. When the gods have first been praised, *then* comes the right time, the *kairos*, to praise human beings (P, 856–7).

Of course, like other mortals, poets are fallible; and in *Olympian* 10 a major theme of the poem is that it seeks to make good on the poet's previous failure to deliver a promised Ode. Picking up on themes we have considered earlier, this provides occasion both to appeal to time as revealing truth (that the poem is, in the end, produced reveals the truth of the poet's original promise) and implicates the poet himself in the narratives of guilt and deliverance that the Ode narrates. As in other instances of liberation from the negative aspects of time, the possibilities of forgetfulness are acknowledged—but the poem succeeds *despite* them. And, recalling that human life should always aim at self-governance, the fault to be amended in this case is a fault in the poet's relation to himself (and not only to his patron), so that to make it good is also to transform his own self-relationship (P, 609). Yet (in what seems an implicit criticism of Heidegger), the story of the unfulfilled promise shows that the horizons of being-with-one-another and of temporality are inseparably fused

---

[45] Referring specifically to *Olympian* 3.48: 'it is best to recognize [*noēsai*] what is fitting'.

in human *Dasein*, and the recognition of his friend and patron's claim and the realization of being delivered over to the contingencies of time factually coincide. That the story can be turned round and made good requires the poet to act, but his action is also possible only by virtue of a power that comes over him, figured at several points in the flow of living water (P, 624[46]), an instance (this time with explicit reference to Heidegger) of 'time temporalizing itself', that is, time releasing its possibilities for transformation and salvation.

When, as it should be, it is focused on telling what is truly the case, the poet's task is mimetic. But what does this entail? The theme of poetic imitation is most extensively taken up in *Pythian* 12, in praise of the winner of the flute-playing competition. Here the poet describes how Athena herself invented flute-playing as a (perhaps to our minds somewhat bizarre) imitation of the Gorgons weeping for their sister Medusa, beheaded by the hero Perseus. In this act of transformative mimesis, the goddess turns terror to a thing of beauty (P, 470–1). But the situation, Theunissen suggests, is more complicated than art beautifying life or sublimating suffering. For, despite their terrifying appearance, the Gorgons are also acknowledged to have had a seductive allure and a strange beauty. What Athena does, then, is not simply to invert a situation of terror, but, where Perseus had transformed the beauty of the Gorgons into the terror of death and loss, the goddess acts 'poetically' in a second transformation—a pattern that can be replicated as the inverse mutual mirroring of art/life and life/death (P, 477). But this pattern is extended even further when Perseus, after his terrible act, uses Medusa's beautiful/terrible head to release his mother, Danae, from her captivity, turning death to life (P, 478). All of this, however, belongs to the goddess's music, of which the poetic myth-telling is itself a secondary imitation, so that the truth of poetic art is seen to rest on its truthful imitation of and therefore its dependence on the divine. As Theunissen puts it elsewhere (with reference to *Olympian* 10.79), it is not the natural thunder that the poet imitates but the divine thunder of Zeus (P, 680).

As the myth of the invention of flute-playing suggests, the ultimate achievement of poetry is a kind of triumph over death, which, in turn, is an eminent instance of time's negative sovereignty over human existence. The happiness accruing to the victor on account of his triumph will be washed away into forgetfulness if it is not kept alive by the poet's song (P, 252), but death is not the end if praise outlives it (P, 156).[47] Under the inspiration of the god, poetic song itself effects the transformation of negative time, the ephemeral, into saving time, a movement marked also by the multiplication

[46] The references to Pindar are to *Isthmian* 7.19 and *Nemean* 7.12.

[47] Again we might see an analogy with the Kierkegaardian thought, articulated by the pseudonym Johannes de silentio, that life would be a meaningless cycle of empty recurrence if there were no hero and no poet to sing the hero (SKS4, 112–3/FT, 15–16) and that it is precisely the poet who therefore becomes the chief witness to 'an eternal consciousness' in human being.

of time dimensions in the Odes themselves, as Pindar moves freely between past, present, and future, exploring and celebrating their mutual interdependence. Yet, again, we must remember that all of this must keep to the law of just measure and to the temporal and material conditions of human life on earth. But, when this is observed, what the poem achieves is a re-envisioning of life on earth itself. It is no longer a matter of being suspended in an alien and annihilating fog of ephemeral time, time that turns all our hopes to illusions. When *Pythian* 4 tells the story of the Argonauts' homecoming, this is the ultimate story of poetic vocation itself: 'As we follow the course of the Ode and mentally re-enact it, we accompany the poet as he raises himself above the limits of human *Dasein* and makes his journey home. Having transcended the ephemeral, Pindar returns to the ephemeral and from poetic hyperbole he returns to modest words and from the vision of a beyond comes back to this world' (P, 922). Thus returning home to the material, concrete, and particular reality of everyday life, the poet's song can provide a point of rest that, at the same time, incorporates the infinity of the journey, providing 'a temporary refuge in the world as it is' (P, 922).

The achievement is avowedly modest. The poet's public role and the view of his poetic art as offering a mimetic transformation of divine reality add up to a stronger claim than Jankélévitch's nostalgic and essentially private musical soirées, but what authorizes this claim? Is there anything beyond the charm of the poetry itself? After all, there are innumerable forgotten heroes whose poets are as forgotten as they are. Pindar's heroes might be seen as just plain lucky to have found a poet who really could sing their names into the collective memory of humanity! Of course, the same objection could be extended to any of our attempts to resist oblivion, such as Berdyaev's creative memory. Surely its reach is only as far as those who are doing the remembering are prepared to take it. And at one level this would seem to be simply and obviously true. If we all just gave up on religion and philosophy and watched TV instead, both religion and philosophy would perish from the world. Both necessarily involve human commitment and human activity. But most believers and most philosophers believe that, nevertheless, their faith and their work is a response to something more than their own fancies. Theunissen's Pindar is a poet who speaks the truth about time and about human life in time and this poetically articulated truth is offered as a mimetic reproduction of the thunder of Zeus. And if he proposes that we take a hopeful view of life, this is because he also believes that this is a right attitude to take to life—a matter of what Theunissen calls *Zeitgerechtigkeit*. We *ought* to be hopeful. But what is going to make us think that his words are really anything more than a poetic rendering of the thunder of Olympus? Perhaps it is simply that the poet shows himself to be an especially careful observer of the truth of the human condition, in which case his claims might be better seen as a kind of proto-philosophy, that is, a judgement on what is actually the case. But then it seems that there is nothing the poet can promise beyond what we are all already capable of knowing if only we cared to look more carefully at how things are with us. So just how much can the poet qua poet promise? *Olympian* 10 is written as making good on a promise (the poet's promise to write in praise of the victor), but this is a promise between one human being and another. Can the poet promise anything more than a mortal word concerning the meaning of mortality or point us towards anything greater than the effective power of his own genius? Or is it all just 'poetry' (in Kierkegaard's negative sense of the aesthetic, perhaps)?

HEIDEGGER, HÖLDERLIN, AND THE GREEKS

In an appendix to *Pindar* (although one is also tempted to see it as philosophically decisive for the intention of the work as a whole), Theunissen turns to consider 'Heidegger, Hölderlin and the Greeks'.

The discussion of Heidegger focuses on a 1946 paper on a saying of Anaximander that speaks of all that is being judged 'according to the order of time'.[48] This gives Heidegger occasion to develop what will be a hallmark of his later philosophy, namely, the reinterpretation of Being in terms of 'the presence of what presences' ('*das Anwesen des Anwesenden*'[49]). But if *Anwesen* 'is' Being (and not merely a means by which Being manifests itself), it is also intrinsically connected with truth, time, and *logos*. It is connected with truth in the specifically Heideggerian sense of 'unconcealment': the coming into view of what is unconcealed and is 'there' for us to see is precisely what is meant by presencing. It is also indissociable from *logos* since thinking (assumed to occur eminently in and as *logos*, discourse) belongs to Being as a necessary condition of presence becoming present. And, finally, both truth and *logos* as the presencing of presence are time-words. In combination, the outcome is the abiding of human thought with what shows itself, in *logos*, as present to it.

---

[48] Martin Heidegger, 'Der Spruch des Anaximander', in *Gesamtausgabe*, v: *Holzwege* (Frankfurt am Main: Klostermann, 1977), 321–73; Eng. 'Anaximander's Saying', in Martin Heidegger, *Off the Beaten Track*, ed. and tr. Julian Young and Kenneth Haynes (Cambridge: Cambridge University Press, 2002), 242–81.

[49] The term '*Anwesen*' has rich connotations for Heidegger. '*Wesen*', customarily translated as 'essence', is reconnected to its verbal root by Heidegger, as when it occurs in the past participle of *Sein: gewesen*, allowing him elsewhere to say '*Wesen ist was gewesen ist*', that is 'Essence is what has become', thus resisting thinking in terms of of timeless essences. But he also coins his own verbal form '*wesen*' ('to essentialize', perhaps?), giving '*das Wesen west*', 'essence essences'—where, again, the point is to highlight the fundamental temporal and processual character of Being. And, as Theunissen comments, Heidegger also avoids '*Anwesenheit*' as a substantive form, since this would again conceal the verbal and temporal root-meaning.

Yet—and in a departure from the usage of *Being and Time*[50]—Heidegger also defines 'presence' in terms of *Gegenwart*, which also has the dictionary meaning of 'the present' but which, as Theunissen points out, is essentially a spatial rather than a temporal word, suggesting the kind of being proper to a localized entity. By thus privileging the spatial and non-temporal term *Gegenwart* as the decisive instance of presence, Theunissen suggests that Heidegger undermines his own thesis that time is the horizon of Being. Indeed, Theunissen suspects this is already the case in *Being and Time*, since despite the closing rhetorical question suggesting that time is the all-encompassing horizon for the meaning of Being, time itself has been consistently interpreted through the experiences of the specific being of human *Dasein*, that is *Da-sein*, being-there—again an essentially spatial expression.

Through the post-War period Heidegger's own use of '*das Anwesen des Anwesende*' is variable, and Theunissen discerns four contexts that nuance Heidegger's meaning: (1) in relation to early Greek thought, (2) as the key to classical Greek philosophy, (3) as the hallmark of the entire tradition of Western metaphysics, but also, finally, (4) as the movement that will overcome metaphysics. Such diversity of usage generates what we might call an internal conflict of interest in the expression, according as to whether it is the processual or the spatial element that is most to the fore. But, even where Heidegger stresses the processual (which he—unlike Theunissen—finds in Parmenides), this is reduced to a present devoid of past and future. This becomes particularly clear when Heidegger calls on the testimony of the ancient Greek poets (though never on Pindar), as in his discussion of the Homeric seer, Calchas. Heidegger interprets the kind of vision ascribed to Calchas as anticipating the philosophical vision of the 'all of what presences' and therefore a knowing that connects to both Anaximander and, ultimately, Parmenides. In the process of this 'homogenization' the seer's relation to both future and past is obscured in favour of a focus on the present qua *Gegenwart*. It is not (as Heidegger assumes) that past and future are taken up into presence but, according to Theunissen, that Calchas' authority rests precisely on his ability to anticipate the future and his knowledge of the past, neither of which are reducible to each other or to the present; 'To this extent the present, as he sees it, stands in the shadow of what is to come and what is past' (P, 940–1). But this homogenization also reveals that Heidegger overlooks the way in which Parmenides' 'No' to time is also a 'No' to the poetic world-view.

For Heidegger, deliverance from the errancy of metaphysics by a 'new' or 'other' beginning of philosophy is simultaneously a return to the Presocratics and a turn to the poetic world of Hölderlin who, within the era of metaphysics, has already thought—in connection with his own dialogue with the

---

[50] Where it relates to the (inauthentic) perceiving of beings as present in the manner of what Heidegger calls presence-at-hand (P, 933).

Greeks—beyond metaphysics. In continuation of the theme of the presence of presencing, Heidegger takes this as the key to Hölderlin's poem '*Wie wenn am Feiertage...*' ('As on a Feast-day'). As read by Heidegger, the poem celebrates the 'gathering' of nature in the poetic word and nature in turn is understood in the light of the Greek *physis* as what gives the power of presencing to all that presences. This power is also named by Hölderlin as 'the Holy'. On Heidegger's interpretation, however, 'the Holy' does not refer to anything divine but is instead a divinization of nature. As Theunissen puts it, 'The Holy is not holy because it is divine, but the Divine is divine because, in its own way, it is "holy"' (P, 946). Following Heidegger's logic, history is grounded in nature and nature, as presence, is grounded in Being.

In Theunissen's eyes this turns Hölderlin's poem on its head, since Hölderlin grounds nature in history, not Being, and, even more specifically, in the historical moment of his own poetic writing. When he writes '*Jetzt aber tagts*' ('Now day is breaking') it is the French Revolution and not Being that he has in sight (P, 947)! Even as a translator of Pindar, Hölderlin does not aim at reviving some kind of Hellenism and is clear as to the difference dividing the ancient and modern worlds. Like the tragedians, he knows of time's lacerating power and its futility, which he too sees as 'the spirit of nature and time' (P, 965) and as sovereign over human life to such a degree that he can also say that the human being is 'nothing but time' (P, 965). But even in finding ourselves thus handed over without reserve to time, he nevertheless also sees time as a medium in which humans and gods can encounter one another, as when Danae turns in prayer to Zeus and, implicitly, as the poet also does when he imitates her action in his verse (P, 965). Even if moments of dereliction may lead to talk of a *deus absconditus*, an absent god, the elementary possibility of some kind of relationship between humans and gods is also the possibility of humans and gods developing a common history. Where the tragedians saw the immediate presence of the god as annihilating the human, Hölderlin, as a modern, no longer looks for such immediacy. But nor does he look for some kind of Hegelian mediation. Rather he looks to the culture in which a historical people encounters itself (P, 977).[51] The liberation offered by such cultural life is therefore also a temporal liberation and the God who makes himself known in it as a liberating power is also therefore a temporal God, a 'God of time' who can be known only in the goal, the measure, and the moment (the *kairos*) (P, 970). Recalling what was said about Heidegger's suggestion, commenting on Hölderlin, that the new kind of time-experience voiced in the poem might enable a new thinking of the eternal, Theunissen's criticism, if correct, would suggest that Heidegger himself cannot attain to such new thinking. At most, a

---

[51] Hölderlin's '*Zucht*' is stronger than '*Bildung*', often also translated as 'culture' (or 'formation'). *Zucht* suggests also breeding and discipline—a *Zuchthaus* is a prison or penitentiary, implying cultivation through rigorous and painful learning and self-amendment.

Heideggerian eternity might be an abiding in a constant presentness that, if it is not a *nunc stans* (i.e., if it is not perpetually the same), is nevertheless always immanent to the moment that now is. As such, it is a moment defined in terms of Being, and cuts off the possibility of time becoming radically other to itself through an irruptive and transformative power from beyond the present, a time coming from beyond any time that we can measure. The word of the poet is, in the last resort, unable to break the spell of ontology.

In terms of Theunissen's discussion, however, we now ask again: what kind of promise is Pindar's poetic hope able to give if human beings are 'nothing but time'? If Theunissen points us towards a saving time whose advent cannot be measured by the chronology of historical time, this is still a time that can only be experienced in time, as time, temporally, and it is certainly not the time of a timeless eternity. Is this enough for it to count as 'saving time', that is, time that saves and that saves time from its own ceaseless vanishing? But how much are we asking for here? How bad must a situation be to be bad enough to need saving from, and how good must a turnaround in our experience of time be for us to call it 'saving'? And if we find Theunissen's saving God too modest, does that say more about us than about such a saving God and, if so, is it entirely to our credit? Should we want more than a *kairos* that gives good measure and that is 'fitting'—the right time—for all our purposes under heaven? Wouldn't such 'wanting' be a paradigmatic case of hubris? If time is indeed the horizon of our being, how might we look for a salvation from all that oppresses us in time elsewhere than in what comes to us as an event in time? Would we ever *really* want to?

And whatever we might make of the modesty of Theunissen's claims, it is important to underline that the kind of hope he describes is not just a matter of an individual surmounting of fate but, in the complex interplay of poet, dedicatee, and community, is also significant for the wider society. In this regard, the triumph of the hero is also a kind of pledge for the future well-being of his city. And that his success is also success in a competition amongst friendly Hellenic states suggests that it may also hold a certain promise for the future relationships between as well as within communities. Probably this falls short of any kind of Messianism, even the philosophers' weak Messianism. Yet it might also provide a context in which the *kairos* of a Messianic hope 'for all' might first gestate.

Theunissen does not discuss the kind of experiences of 'timeless moments' or of 'the eternal now' that we considered briefly in Chapter 3.[52] Yet he provides a perspective in which these receive a new significance. As, this side of the eschaton, we take up the task that each day or each hour lays upon us, such experiences may serve less as evidence of our immediate relation to a timeless

---

[52] See the section 'The Eternal Now' in Chapter 3.

absolute and more as awakening us to time's depth and complexity, to the truth that time is more than 'just' time. Disrupting the hegemony of linearity, they open the way to a greater fullness of time, to what the title of this study refers to as 'saving time'. But if, however minimally, we experience such saving time as, specifically, eternal time, then, conversely, we establish the basis for how the eternal God may also be, for us, a saving God.

Many questions remain hanging. We need to probe further the mutual limits of poetic and religious promise as well as to examine how this might be effective in both individual and communal life. It is it is in the cause of such further exploration that we now move on to Kierkegaard and Rosenzweig, each of whom also straddles the complex boundary area between philosophy, the poetic, and religious life.

8

Kierkegaard: The Eternal Gift of Time

In Chapter 3 we encountered Kierkegaard as anticipating the polemical contrast of time and eternity that found expression in Barth's commentary on Romans, and this contrastive view of time and eternity has often been taken as 'the' Kierkegaardian view. In reality, however, it is only part of a complex dialectical picture: Kierkegaard may also be read as an example of what it might mean to follow up on Heidegger's hint that to think of time as the horizon of the meaning of being does not immediately render all talk of eternity meaningless but simply requires us to start thinking of eternity in terms other than either timelessness or sempiternity. In this chapter I shall provide evidence for the view that this is just what Kierkegaard does and that the consequences of reading him in these terms is crucial to our entire conception of his thought. This has not always been recognized. As Arne Grøn has written 'Time and history do not figure prominently in the literature on Kierkegaard'.[1] Yet, as he also goes on to say, 'Kierkegaard's key concepts can only be unfolded through the questions of time and history', which are therefore 'of critical importance for understanding what is at issue in Kierkegaard's key concepts as, for example, selfhood, existence, the ethical, faith, and love'.[2] As this brief list implies (and, as we shall see, it is not exhaustive), time—and therefore also eternity—is not one theme amongst others but a unifying and decisive one. In support of this last claim, I shall in this chapter draw on both pseudonymous and signed, early and late works, arguing that whatever the difference between genre and periodization, Kierkegaard has, in this respect at least, a coherent and consistent teaching. With the critical account of German Idealism's conceptualization of existence *sub specie aeternitatis* in the background, I begin with a passage that has been amongst the most influential of all his writings on the development of both philosophical and theological conceptions of time, from *The Concept of*

---

[1] Arne Grøn, 'Time and History', in John Lippitt and George Pattison (eds), *The Oxford Handbook of Kierkegaard* (Oxford: Oxford University Press, 2013), 273.
[2] Grøn, 'Time and History', 273–4.

Anxiety, before going on to show how the position developed there is reflected in other pseudonymous works, both early and late.[3]

THE ETERNAL AND THE SELF

In Chapter 4 we saw how Kierkegaard set out time and eternity as mutually exclusive, mutually negating powers. Yet if Kierkegaard insists that God/the Eternal is in some sense external to the human self, there is also a sense in which he supposes that a relation to the eternal is also internal to the self, a defining factor in the possibility of our coming to be selves at all.[4] We shall now explore what this means by examining the account of 'the moment' in *The Concept of Anxiety*, the notion of repetition, and, *ex negativo*, a selection of accounts of despair found in both pseudonymous and upbuilding works.

The key points are outlined in just a short section of Chapter 3 of *The Concept of Anxiety*, 'Anxiety as that Consequence of Sin which is the Absence of a Consciousness of Sin'. In the previous two chapters, Kierkegaard has offered a revisionary interpretation of the Christian narrative of Adam's fall and of the consequences of this fall—'original' or 'inherited sin'—for subsequent generations.[5] This has involved locating the fall in the emergence of human freedom from the indeterminate continuum of animal life, a freedom that nevertheless characteristically 'falls' by failing to sustain its own movement towards becoming capable of governing and directing the whole of its life, inclusive of both body and soul. In short, it fails to become the spiritual existent it has the potential to be. But this failure is not something transmitted quasi-biologically from Adam to subsequent generations. Rather, it is a failure grounded in the experienced anxiety evoked by the fact that emergent freedom is, in an objective perspective, 'nothing', since what it is directed towards are only possibilities—what it may do or might become but not what it already 'is'. The only difference between Adam and subsequent generations is that Adam's is a purer

---

[3] I am more than usually aware in this chapter of by-passing much important secondary literature, although my reading is, of course (and with gratitude) strongly informed by what I have learned from others. But as this is not a book about Kierkegaard or for Kierkegaard scholars, I have had to limit myself to what is strictly essential.

[4] A good discussion of the founding role of eternity in the constitution of the self remains Louis Dupré's article 'Of Time and Eternity in Kierkegaard's *Concept of Anxiety*' in *Faith and Philosophy* 1/2 (April 1984), 160–76. Dupré especially highlights the connection between Kierkegaard's treatment of the self and Schelling's *Freiheitsschrift*, discussed in the section 'Ur-time and Eternity' in Chapter 2 of this volume.

[5] For discussion of Kierkegaard's view of the fall see Chapter 6 'Sin', in my *Kierkegaard and the Theology of the Nineteenth Century* (Cambridge: Cambridge University Press, 2012), 124–49. This is itself a 'revisionary' reading in that it goes against the grain of much Kierkegaard-interpretation, for which I there provide references.

anxiety, relating only to his own unique possibilities, whereas for others it is an anxiety quantitatively induced by the knowledge that we stand in a history of misdirected, abused, or failed freedom. Even so, our historical inheritance does not condemn us to fail. It can at most provoke an anxiety *like* Adam's and, *like* Adam, each of us falls, if we fall, through our own failure in freedom. In this chapter, however, Kierkegaard turns to the self that has not yet identified itself as a sinner or brought the specifically Christian account of sin to bear on its own situation. This is important, amongst other reasons, because, as we shall see, it directs us towards a dimension of selfhood that lies below the threshold of articulate self-consciousness and it is in this dimension that the interplay of time and eternity first makes itself manifest.

Kierkegaard presents the human being as a synthesis of body and soul (where 'soul' indicates the affective or emotional aspect of the self that is shared with non-human animals) sustained (or, in the situation of sin, not sustained) by spirit or freedom. But how might such an entity or self change its state? How might it 'become' a sinner—or 'become' anything? Hegelian philosophy answers this question by using the notions of negation, transition, and mediation to establish a principle of movement in the basic logical structures that underlie all appearances, although Kierkegaard thinks this is illegitimate and that Aristotle is closer to the mark when he speaks of kinesis as the change from possibility to actuality.[6] However, in distinction from Aristotle, Kierkegaard is focused on existential relations and on the kind of transition involved in causality through freedom. In this case the transition from one state to another is never a matter of pure automatism but, as he puts it, a leap (SKS5, 388/CA, 85)—just as Adam's fall qua free act was also a leap. This must be so if what is to emerge as the result of a free action is genuinely novel. If we do not hold on to this, he suggests, 'then the transition [at issue] becomes predominantly quantitative and loses the elasticity of the leap' (SKS, 388/CA, 85). So, how might this 'elasticity' be preserved?

Kierkegaard's presentation is almost aphoristic, yet it is possible to reconstruct a clear line of argument running through it. Thus, when he abruptly starts speaking about a new synthesis, the synthesis of the temporal and the eternal, I take this to be offered as a way of understanding freedom's leap in such a way as not to lose its 'elasticity'. How? Because when it is considered under the aspect of time/eternity, freedom cannot congeal into a cipher of Being.[7]

But what of the objection that time does not provide an adequate basis on which to establish any lasting truths, including lasting truths about the human condition? Kierkegaard acknowledges this when he says that time has been

---

[6] Kierkegaard discusses these issues more extensively in *Concluding Unscientific Postscript*. See especially SKS7, 274–314/CUP, 300–42.

[7] As, in a journal entry from 1850, he says happens in Schleiermacher—cf. SKS23, 58 (NB15:83)/JP4, 3853.

regarded as an infinite succession of vanishing moments that cannot even provide a basis for the division of time itself into past, present, and future. Time is
sheer flux, endlessly flowing away into nothingness. How, then, can we establish a point within time—a moment of presence—that might enable us to start
structuring this structureless flow? One answer is that of Christian Platonism
(as in Boethius), namely, to propose an idea of the eternal as an absolute and
fulfilled present that gives meaning to time as a whole (SKS5, 389/CA, 86).
The pure presence of God in the moment is a presence without past or future
(SKS5, 390/CA, 86). But if this Christian Platonic solution marks an advance
on that of Hegel (since it is at least speaking in existential terms), it cannot
be the ultimate solution. This is because—or so Kierkegaard now suggests—it
perpetuates the perception that time, purely as such, lacks any specific content.
If, however, the eternal is to be exerienced together with time, as it must be if
it is to be known by temporal beings such as we are, this will be in the kind of
intuition that Kierkegaard calls the 'moment of vision'. 'If time and eternity are
to be in contact with one another then it must be in time and this brings us to
the moment of vision' (SKS5, 390/CA, 87). In a significant shift of terminology,
Kierkegaard at this point moves away from the Latin 'moment' he has been
using to the Danish *Øjeblikket* (cf. German *Augenblick*), literally 'the glance of
an eye'.[8] This will prove to be his preferred term for the 'moment of vision' that
does justice to the congruence of the temporal and the eternal.

Kierkegaard immediately concedes that the 'moment of vision' is a highly
figurative term—and not only concedes but also exploits its metaphoric force.
He offers a poetic example, familiar to his first readers though less so to us,
of a scene from Frithiof's Saga in which the heroine, Ingeborg, watches her
beloved being carried away across the sea and knowing—as he does not yet
know—that he will be assassinated and their separation be forever.[9]

> When Ingeborg thus looks out at sea after Frithiof, then this is a figure of what
> the figurative saying means. If her feelings were to break out in a sigh or a word
> the sound [as opposed to the purely pictorial quality of the image] would already
> have more of the character of time as a vanishing present. Similarly a sigh or a
> word, etc. has the power to help rid the soul of what weighs upon it but precisely
> because the weight is uttered it already begins to belong to the past. The glance is
> therefore a characterization of time, but, n.b., of time [as seen] in a fateful conflict,
> when it is touched by eternity (SKS5, 390–1/CA, 87).

By not being spoken the burden of the glance maintains the tension of the
spring, something Kierkegaard also sees in the Platonic *exaiphnēs* (SKS5,

---

[8] He defines the Latin 'momentum' as relating only to time as vanishing and insubstantial.
[9] For a full discussion of this passage and an exceptionally insightful exposition of the treatment of time in *The Concept of Anxiety* that also makes important connections to Heidegger
and to Lévinas, see Niels Nymann Eriksen, *Kierkegaard's Category of Repetition: A Reconstruction*
(Berlin: de Gruyter, 2000), 65–81.

901/CA, 88), saying too that 'Nothing is as rapid as the glance of an eye and yet it is commensurable with the fullness of the eternal' (SKS5, 390/CA, 87).[10]

In one of the many aphoristic turns of phrase that are typical of this chapter, Kierkegaard declares that 'Understood like this, the moment is not really an atom of time, but an atom of eternity. It is eternity's first reflection in time, its first attempt to, as it were, bring time to a halt' (SKS5, 391/CA, 88). To the limited extent that the Greeks did grasp this, he says, they thought of it retrospectively, not prospectively, as a moment of recollecting eternal truth, so that, Kierkegaard adds, 'justice is done neither to time nor to eternity' (SKS5, 392/ CA, 88).

We are now in a position to see why Heidegger might have concluded that Kierkegaard's concept of the moment of vision is defined in terms of the 'now' and 'eternity'.[11] However, we can also see how Heidegger may be misreading Kierkegaard at this point, since, albeit somewhat convolutedly, Kierkegaard seems to be taking pains to distance himself from any concept of the 'now' as a vanishing moment and from an 'eternity' that would be timelessly available for recollection. Although he speaks of eternity as attempting to 'bring time to a halt', what happens in the moment of vision does not strictly mean time being brought to a halt but, more precisely, becoming inflected otherwise than as pure vanishing. As Kierkegaard goes on to say, 'The moment of vision is that twofold [state] in which time and eternity touch one another and whereby the concept of temporality is established, in which time constantly cuts short eternity and eternity constantly permeates time' (SKS5, 392/CA, 89). It is therefore not as if, prior to the moment of vision, we already knew what time and eternity were and then merely put them together in the moment of vision; rather, the moment of vision is precisely the productive centre from which time is set free to become time and eternity to be eternity. And where the Greeks (as understood by Kierkegaard) represented eternity as in the past and therefore the object of recollection, the logic of the moment of vision must be thought prospectively, as futural. 'For in a sense the future is the whole of which the past is a part and the future can in a sense signify the whole. This is because in the first instance the eternal means what is to come, or, what is to come, the future, is the incognito in which the eternal, as incommensurable with time, nevertheless seeks to preserve its intercourse with time' (SKS5, 392/CA, 89). In phrases reminiscent of that most-quoted of Kierkegaardian lines, that life is lived forwards and understood backwards, he here suggests that the Greek eternity is only attainable backwards whereas his own concept, repetition,

---

[10] We recall the emphasis on the *speed* of the eternal's appearance in time in Theunissen's interpretation of Pindar, again with reference to the Platonic *exaiphnēs*.

[11] Martin Heidegger, *Gesamtausgabe*, II: *Sein und Zeit* (Frankfurt am Main: Klostermann, 1977), 497 (Eng. *Being and Time*, tr. John Mcquarrie and Edward Robinson (Oxford: Blackwell, 1962), 338).

signifies that we are to approach eternity forwards, through our own temporal existence (SKS5, 393/CA, 89–90). This shortcoming in the Greek conception is associated both with the sightlessness of Greek sculpture, a plastic form that lacked the glance (SKS5, 391/CA, 87),[12] and with the Greeks' identification of temporal life with sensuousness, that is, with a life that is not ordered by or subject to a spiritually internalized understanding of the relationship of time and eternity.

As we have seen, Kierkegaard views human life as a synthesis of body and soul united by Spirit. But how does this threefold synthesis relate to the synthesis of time and eternity in the moment of vision? The latter is not another synthesis, Kierkegaard says, but coincides with the first in such a way that 'as soon as Spirit is posited the moment of vision is there' (SKS5, 392/CA, 88): Spirit and temporality are interdependent. To exist as Spirit is to exist as temporal, in other words as having a history—and to be able to have a history is to exist as Spirit. And to exist as Spirit is also to exist as open to the future, in relation to the eternal, and as having significant existential possibilities: 'the future is the eternal's (freedom's) possibility in the individual experienced in the mode of anxiety' (SKS5, 394/CA, 91). In this aspect, to be anxiously expectant has a positive meaning that Kierkegaard calls 'anxiety as saving through faith'.

THE TIME OF THE AESTHETIC

Plainly, not all succeed in living by the light of this positive anxiety, otherwise Kierkegaard's assessment of sin as the typical human condition would be misplaced. But if Kierkegaard can use the traditional Christian vocabulary of sin to describe lives that fail to seize the moment of the eternal in time, he also interprets the prevalent contemporary form of sin in terms of his distinctive category of the aesthetic. In *The Concept of Anxiety* Kierkegaard alluded back to his earlier work *Repetition*, and this sometimes humorous, sometimes tragic take on the problem of temporality and of movement in time may be a good place to start looking at how aesthetic existence arises as a failure to exist in time as open to the eternal.

Repetition starts by citing Diogenes' practical refutation of the Eleatic philosophers' denial of movement by simply walking up and down, thus indicating the limits of any purely theoretical response to the problem. Against the Greek privileging of recollection (*anamnēsis*), Kierkegaard declares that '[repetition] will come to play as important a role in modern philosophy since repetition is a decisive expression for what "recollection" was for the Greeks' (SKS5, 9/FTR,

---

[12] It was not generally recognized at that time that the sculptures of the ancient world were in fact painted and would therefore not have appeared eyeless, as they do to us.

131). In the light of the subsequent development of modern philosophy this may seem like a misplaced expectation, and even in Kierkegaard's own writings the term itself only rarely appears again. However, if we look beyond terminology, it is clear that 'repetition' plays into the central exploration of relations of time, existence, and eternity that we have already encountered variously in the *Postscript* and in *Anxiety*; and as for its later role in modern philosophy, the issues associated with the concept 'repetition' will also prove decisive—with or without direct reference to Kierkegaard—for Nietzsche, Heidegger, and those whose philosophical reflections are guided by them.[13]

As *Repetition* develops it works through a variety of situations that illustrate different ways of understanding what it might mean to base life on repetition—and what it would mean for existence if repetition is not possible. The pseudonymous author Constantin Constantius takes a journey back to Berlin to see if he can recapture the pleasures of a previous visit—but fails. A young man falls in love, breaks off the engagement, and flees, but, in his self-imposed exile he begins to wonder whether he might, after all, take up the relationship again and overcome the problems that prevented him from becoming a husband. He meditates on the example of Job, who lost everything, but then got everything back. His own plans, however, are frustrated by the simple fact that his former fiancée is now married to another. Here too, it seems, is failure, although he maintains that he has found repetition by once more becoming the free-spirited poet he was before he fell in love.

In an unpublished response to a review that mistakenly assumed that the idea of repetition referred to the cycles of nature and the celestial bodies, Kierkegaard spells out that repetition is not at all intended to be understood as or in relation to any natural phenomenon: it is a phenomenon of freedom and does not just happen to us, as growing up and growing old happen to us (SKS15, 61–88/FTR, 281–325). To find repetition is to find, through freedom, the continuity and consistency that are the foundation of personal life: it is to be able *to be*, in time, to be someone, a person who is irreducible to the products of biology or environment. It is what another pseudonym calls acquiring an inner history (SKS3, 136–7/EO2, 137–8). But if, as the book *Repetition* suggests, repetition is not possible, then life is delivered over to sheer flux, meaninglessness, and, ultimately, death. In a passage of exuberant pessimism at the end of Part 1 of the book Constantin seizes upon the coach-horn as a symbol of existence that might serve the modern world in the same way that a death's head served the ascetics of the Middle Ages, that is, as a memento mori, since the 'virtue' of a coach-horn is precisely that it is impossible to play

---

[13] See the section 'God is in Heaven, We are on Earth' in Chapter 4. Clare Carlisle has argued that not only is repetition Kierkegaard's own single central thought but that it is also Heidegger's. See Clare Carlisle, 'Kierkegaard and Heidegger' in Lippitt and Pattison (eds), *The Oxford Handbook of Kierkegaard*, 421–39.

the same note twice: you can't step twice into the same stream! It is therefore the epitome of a life that is handed over to arbitrary and unconnected events, life that is sheer flux, and time as infinite vanishing and thrownness towards death. And this is also why, for Constantin, it seems that repetition must, in the end, also be a *religious* concept, since it is a concept that bespeaks the possibility of transcending the worlds of nature and death—of being able to exist in and through a relation to the eternal.

The failures of repetition in *Repetition* can be read as offering powerful illustrations of what Theunissen described as the tyranny of time over human life, and this same tyranny is similarly illustrated and explored in many other passages from the early pseudonymous writings dedicated to expounding (and exposing) the aesthetic view of life, including the famous Kierkegaardian melancholy. That this is essentially a form of time-sickness is forcefully revealed in the opening aphorisms of *Either/Or* 1, which, in turn, suggests that the relation to time is key to the whole domain of what Kierkegaard calls 'aesthetic' existence. The aesthete, epitomized for Kierkegaard in some of the writers and theorists of Early Romanticism, seeks to create and inhabit a world of beautiful images that mirror back to him a view of reality as conforming to his own desires and passions. But this is a project that is doomed to fail. As he puts it in his criticism of Romantic irony, 'It is through the negation of the imperfect reality that poetry [as conceived by the Romantics] opens up a higher reality, extends and transfigures the imperfect into the perfect, and thereby reconciles the deep pain which seeks to darken all things. Thus far poetry is a kind of reconciliation but it is not the true reconciliation for it does not reconcile me to the reality in which I live'.[14] And a salient feature of 'the reality in which I live' is precisely time. If Romantics are always eager to talk about eternity or about eternal truths, eternal values, and eternal images, their eternity is precisely an eternity abstracted from and kept, so to speak, at arm's length from the reality of everyday life in time. This, Kierkegaard's ethical persona will say, is why such aesthetes disdain marriage, since marriage requires commitment to the everyday and faithfulness to a promise that binds the self to the promised future.[15]

That the aesthete's problem involves a pathological relation to time is clear from the opening aphorisms that epitomize his 'life-view'. Indeed, his attraction to the aphoristic form already indicates his inability to develop a coherent trans-temporal identity and his limitation to what is fragmentary and disconnected. As the sub-title of another of the essays in *Either/Or* 1 will say, it is 'An Attempt at a Fragmentary Effort' (SKS, 137/EO1, 137). Such fragmentation manifests itself in diverse and sometimes seemingly contradictory ways.

---

[14] SKS1 330–1. Cf. my *Kierkegaard: The Aesthetic and the Religious* (2nd edn, London: SCM Press, 1999), 35–62.

[15] This is a central theme of the first part of *Either/Or* 2 'The Aesthetic Validity of Marriage'.

The aesthete cannot move, yet rushes through life too quickly to take account of what is happening. 'I feel like a chess-piece must feel when the opponent says: that piece cannot be moved' (SKS2, 30/EO1, 22) and 'People say that time passes and life is a stream, etc. I don't see this: time stands still and so do I' (SKS2, 34/EO1, 26) are aphorisms that illustrate the loss of motion in time. But the aesthete also confesses to the opposite affliction: 'I entirely lack patience for living. I can't see the grass grow and since I cannot I just don't care to look at it at all. My observations are fleeting remarks of a "wandering scholar" who rushes through life in the greatest haste. They say that our Lord satisfies the stomach before he satisfies the eyes, but I don't see it—my eyes are full up and weary of everything and yet I'm still hungry' (SKS2, 33–4/EO2, 25). And, in an image that perhaps draws both aspects of the loss of temporal location into one, the aesthete describes himself as follows:

> What is to come? What will the future bring? I don't know and have no idea. When a spider plunges out from a fixed point down into whatever follows, all it sees before it is an empty space in which it cannot get a foothold, no matter how much it stretches. That is how it is with me. All I ever see before me is an empty space and what drives me forward is determined by what lies behind me. Such a life is back-to-front and unendurable. (SKS2, 33–4/EO1, 24).

The more fully developed figures of despair that follow in the subsequent essays therefore span a range of existential possibilities, from Don Giovanni, restlessly driven on by his consuming sexual passion to seek one conquest after another to the essay on boredom, 'The Rotation of Crops', which includes a virtuoso parody of biblical history of the world as a history of boredom.[16] The essay concludes with the advice that we should refrain from any form of friendship or commitment so as to maximally vary our social experience and that we should randomize our activities and observations, so as to avoid doing, seeing, or reading the same thing twice (precisely *avoiding* repetition), focusing on what is arbitrary and beside the point, such as watching the sweat dripping from the face of an acquaintance holding forth on some philosophical subject instead of listening to what he is saying (SKS2, 288/EO1, 299).[17]

But if the writer of 'The Rotation of Crops' can turn his boredom into an entertaining, if malicious oration, 'the unhappiest man' of another of Kierkegaard's aesthetic essays is without even this consolation. Picking up on Hegel's category of 'the unhappy consciousness', the essay remarks that the unhappy man 'is always absent from himself and never present to himself', but it is clear that one can be absent either by being in the past or by being in the

---

[16] This is already anticipated in the aphorism beginning 'I don't care at all...' (SKS2, 28/EO1, 20).

[17] For a full discussion of Kierkegaard's theory of boredom see Chapter 3 'Boredom', in my *Kierkegaard and the Quest for Unambiguous Life* (Oxford: Oxford University Press, 2013), 58–85.

future' (SKS2, 216/EO1, 222). However, as he adds, with reference to grammatical tenses, there is a kind of past in which one is present and a kind of
future in which one is also present but there is also a kind of past, the pluper-
fect, in which there is no presence and a future, the future perfect, in which
there is likewise no presence. Kierkegaard proceeds to explore psychological
situations in which such lack of presence might be exemplified before finally
coming to a situation that combines absence from both past and future. Such
a one 'is continually recollecting what he should hope for, since he has already
encompassed the future in thought and lived it in thought and it is this experience in thought that he remembers instead of hoping for it. So, what he
thus hopes for lies behind him and what he recollects, lies ahead. His life is
not backward-looking but is back-to-front in a twofold way' (SKS2, 218–19/
EO1, 225).

What is at issue here is a despair that is essentially connected to the loss
of or failure to attain the 'moment of vision', which, if found, would provide
such an unhappy one with the possibility of reintegrating his fragmented
temporal relationships. Without this, time shatters into a shapeless multiplicity of unrelated pieces that are neither separately nor together capable of
giving him a sense of who he is or what he is living for. As Kierkegaard will
say of a comparable case-study analysed in one of his *Christian Discourses*,
such a person is in a worse situation than those of whom the doctors say
that they will not live until the next day, since at least they are still living
today. But those who are consciously or unconsciously consumed by anxiety about the next day 'do not live today, let alone live to tomorrow, even
though they are still alive—yet to live to tomorrow one must nevertheless
live today' (SKS10, 87/CD,79). By worrying about a future constructed out
of their own imaginations those Kierkegaard here calls the 'pagans' repli-
cate the misfortune of the unhappiest man and effectively cease to *be*. Not
having the eternal, they also lose time, since time itself is the gift of the
eternal.

The same logic also applies to any would-be speculative philosophy that
views human existence *sub specie aeternitatis*, since its 'eternity' is conceived
of in entire abstraction from temporal life. Like the pagans' 'next day' the
philosophers' eternity prevents them from really living the time that now is.
And this is a kind of distraction or self-forgetfulness: 'Someone who exists
but forgets that he exists will become more and more *distrait*, and as people
sometimes deposit the fruits of their *otium* [leisure] in literary works, so might
we venture to expect as the fruits of his distraction the expected existential
system—well, perhaps not all of us, but only those who are almost as *distrait* as
he is' (SKS7, 116/CUP1, 121). Speculation is a flight beyond the conditions of
earthly life, something proper, Kierkegaard thinks, only to winged creatures or
moon-dwellers (SKS7, 119/CUP1, 124). And, as we saw in Chapter 4, the forgetfulness of aesthetes and philosophers extends to the age as a whole—which

is why Kierkegaard can use the adjective 'aesthetic' to describe the basic cultural orientation of his time.

Yet, in a figure that recurs throughout Kierkegaard's authorship, the point of danger is also the point at which the possibility of transformation may open up. The realization that one is in despair and the acceptance that one is in despair may turn one towards that which can transform despair. The advice to those in despair is: despair, really despair and see yourself for how you are—and then you might change (SKS3, 203/EO2, 211; SKS11, 130–3/SUD, 14–17).

In *The Sickness unto Death* the kind of despair seen in the unhappiest man and amongst the pagans achieves a level of consciousness of itself and becomes 'despair of the eternal', which Kierkegaard classifies as a form of the despair of weakness, which is in turn defined as not willing to be the self that one is, namely a self capable of existing as Spirit, in relation to the eternal, to the future, and to possibility. Such despair is the belief that one cannot make of one's life what one wants it to be or that one cannot grasp one's life as a whole because one is bound to the passage of time that threatens, dulls, or evacuates whatever one strives for or even achieves. Life somehow disappoints, and whilst we know that there should be more to it than there is we can never grasp what this 'more' is or how it might be realized. Those who feel like this may well be well-educated, married with a family, competent in their job, occasional churchgoers, and probably lovers of solitude. Yet, as Kierkegaard puts it, they remain shut in on themselves, keeping to themselves the nagging thought that they are less than they ought to be and that 'there must be something more'. So, Kierkegaard says, 'This self-enclosed despairer goes along in life, *horis succesivis*, in hours that if not lived for eternity nevertheless have some business with the eternal, being occupied with the self's relation to itself—but he doesn't really get any further than that' (SKS11, 179/SUD, 63–4). Such people never identify themselves so completely with their despair as to reach the point of really looking beyond themselves for help. They sense something missing, but don't pray for it and can't believe it could ever be given. In the Christian terminology towards which *The Sickness unto Death* moves, they cannot believe in forgiveness or that a person can be accepted 'just as I am'.

Alongside the despair of weakness, not willing to be the self that one is, Kierkegaard describes a despair he calls the despair of defiance, which is willing to be oneself without reference to any human or transcendent other. Again, Kierkegaard doesn't use the word, but it is a kind of extreme version of autonomy that could be seen as characteristically modern, in a line running from Byron to Sartre and down to the present. It is the stance of 'the rebel' who believes he owes the world nothing and who, when he is consistent, doesn't believe that the world owes him anything either. Following Kierkegaard's own analysis, commentary on these forms of despair typically emphasizes the element of will in their respective definitions of weakness and defiance, of willing and not willing to be oneself. However, just as the despair of weakness is also characterized by the lack of a proper relation to time and to eternity, so too the despair of defiance involves a misuse of the relation to the eternal and therefore also a distortion of the self's time-experience. Because the eternal is what gives the power of extending the self over or through time, it is the power of the eternal itself—though unacknowledged—that gives the defiant self the illusion of being able to establish its own continuity and identity through time. As related to the eternal, it does always have a certain continuity, whether or not this is acknowledged. But the possibility of such continuity is not a possibility it gives itself. It is the eternal, as the power of what is not-self in and at the basis of self, that grounds this possibility and not the autonomous will (see SKS11, 181/SUD, 67).

Hegel had used the etymological root of the German *Verzweifelung* to equate 'despair' with the condition of the divided self and, as such, intensifying the duality that is put in motion in Cartesian doubt. Kierkegaard similarly uses the Danish *Fortvivlelse* to explore despair as the division and separation of the polar structures of finitude and infinity, necessity and possibility that he sees as integral to a well-grounded self. This approach perhaps obscures how Kierkegaardian despair is also a failure in the human time-relationship. If we are attentive to the role of the time/eternity dialectic that we have been considering here, we may come to agree with Michael Theunissen's conclusion that, in purely formal terms, it is *desperatio*, a lack of hope, that lies at the root of the self's inability to constitute itself as a whole and not vice versa.[18]

CHRIST: THE ETERNAL IN TIME

We have established the general shape of Kierkegaard's conception of the relations of time and eternity and seen how this is both more affirmative and more subtle than the readily anthologized excerpts that are reflected in Barth's astringent iteration of the 'acid difference' between time and eternity. Especially, we have seen how the self is itself constituted as a synthesis of time and eternity such that the loss of a proper relation to the eternal entails the dissolution or disintegration of our own self. But how might a self that has lost this relation and therefore lost itself (or its self) be restored to its original potential for self-integration on the ground of a true God-relationship? *The Sickness unto Death* points towards the need to be always ready to believe in the possibility of forgiveness and reconciliation, in other words, to trust and to hope in all that the gospel promises. But what, in turn, grounds these promises and what is involved in taking them seriously?

---

[18] See Michael Theunissen, *Der Begriff Verzweiflung: Korrekturen an Kierkegaard* (Frankfurt am Main: Suhrkamp, 1993), 132.

At its simplest, Kierkegaard's answer is that the manifestation of Jesus Christ as the God-man or 'the God in time' reveals the eternal as entering into and redeeming time and therefore showing time to be both the bearer and revelation of eternal truth and a fully appropriate medium in which human beings can work out their relation to the one, eternal God. This, however, leaves open the question as to whether it is simply one instance—even if an eminent instance—of a general structure of temporal–eternal relatedness or whether it is, in fact, only in the light of and on the basis of the Incarnation that the more general possibility of time becoming the 'stairway to heaven' is established, that is, whether we are to read Kierkegaard in a Tillichian or a Barthian way. We shall return to this important interpretative question, but first to the texts themselves.

The most referenced text with regard to the relationship between time, eternity and the Incarnation is, rightly, *Philosophical Fragments*, and if this is not Kierkegaard's last word on the topic it certainly puts down a number of markers that remain definitive for his developing meditations on the meaning of Christ for human life. In the twentieth century the discussion of the issues raised by this work generally occurred within the horizons of debates about faith and history—indeed, this whole debate, in the shape it took in the middle of the twentieth century, was decisively influenced by Kierkegaard. A crucial factor was the effective conclusion—at least for a while—of the 'Life of Jesus' movement, as marked by Schweitzer's *The Quest for the Historical Jesus*, which, for reasons very different from those put forward in Kierkegaard's time by D. F. Strauss, challenged the relevance to faith of the historical investigation of the life of Jesus. Those who reacted positively to Schweitzer found Kierkegaard's *Philosophical Fragments* a significant further aid in reflecting on what this could mean for theology. Often they would be accused of a 'flight from history', in that they accepted the view that the Christ of faith could not be based upon the testimony of the Jesus of history. However, as we shall see in relation to Kierkegaard, acknowledging the very limited role of historical reconstruction in supporting the claims of faith should by no means be confused with denying the historicity, that is, the all-pervasive and ineluctable temporality, of the historically existing human person. On the contrary, it is precisely the seriousness with which Kierkegaard and others took temporality that, I suggest, led them to recognize the limitations of a positivistic reconstruction of historical 'facts'.[19]

Philosophical Fragments does not directly respond to Strauss, and the stage-setting questions refer more directly back to Lessing's publication of H. S. Reimarus' attempt to reveal a historical Jesus devoid of supernatural endowments. The questions run: 'Can there be a historical point of departure

---

[19] As when Heidegger sees, e.g. Harnack and Troeltsch as still caught in the basic a-temporality of a certain kind of inherited Platonism (see Chapter 4).

for an eternal consciousness? How can such [a historical point of departure] be of more than historical interest? Can one build an eternal happiness on historical knowledge?' (SKS4, 213/PF, 1).

Kierkegaard addresses these questions via a consideration of the nature of truth and how truth can be learned.[20] On the Socratic view, truth is recollection. We already have an implicit knowledge of the truth such that the role of the teacher is merely to remind the learner of what he already knows, as in the classical instance of Socrates eliciting the principles of geometry from an uneducated slave boy. The teacher is a kind of midwife, bringing to birth what is already 'in' the learner. But if this is identified in the text as 'the Socratic view' it is clear that Kierkegaard elsewhere associates recollection with idealism in general, and Hegelianism in particular. The justification of this association is, of course, summarized in Hegel's own reflections on the Owl of Minerva flying at dusk, at the end of the era of historical becoming that, having become actual, is now to be recapitulated in thought and brought into the form of the concept. At the same time—on the interpretation that makes Hegel the teacher of a doctrine of timeless eternal truths—the historical development is itself merely the manifestation of what was always already timelessly true in the realm of pure thought.

But if Socrates and Hegel are right, then this not only means that the teacher is limited to the role of a midwife but also that no historical event or complex of events can have a more than occasional significance. What happens in time doesn't make what is true true and cannot make it untrue: it merely provides us with the opportunity for knowing what is timelessly true. If what Christianity teaches about the Incarnation is true, however, then this Socratic-Hegelian view must be rejected, since the Christian teaching makes what happens in time decisive for the overall configuration of human beings' God-relationship and therewith their relation to the truth of their own being. What God does in Christ is, on Christianity's premises, a truly new thing—and even if Christianity also teaches that there were prophetic and typological prefigurations of this event these were only foreshadowings and not the thing itself.

Such a Christian view of the relationship between truth and history is curiously ambiguous. It makes what happens in time, at a unique, unrepeatable point in time, absolutely decisive. Yet it simultaneously asserts that this truth comes to us from beyond the horizon of our historical possibilities. As Chapter 3 of *Fragments* provocatively states, such a happening is not just puzzling, it is 'the absolute paradox'. However, and despite the questions about history

---

[20] Marilyn G. Piety's translation suggests the question is how the truth can be taught. This is certainly a thought-provoking alternative to the conventional 'taught', although it has no major implications for the question we are pursuing here, which is that of the relationship between time and eternity, since the critical point of identity/difference remains the same, whether we approach it from the one side (that of the learner) or the other (the teacher).

and the eternal consciousness on the title page, questions of time and eternity are not significantly emphasized in the first three chapters, which speak more generally of what we might call the realms of immanence and transcendence. However, when Chapters 4 and 5 turn, respectively, to the kind of faith of those contemporary with Christ and that of those 'at second-hand' (at a historical distance from the events described in the New Testament), these questions start to come into their own. Their importance is underlined by the so-called 'Interlude' addressing the questions: 'Is the past more necessary than the future? Or by becoming actual does the possible thereby become more necessary than it was?'[21]

The Interlude is written in a style not unlike that of the discussion of time and eternity in *The Concept of Anxiety*, that is to say, it reads more like a concatenation of aphoristic assertions than an argument. However, if we have the patience to weigh Kierkegaard's terms we can see that it offers a coherent account of historical knowledge that makes clear the differences between Kierkegaard's own position and the positions of his contemporaries and predecessors. He starts by distinguishing between simple alteration and coming into existence. Although this is not spelled out, the context suggests that an example of the former might be the movements of the celestial bodies or the different phases of plant or animal life—paradigmatic instances of change in Aristotelian philosophy. As Kierkegaard will make clear, the latter, coming into existence, is the kind of change that occurs as the result of a free action, 'doing a new thing', and it is this kind of change that most preoccupies him in the Interlude. But how might we understand this kind of change? As the coming into existence of a new thing it cannot be explained merely as the outcome of a sequence or constellation of preceding causes since it would then be no more than a case of alteration—the reconfiguring of what already existed but in a different state. Furthermore, it cannot be necessary. If, as Hegel asserts, necessity is the unity of possibility and actuality then possibility will be subordinated to necessity in such a way as to be removed from any possible relation to existence, since, on Kierkegaard's view, necessity neither brings anything into existence nor does anything that comes into existence thereby become necessary. The necessary is entirely distinct from both actuality and possibility and simply is what it is, immutably. It neither undergoes nor brings about any kind of change (examples are, presumably, mathematical or logical truths). Not even change in nature is strictly necessary since although natural change may seem to be necessitated by a heavy layering

---

[21] SKS4, 272–86/PF, 72–88. This section is amongst the least-commented on in Kierkegaard's authorship, not only because it is exceptionally abstract but also because its relevance to the argument of the whole is not, perhaps, immediately obvious. Both explicit and implicit references to Hegel, Schelling, Franz von Baader, and Karl Daub in this passage confirm that Kierkegaard is engaging primarily with the generation of his teachers and not indulging a purely fanciful and unhistorical meditation on 'the Socratic'.

of intermediate causes, its ultimate ground is a free act.[22] But when we are speaking of what Kierkegaard calls 'the historical, in a stricter sense' (SKS4, 276/PF, 76) there is what he refers to as a 'doubling' of coming into existence in that what comes into existence as a natural phenomenon—a human being qua animal life-form—also contains a possibility for coming into existence within its coming into existence (SKS4, 276/PF, 76). I take this to mean not that a human being qua animal has the power simply to reproduce itself qua animal, which would be a purely natural form of causing something to come into existence (the relationship of paternity, for example), but that, animal that it is, it also has the power to exercise another level of free causality, in other words, to act through choice or will.

An important consequence of these basic definitional steps is that even though what has came into existence in the past cannot be undone or be made otherwise than it was, *it will never have been necessary*. Historical action, the subject of genuine historical enquiry, is what has been done in and through freedom, and this freedom preserves its character of freedom even when past. This assertion not only wrecks the Hegelian project of gaining insight into the necessity of the historical sequence but also indicates that we can only understand a free action in time past if we too relate to it freely. As Kierkegaard quotes the idealist philosopher Karl Daub, the historical philosopher is a 'backwards prophet', meaning that the past, just as much as the future, is present to us in the mode of possibility and, as such, is accompanied by a margin of uncertainty that requires all interpretation to accept the need for judgement and decision (SKS4, 279/PF, 80). One who understands this will therefore always go towards the encounter with the past with an attitude of passion and wonder that befits the mysterious reality of freedom. A chain of historical causes is never self-contained or teleologically determined but is always open to what is not encompassed within the causal sequence itself.[23] Knowledge of a genuinely historical event or act is therefore a kind of faith; it is not cognition in the conventional sense but a determination of the will: what we see depends on how we choose to see. As we saw in reflecting on hope and love, we can

---

[22] Kierkegaard's commentators explain this by reference to a paraphrase of Aristotle found in Tennemann's history of philosophy: 'A stone is moved by a stick, this is moved by the hand and this in turn by a man. But in the case of this last the reason for the movement is in itself, which, like every being that has soul is a self-moving being' (SKSK4, 270). This suggests that the point is not to trace all movement back to God as the prime (and free) mover but to see all movement in animal life as having a kind of freedom. Dogs chase cats because they want to, not just because they have to, seems to be the point—a view which, arguably biologically unsound, is perhaps endorsed by dog-owners who call on their dogs to 'Stop!', with the implication that the dog has the power to cease its action and therefore some degree of control over it.

[23] Kierkegaard also emphasizes that what is at issue here is not a matter of empirical knowledge either, since sensuous knowledge is no more 'historical' than is knowledge of what is necessary. Sense-certainty is immediate and irrefutable, but historical action on the basis of freedom is always open to alternative interpretations.

always choose to see others' actions in the belittling mirror of our own anxiety or in the expansive light of eternal charity.

Returning to the question of the Incarnation, this then leads Kierkegaard to the position that there can be no essential difference between those who walked and talked with Jesus in Galilee and those who believe in him two thousand years later on the basis of a distant historical testimony. In each case we are dealing with a relationship between two freedoms: the freedom enacted in the God-man's own words and actions and the human freedom to respond to them with (or without) faith. Jesus' contemporaries were exposed to manifold possibilities of misinterpreting who he was on account of his lowly and humble appearance. We, for our part, may be exposed to the scandal of claims resting on questionable historical sources. But both past and present encounters with Christ come down to a relationship between two freedoms, divine and human. By virtue of their being grounded in freedom, both revelation and faith can therefore only come into existence in, with, and under time. Time is the medium in which time and eternity meet and mesh.

This leaves many questions. Firstly, it is clear that those who see Kierkegaard as some kind of fideist may think that the way in which he describes faith as an act of will fully justifies their interpretation. It might also seem to bring Kierkegaard into a certain proximity to the position of the later Schelling, namely, that the fusion of time and eternity is possible only as an act of supra-temporal volition. In Kierkegaard's case, however, it does not seem that there is anything supra-temporal going on here. Whatever else Kierkegaard is talking about, it is essentially a relationship between one historically situated event (the life of Jesus) and another historically situated event (the believer's act of faith). But is this, then, to hand faith over to the arbitrariness of sheer wilfulness?

In terms of Kierkegaard's own writings, it is crucial at this point to also take into account how he shows the decision of faith to arise out of and feed back into the formation and reformation of the self that has been fragmented through its negative experiences of time. And here it is just as important to take note of what his many 'upbuilding' and 'Christian' discourses have to say about the preparation of the self for faith, and about what the reception and practice of faith mean for the self, as it is to follow his analyses of the sick and divided self looked at in the previous section. I shall turn to these in the following two sections, but, first, we are still left with the question posed at the start of the present section, namely, whether all this means that Kierkegaard makes the possibility of the moment of vision in the individual dependent on faith in the historically particular revelation of Jesus as the Christ or whether this is, as it were, simply a Christian colouring of a universal structure of time–eternity relationship.[24] Does Kierkegaard mean that only Christians

---

[24] Very simplistically, we may once again say that this is a choice between a Barthian and a Tillichian reading of Kierkegaard.

who have an explicit faith in the dogmatic claim that Jesus is the Christ, the Son of God, can know the eternal in time and time as revelatory of eternity? The teaching of the Interlude suggests that an answer will not be a simple yes/ no. For there, as elsewhere, we have seen that Kierkegaard resists seeing the eternal in time as a past event that has become actual (still less necessary). As an event still characterized by possibility it is an event of which we can say, with Faulkner (but against Jankélévitch!), that the past is not only not dead, it isn't even past. Insofar as it shares with the future the defining character of possibility, the historical past of Jesus Christ—as of any genuinely free histori-cal agent—remains, in a sense, still future and still to come. Consequently, the inability to experience him in the historical record of his past appearing leaves open the possibility of experiencing him otherwise than as an event of past history. As Schweitzer put it, in reflecting on the dead-end of the nine-teenth century 'Life of Jesus' movement, we cannot come to know him in such categories of the past as Messiah, Son of Man, or Son of God, which, he says, are 'historical parables'. It is only in following him as one unknown that we can today learn who he is for us.[25]

For Kierkegaard himself, the gospel narratives, read otherwise than in the perspective of historical research, would always remain the paradigm of a human life lived in the light of the eternal, and the highest possibility for a contemporary human being is always to aspire to the imitation of that paradig-matic life. But as the release into history of possibilities of hope and love that overspill the boundaries not only of cultural but also of cognitive limits, the meaning of the life of Jesus is meaningful in terms of the twofold possibilities of hope and love. Certainly, on Kierkegaardian premises, no Christian would ever have any justification for regarding any other human being, Christian or non-Christian, with anything other than the loving hope that sees them in the light of the eternal.

BEING BUILT UP IN THE ETERNAL

To fill in the picture of how Kierekgaard sees the act of faith as grounded in the whole life of the individual self, I shall now summarize what we might call the theological anthropology set out in his many upbuilding or devotional writ-ings. I have discussed this extensively elsewhere and what follows aims only to pick out the main relevant points. It should also be stressed that these are neither systematic works nor do they have the kind of intensely concentrated argumentation we have been seeing in some of the pseudonymous works.

---

[25] Albert Schweitzer, *The Quest of the Historical Jesus*, tr. William Montgomery (London: A. & C. Black, 1954), 401.

Instead, they are works of a looser, more discursive, and even conversational character, with many digressions and illustrations. Inevitably, then, the following discussion cannot avoid giving a more 'systematic' impression than the texts themselves suggest. These cautions in place, let us begin.[26]

Kierkegaard supposes that the initial or immediate state of human beings (i.e. in infancy) is to be 'carried along by life, one link in the chain that joins past and future, unconcerned as to how it happens, they are carried along on the wave of the present' (SKS5, 42/EUD, 33). For some people, this state seems to continue into adult life, such as the person Kierkegaard calls 'the fortunate man', 'whom fortune seems to delight in all things… [his] every dream is fulfilled, his eye is satisfied more quickly than it is aroused to desire, his heart hides no secret wish, his longing has learned no limit' (SKS5, 94/EUD, 89). We might be tempted to envy such persons, but, Kierkegaard suggests, there is something missing in their lives. Although they seem to be not merely content but enjoying each new day that comes, their good fortune is just that: fortune. Their life is not their own in the sense that it is not what they have actively chosen and is not the result of anything they have set in motion. Because they have not actually chosen to be who they are, there is therefore a sense in which their lives are not really their own. So far from being the subjects of their lives they are 'lost' in them (SKS5, 164/EUD, 164) or are mere 'instrument[s] in the service of obscure drives… a mirror that reflects the world, or better, in which the world reflects itself' (SKS5, 301/EUD, 308). Rather than existing in the world or fully living it they are, as in the first quotation in this section, merely 'carried along' by it.

But this is enviable only as long as fortune continues to be good fortune. If life changes and experience turns sour—as Kierkegaard supposes to be the case for most people in some degree—the simple recipe of taking life as it comes seems less effective and for some it will become a recipe for downright despair. The task that Kierkegaard sets himself and his readers, then, is to see how to become responsible for our existence in such a way that the changes and chances of fortune, good or bad, do not carry us away from what is truly important in life. An immediate problem is that we typically don't want to wait for results, we want them now, a trait which Kierkegaard illustrates by describing how spectators will be full of admiration for a diver who performs a feat of daring that is all over in just a few seconds but will scarcely pay attention to one whose work and achievement is spread over many years. We want stars, fireworks, and something to astonish us, but, he says (reflecting the notion of 'inner history' in *Either/Or*), 'one must learn to crawl before one can walk, and to want to fly is always suspicious' (SKS5, 337/EUD, 349). Kierkegaard is at his most inventive in describing how such impatience manifests itself: we have

---

[26] A fuller account of the materials dealt with in this section can be found in my *Kierkegaard's Upbuilding Discourses: Philosophy, Literature and Theology* (London: Routledge: 2002), 39–63.

already seen an example of this in his account of the pagan who falls prey to 'the next day'.

Impatience emerges as one of the dominant forms of human time-sickness in these analyses, and the cure for it is, of course, to learn patience. Kierkegaard understands patience in terms of the literal meaning of the Danish term '*Taalmod*', 'courage to bear', and, principally, courage to bear the burden of time itself. In this sense patience is not just one of many virtues that the self would do well to acquire, but, as the titles of two of the discourses indicate, is the way in which we both acquire a self and sustain it in being.[27] Examples from various fields of life teach us that patience is a necessary condition of succeeding in any significant undertaking—walking, farming, business, and parenting are amongst the examples Kierkegaard cites—but there is a difference between using patience and being patient. Precisely because the soul is 'the contradiction of the temporal and the eternal' (SKS5, 163/EUD, 163), the gratification of temporal wishes or projects will never entirely fill the soul and, fully to be who we are, we must learn to live with what is not exhausted in any particular temporal goal or combination of goals. Becoming patient is how we combine the temporal and the eternal in existence, humbling ourselves under the incompleteness of time by virtue of the power of the eternal, which is, again, to understand the eternal as the power that allows us fully to be in time, to bear time, and to endure time and not as time's negation. It is how we live the prospect of the self's unfinished distension in time.

But Kierkegaardian patience is also qualified by a future-directed orientation that he calls 'expectation', and again some of the discourse titles are telling: 'The Expectation of Faith', 'Patience in Expectation', and 'The Expectation of an Eternal Happiness'. Expectation is distinct from wishing in that it involves real commitment on the part of the one who expects and is itself a response to a kind of time-experience that enables the self to rise above the transient moment of immediate time. In the very first discourse of all, Kierkegaard already identifies this in terms of the eternal. Drawing an analogy with the seafarer who finds his course through the inconstant and featureless flux of the ocean by looking up at the stars, he sees confidence in the future—which he here calls 'victory'—as coming from a relation to the eternal:

> By means of the eternal one can overcome the future, because the eternal is the ground of the future, therefore one can use it to provide a basis for the future. What is the eternal power in a human being? It is faith. And what is the expectation of faith? It is victory, or, as Scripture so seriously and feelingly teaches us, that all things work together for good for those who love God' (SKS4, 26/EUD, 19).

---

[27] The titles are 'To Acquire one's Soul in Patience' and 'To Preserve one's Soul in Patience'. See Clare Carlisle's discussion of 'humble courage' in *Kierkegaard's Fear and Trembling: A Reader's Guide* (London: Continuum, 2010), 193–9.

And, as we have seen in the diagnosis of the pagan's preoccupation with 'the next day', the relationship to the future is also a form of self-relationship: 'He who struggles with the future . . . struggles with himself, [since] the future does not exist, it borrows its strength from the self . . .' (SKS4, 24/EUD, 18). But precisely because this kind of orientation to the future is grounded in the eternal it does not result in impatience or the loss of the present, actual time. As 'Patience in Expectation' shows, taking the case of the biblical prophetess Anna (who is reported to have waited in the Temple through many years after her widowhood in the hope of seeing the promised Messiah (Luke 2:36–8)) Kierkegaard suggests that a truly powerful expectation such as hers, an expectation directed towards the eternal, can endure a lifetime of waiting. In a sense, even the absence of the promised fulfilment cannot disappoint: in waiting patiently and expectantly, Anna already has the spiritual life that the Messiah is to bring. In a word-play on the Danish '*maa-ske*', 'maybe' but also 'must be', Kierkegaard writes that

> We do not know whether there has ever been an age more serious than ours that did not know this word [maybe] but rested in eternity's assurance that 'it must be so'; we do not know of a more impatient generation that by more and more rapidly repeating this expression for eternal expectation actually produced impatience's short, hasty, hurried, light-minded, superior, smart, comfortless 'maybe'. But good for those who, when earthly expectations disappointed them, [spoke,] like Anna, with a mind devoted to God and in the ceremonious language that befits eternity, when she said: It must be. Good for those who, having lived many days, can still, in their eighty-fourth year say: It must be. (SKS4, 217/ EUD, 217)

Perhaps this is already an anticipation of deconstructive messianicity—a messianic faith that requires no Messiah and perhaps even demands that the Messiah should not come *yet*? In any case, a striking instance of how expectation serves not to distract the self from the present but to give it consistency in and through time.

But how might such expectant hope arise in any human heart? Is the idea of an eternal happiness just a kind of wish-fulfilment, a fantasized transformation of an existential lack? The first step in Kierkegaard's response is to suggest that it is rather the consistent extension of a movement that begins when the self, carried along on the flow of time, starts seriously to question itself about the meaning of its life in time and how it should make use of the time allotted it. In such a moment of concern for who it is and what it is to be, the self, as Kierkegaard puts it, becomes 'older than the moment and [this concern] lets him grasp the eternal' (SKS4, 93/EUD, 86). The second, and more basic, is the realization that the self is, so to speak, dialogically grounded, that is, that its relation to itself is inseparable from its relation to a larger reality—of which the eternal is the ultimate term. And, more precisely, it is the realization that its life is not just constrained by a certain existential or ontological structure but is ordered and called to exist by and as love.

ETERNAL LOVE

Many of Kierkegaard's discourses are devoted to the theme of love, including the collection entitled *Works of Love*.[28] That this theme has an important and not merely circumstantial connection to matters of time and eternity is signalled in the rhapsodic introduction to the first discourse on this theme, from the *Three Upbuilding Discourses* of 1843.

> What is it that makes a person great, admirable amongst creatures, and well-pleasing in God's eyes? What is it that makes a person strong, stronger than the whole world or so weak as to be weaker than a child? What is it that makes a person firm, firmer than a cliff or yet so soft as to be softer than wax? It is love. What is older than everything? It is love. What is it that outlives everything? It is love. What is it that cannot be taken away but itself takes it all? It is love. What is it that cannot be given, but itself gives everything? It is love. What is it that stands fast when everything falters? It is love. What is it that comforts, when other comforts fail? It is love. What is it that remains, when everything is changed? It is love. What is it that abides when what is imperfect is done away with? It is love. What is it that bears witness when prophecy is dumb? It is love. What is it that does not cease when visions come to an end? It is love. What is it that makes everything clear when the dark saying has been spoken? It is love. What is it that bestows a blessing on the excess of the gift? It is love. What is it that gives pith to the angel's speech? It is love. What is it that makes the widow's mite more than enough? It is love. What is it that makes the speech of the simple person wise? It is love. What is it that never alters, even if all things alter? It is love. (SKS5, 65-6/EUD, 55)

Of the eighteen questions structuring this passage, nine are related to love's power to endure and even to outlast the passage of time. Thus, the presence of the eternal in human existence is not just a matter of the self being empowered to become older than the moment, grasping itself in the continuity and coherence of its temporal life as a whole, but also as coming to be in relation to others. And it can come to be in relation to other precisely by virtue of the word—the 'angel's speech' or the 'wise' 'speech of the simple person'—that assures us that we are loved.

We have already heard extensive testimony to this *ex negativo*: the aesthete is depicted as a solitary and asocial being, whose relations to others are

---

[28] The discourses on love not included in *Works of Love* are collected in Part III of my selection of Kierkegaard's *Spiritual Writings: Gift, Creation, Love: Selections from the Upbuilding Discourses* (New York: Harper, 2010), 227–98.

contemptuous, indifferent, or, in the case of his alter ego Johannes the Seducer, abusive. Likewise the despairer who lacks faith in the eternal is, Kierkegaard says, likely to be a solitary person, anxious to conceal from others his sense of failing to be who he wanted to be in life. The connection of love and the eternal gives us the positive counterpoint to this and, structurally, it is clear that it is this positive counterpoint that is the more properly basic: eternal love gives the possibility of loving relationships from which those who do not love diverge.

Perhaps the most developed statement of the interconnection between love and the eternal is in the discourse (from *Works of Love*) 'Love Hopes all Things yet is Never Put to Shame'. As Kierkegaard says in this discourse, 'Only an earthly mind, that understands neither what love is nor what hope is thinks that to hope for oneself and to hope for others are two separate things and that love is yet a third thing on its own'. Instead, as he explains, 'Love is not a third thing on its own but is an intermediate determination: without love there is no hope for oneself and with love there is hope for all others: to the same degree that one hopes for oneself one hopes for others, for one is loving in exactly that same degree' (SKS9, 259/WL, 260).

In the opening paragraph of the discourse, Kierkegaard speaks of how time that has become separated from the eternal is experienced as simultaneously 'creeping' and yet rushing past or relapsing into immobility, recalling not only his own analyses of aesthetic existence but also the kind of tyranny of time over existence Theunissen found in the pre-Pindaric poets. In the grip of such degenerate forms of time, it feels as if the air around us becomes stuffy and poisonous and, bound to the transient moment, causes us to lose all perspective on our lives. But this is just what the eternal offers us—fresh air and a perspective on life. 'Although Christianity knows only one way and one way of escape it nevertheless does always know of one way and one way of escape, and it is with the help of the eternal that Christianity creates space and perspective in each moment' (SKS9, 246–7/WL, 246). Thrown back upon itself, without relation to the eternal, the moment is reduced to a vortex, that is, to a movement that cannot move forward. Against such unmoving movement Christianity metaphorically re-envisions worldly time as the time of sowing preceding eternity's time of harvest, teaching that the moment is only a passing moment of strife and sorrow that will yield to the victory of the eternal. Yet if, as Kierkegaard also goes on to say, eternity is appropriately figured as a ceremonial hall in which the walls are covered with mirrors that infinitely reflect even the merest trace of unworthiness, surely we are even more trapped in an infinitely reflected present from which we cannot ever escape? Yet Kierkegaard, nevertheless, counsels us to hope—but how can we hope? What is it to hope? And what is it to hope 'all things', as Paul puts it in the text on which the discourse is based?

Commenting on this expression, Kierkegaard distinguishes between hoping all things and always hoping. The former sounds as if it is something we might do once and for all, as if a single 'eternal moment of vision' might empower us to take in the whole compass of our lives and inscribe it under the rubric of hope. Always hoping, however, is hoping that goes on year after year, day after day, hour after hour; it is hoping extended through time. The one, hope that encompasses 'all things' in a moment of vision, is the hope that most directly manifests the eternal; hope that is always hoping is the kind of hope characteristic of the temporal. But, Kierkegaard says, these are so far from being opposed to each other that the one becomes untrue if separated from the other. The aim is 'in each moment always to hope for everything' (SKS9, 249/ WL, 249). This, however, is precisely to enact the unity of the temporal and the eternal or to reveal the unity of the temporal and the eternal in the manner of our comportment towards life.

Whereas the eternal 'is' without further qualification, it is nevertheless not in the present that time and the eternal come together since the 'now' is a vanishing moment. Nor can this conjunction be in the past, since the past no longer exists. 'If, then, the eternal is in the temporal then it is in time to come [the à-venir]...or in possibility' (SKS9, 249/WL, 249). Or, as Kierkegaard immediately parses this assertion, 'What is past is actual, what is to come is the possible; the eternal is eternally the eternal, [but] in time the eternal is the possible, what is to come' (SKS9, 249/WL, 249). So, whereas the 'classical' timeless theories we considered in Chapter 1 locate the presence of the eternal in time primarily in the present, Kierkegaard here suggests that temporal beings such as we are will best relate to the eternal as it is revealed in our concern for what is to come, since it is in our concern for what we are to be or to become that we raise ourselves above the merely transient moment. Nevertheless, if the future thus becomes a possible site of the revelation of the eternal, this is strictly in the mode of possibility and not actuality. The eternal is not a neutral fact waiting to be discovered but is disclosed in and through our relation to our own existential possibilities. As Kierkegaard puts it, this means that how we relate to the future will decide whether or not we get to know the eternal and this in turn will depend on whether we choose good or evil. If we choose evil, then the future will be experienced in terms of fear; if good, then hope. To choose hope is to choose the possibility of knowing the future as the future of the eternal. This is why those who chose hope are, as the title of the discourse has it, 'never deceived'.

But surely they are, often, deceived? The person we lent to doesn't repay the loan and the rebel's belief that he will never sell out to the establishment is rather frequently proved to have been ill-founded. Here, however, we must distinguish between the kind of hope that is simply the expression of youthful optimism and hope that is based on a genuine existential decision. Precisely because the eternal thus encompasses the whole of human life it is impossible for us to know it in an instant. We can only come to know it by patiently learning to be and to stay true to who we are. It is, Kierkegaard says, as when one sets a child a task. If the task is very large, one doesn't ask the child to do it all at once but breaks it down into easily achievable elements. And this is how the eternal too proceeds. Although it is eternally eternal, it breaks itself down into the temporal dimension of what is to come, the possible, and in this way teaches us to hope so that, through hope, it teaches 'time's child' to know the eternal. 'The eternal, properly understood, puts down only a little piece of itself before us at a time, as possibility. By means of the possible eternity is constantly *near* enough to be at hand and yet *distant* enough to keep a person moving forward towards the eternal, going, going forward. Thus eternity entices and draws a person forward in possibility from the cradle to the grave—if we choose to hope' (SKS9, 252/WL, 253).[29]

And hope, if it is to educate us to knowledge of the eternal, must also hope all things *lovingly*. This returns us to the question as to whether hope can be deceived and put to shame. Don't we see countless examples of enthusiasm growing dim, faith turning to unbelief, love becoming cold or the just unjust or friend foe? All these things are, at least, *possible*! Isn't possibility, then, more likely to instruct us in the vagaries of life in time than to educate us to eternity?

Before Kierkegaard answers this question, however, he makes a further clarification. The hope and the possibility that are at issue here are not just the hope we have for ourselves or the possibilities that await us in our own lives. First and foremost it is a matter of our relations to others. A person who truly despairs is a person who despairs of others. The one who loves, however, has the opposite attitude. Both may think 'All is possible' as they look round at their fellow human beings, but where the person who despairs interprets this in terms of the good becoming bad, the one who loves will see the possibility of good in even the most depraved or perverted. And to see only the possibility of the bad in others is, in fact, to foreshorten and therefore to lose possibility and, in losing possibility, lose the eternal and be delivered over to time and to time's changes and chances. As Kierkegaard puts it

> None can hope unless they are loving [since] we cannot hope for ourselves without also being loving, for all that belongs to the good hangs together infinitely. Therefore those who love hope for others. In the same degree that they hope for themselves, they hope also for others [and] in the same degree that they hope for themselves, they are loving. And in the same degree that they hope for others, they hope [also] for themselves, for this is the infinite precision of the eternal like for like that is in everything eternal. (SKS9, 255/WL, 255)

In this perspective, those who withhold hope from others by writing off their chances of improvement actually damage themselves: to despair over another is to be in despair. To deny the possibility of hope to another—to see

---

[29] This might be read as a more personalistically inflected account of the relationship between the future and eternity that we encountered in McTaggart (see Chapter 2).

the other's life as a life that can't get better—is, morally, to condemn that person to death. Love, on the other hand, hopes all things. With a logic that recalls the theme of a number of other discourses on love, that 'love covers a multitude of sins',[30] the person who loves sees the possibility of the good even in what is bad. And, Kierkegaard suggests, one of the reasons why those who refuse to hope for others do this is because they are frightened of appearing ridiculous or naive in the eyes of others. Privileging the judgement of time, based on likely outcomes of predictable behaviour patterns, over the judgement of eternity, which cannot be measured by the measure of time, their fear of being put to shame overwhelms whatever more generous feeling towards others they may have. Although Kierkegaard does not develop the point here, this comment reflects a major theme in his general treatment of sin, namely, that a primary feature of fallen existence is a kind of negative mimeticism, whereby we are continually trying to make good our sense of not being who we are or could be by identifying ourselves with others and borrowing our identities from them.[31] But others' judgements as to what is or isn't or should or shouldn't be shameful are very far from being incorrigible. To hope for others is therefore also to see them in a perspective that transcends the simple application of social norms.

Of course, if they are not just naive optimists, those who hope do see the bad and even the sin. But they only ever see sin in relation to the possibility of forgiveness. Whatever sin there is or may have been, it can be forgiven. And there is nothing shameful in believing this. Thus, in a characteristic re-telling of a well-known biblical story, Kierkegaard considers what might have happened to the prodigal son if, instead of repenting and returning to the paternal home he had died in his sins. He pictures the father standing at the grave, having hoped for his son until the very last breath and hoping for him still, even beyond death—such a love, he says, is nothing to be ashamed of.

This meditation, then, returns us to the image of eternity as a festal hall into which no one who has cause for shame can enter. Initially, this seemed more threatening than promising. In the light of the argument of the discourse as a whole, however, it emerges that the only real reason for shame is if we have withheld the hope of forgiveness from others. The only 'condition' is the conviction that good can come. This gives us confidence vis-à-vis what is to come and, in this way, what is to come plays its role (in the phrase of *The Concept of Anxiety*) as the incognito of the eternal.[32]

---

[30] See the two discourses with this title from the 'Three Upbuilding Discourses' of 1843 and a further discourse included in *Works of Love* itself, also with this title.

[31] See my *Kierkegaard and the Theology of the Nineteenth Century*, 140–5.

[32] In the period following *Works of Love*, the sinful woman of Luke 7 becomes the eminent figure under which Kierkegaard considers hope's triumph over sin, and her story can be seen as illustrating and deepening the teaching of this discourse. See my *Kierkegaard and the Theology of the Nineteenth Century*, 150–61.

Implicit in the argument of this discourse is the following thought. By means of the future qua possibility time can become a means of revealing the eternal. That it can do so and how it does so also involves a certain relativization of time. But this is not the conventional relativization of temporal becoming in favour of eternal changelessness. Rather it is a relativization of the duality of time/eternity in favour of the dialectic of good and evil or, to be more precise, in favour of the sovereignty of love. As Kierkegaard puts it in another discourse 'the eternal is the difference between right and wrong' (SKS10, 216/CD, 207)—although, in the light of what he also says about hope and love, 'right' and 'wrong' are themselves to be understood not in terms of conformity to conventional social norms but of our capacity to hope for others.

The discourse on 'Love Hopes All Things' ends with the image of the father at the grave of his prodigal son and throughout the discourse, as we have seen elsewhere, Kierkegaard evokes the convergence of the idea of the future as 'time to come' with the idea of 'the life to come', the 'eternal life' of Christian teaching. Often, he will sound as if he is simply restating the familiar doctrines—indeed, it is his claim that this is just what he is doing. However, it is no less striking that he is restating the familiar doctrines in a new light. It is no longer the case for Kierkegaard that human destiny is to be described in terms of an immortal soul. What is decisive, what constitutes a bridge from time to eternity, is simply and solely the possibility of love. We shall consider below how this differentiates Kierkegaard and Kierkegaard's understanding of the eternal from the trajectory that leads through German Idealism and Nietzsche to Heidegger, but it is probably necessary at this point to make a brief comment about another of the discourses in *Works of Love*, the discourse 'The work of love in remembering the dead'.[33]

Adorno famously saw this discourse as the 'culmination' of *Works of Love* as a whole and as revealing Kierkegaard's view of love to be an empty foil for the 'loving' self's narcissistic obsession with its own motives. This suggests to him that, on this account, even our love for our living neighbours must practically treat them as if they were dead. Kierkegaard, in other words, repeats the Goethian view that 'If I love you, what concern is that of yours?' This, however, is to ignore the ways in which Kierkegaard clearly sets up his graveyard excursion as an essentially social event, both in the sense that he pictures it as happening at a time of day when many others will also be paying their familial respects, and as reaffirming the bonds of family and friendship that have been

---

[33] See Theodor W. Adorno, 'Kierkegaards Lehre von der Liebe' in *Kierkegaard: Konstruktion des Ästhetischen* (Frankfurt am Main: Suhrkamp, 1974), 271–2. These comments are more fully justified in my *Heidegger on Death: A Christian Theological Essay* (Farnham: Ashgate, 2013) and in the related article 'Kierkegaard, Metaphysics and Love', in Heiko Schulz, Jon Stewart, and Karl Verstrynge (eds), *Kierkegaard Studies Yearbook 2013* (Berlin: de Gruyter, 2013), 181–96. For present purposes, I shall assume these other discussions and therefore keep my comments to a brief summary of their outcome.

sundered by death. In such ways the aim of 'visiting' the dead is to lead to greater insight into the commonality of human life and our essential solidarity one with another. The very opening lines emphasize that the importance of death is that it is 'life's epitome' and that the value of listening to death is that death knows how to speak well about life and directs us to what is most decisive in it. The point is not to lose ourselves in debilitating melancholic thoughts, but to learn or re-learn what is essential in life and what is being demanded of us in life, in our relationships with the living.[34] In the light of the discourse on 'Love Hopes All Things', the point is not to play off a love for the mythicized dead against the claims of the living, but to show—this time from another angle—how relations amongst the living, when we are faithful to and hopeful for each other (i.e. when we, quite simply, live in love), become the anticipation of eternal life and reveal time to come as opening out onto the life to come—eternal life. And life can become like this when our experience of time is shaped by the promise implicit in the possibility of mutual address between human beings and God.[35] In this respect, Kierkegaard's view of the dead may be assimilated more to Berdyaev's creative memory as anticipatorily enacting resurrection life than to the melancholy twilights of Romantic graveyard scenes.

But if Kierkegaard's view of the human self's relation to the eternal is that this relation is both internally constitutive of the self *and* grounded in the other-oriented relation of love, what about the God who calls us to love and who, through the possibility made actual in the Incarnation, also reveals this same possibility of love as our human future? Does the interdependence of time and eternity in human life reflect also a dynamic in the life of God?

To address these questions, I turn next to the late discourse, 'The Unchangeability of God', a discourse that might seem to answer them with a decided negative, endorsing the view that God doesn't change in any possible sense. A close reading of this discourse, however, suggests a more subtle and complex view.

THE UNCHANGEABILITY OF GOD

The Unchangeability of God is the title of the published version of an address delivered in Copenhagen's Citadel Church on 18 May 1851 and published in

---

[34] SKS9, 339–40/WL, 345–6. This passage is also to be read in conjunction with Kierkegaard's early 'Gilleleie Journal', in which, in semi-fictionalized terms, he describes his own early experiences of bereavement as both intensely individualizing yet also—with regard to the dead themselves—communal.

[35] This is the central argument of Dorothea Glöckner, *Das Versprechen: Studien zur Verbindlichkeit menschlichen Sagens in Søren Kierkegaards Werk Die Taten der Liebe* (Tübingen: Mohr Siebeck, 2009).

September 1855. At the time of publication Kierkegaard had begun a fierce attack on the established Church in a series of fiercely polemical pamphlets, *The Moment*. Like his other discourses, it was dedicated to the memory of his deceased father. The Foreword to the published version remarks that the text, James 1:17–21, was both the first he had used and one to which he had often returned and also draws attention to the fact that the text was that of an address from 1851. Thus we can see that Kierkegaard is taking pains to emphasize the continuity between the discourse and the religious teaching he had been expounding for twelve years, relating both to the early phase of his authorship (1843–6) and to the 'middle period' or 'second authorship' of 1847–51. In an earlier draft of the Foreword this continuity is empha-sized even more strongly, as he writes 'If a human being might be permit-ted to distinguish between biblical texts, I could call this my first love, [the love] to which one usually ("always") returns in temporal life; and I could [also] call this text my only love, [the love] to which one again and again and again and "always" turns back' (SKSK13, 521). In this first draft the date was given as 5 May 1854, the characteristically significant birthday that was also the day on which he signed off the first two upbuilding dis-courses. Everything therefore points to Kierkegaard's own view of the dis-course as repeating and summing up something close to the heart of his religious vision. Interestingly, however, he does not now offer his custom-ary disclaimer that this is not a sermon and in his journal notes he refers quite straightforwardly to his having *preached* it (SKS13K, 524). Since, in his view, the distinction between discourses and sermons has to do with authority (the sermon is authoritative, the discourse not), this suggests that the discourse has significant authoritative status within Kierkegaard's over-all body of religious writings.[36]

It is not hard to guess why the emphasis was important at just this point of the *Kirkestorm*. To those who might be imagining that he had abandoned his earlier positions and was offering some radical new teaching, possibly even running in the direction of a secularist point of view, Kierkegaard couldn't have said more clearly: No, the polemical thrust of *The Moment* is so far from negating the need for upbuilding that it calls for that need to be re-stated; conversely, it is precisely reflection on the unchangeable character of God that most accentuates the need for a polemical reck-oning with a Church that has proved only too changeable—to the point of having abandoned the Christianity of the New Testament. As in the Preface to *Two Communion Discourses* from late summer 1851, in words that themselves 'repeat' the *Postscript*, these are the words of a thinker who 'has nothing new to bring' and 'wishes only to read through the old

---

[36] For a full discussion of the question of authority in the discourses see my *Kierkegaard's Upbuilding Discourses*, 12–34.

foundational text of human existential relationships—those ancient, familiar texts, handed down from the fathers—yet one more time, if possible in
a more inward way' (SKS12, 281/WA, 165). Here, of course, the theme of
divine unchangeability reinforces the emphasis on consistency and same-
ness: what use would it be to say anything different or new about a God who
never changes? But if that might seem to mark a difference from the two
communion discourses, the theme of which is emphatically love, the opening prayer of *The Unchangeability of God* leaves us in no doubt that God's
unchangeability and God's love are two sides of the same coin. If a man is
to make himself unchangeable, Kierkegaard writes, he must resist letting
himself be moved by anything external to him, but in the case of God it is
different. God is moved by everything, precisely because he is infinite Love.
He is, in fact, moved by even the most insignificant aspects of creation and,
unlike human beings, cannot be indifferent to the hunger and need of even
a sparrow. Thus, unchangeable as God is, the prayer asks that he be moved
by the prayer to bless it (the prayer) so that the one who prays may be conformed to His unchangeable will, which is, simply, Love. The relationship
being sought is therefore so far from being one of stasis as to be essentially a
relationship of mutual movement, turning towards the God who is always,
constantly, turning towards us.[37]

If the themes of this discourse thus reprise those of many earlier discourses, they also reach even further back, right to Kierkegaard's 1834 stu-
dent notes on Schleiermacher's *Glaubenslehre*. The final group of excerpts
Kierkegaard made at that time reflect Schleiermacher's discussion of 'The
Divine Attributes which are related to the Religious Self-consciousness so
far as it expresses the General Relationship between God and the World'.[38]
As Schleiermacher immediately glosses this: 'All attributes which we
ascribe to God are to be taken as denoting not something special in God,
but only something special in the manner in which the feeling of absolute
dependence is to be related to him' (SKS27, 13:7[39]). Amongst the attributes
discussed by Schleiermacher and listed in Kierkegaard's notes are inalterability and eternity, and he also noted Schleiermacher's argument that there
is no religious requirement to distinguish between eternity and unchangeability. 'Unchangeability' is merely a way of applying the idea of eternity
in relation to the world and is 'merely . . . a cautionary rule to ensure that
no religious emotion shall be so interpreted, and no statement about God

---

[37] If we take Pike's view that timelessness and immutability are mutually entailing, this clearly
has significant implications for Kierkegaard's position on God's timelessness (see the section 'The
Life and Personality of the Timeless God' in Chapter 1 of this volume).

[38] Friedrich Schleiermacher, *The Christian Faith*, ed. and tr. H. R. Mackintosh and J. S. Stewart
(Edinburgh: T. & T. Clark, 1928), 194. For full discussion of Kierkegaard's Schleiermacher notes
see Chapter 1 of my *Kierkegaard and the Theology of the Nineteenth Century*.

[39] Cf. Schleiermacher, *The Christian Faith*, 194.

so understood, as to make it necessary to assume an alteration in God of any kind'.[40]

Applying these comments to the sermon, we can see that Kierkegaard was from the beginning familiar with the philosophical question as to how concepts drawn from the sphere of human temporal experience might really say anything significant about a God who is not merely 'unchangeable' vis-à-vis creation but eternally self-same.[41] Is Kierkegaard's appeal to divine 'unchangeability' a figurative and, as it were, 'all-too human' representation of an idea of God as non-temporally eternal, or is he thinking it in such a way as to open up the possibility of an interpretation of divine eternity as capable also of entering into time and remaining itself in and through the flux of time as, shall we say, 'abiding love'? To address these questions we move now to a closer exegesis of the text.

It begins, as some but not all of Kierkegaard's other discourses begin, with a prayer. Its opening word of address to God—'Du'—effectively inscribes the whole of the discourse within what Rosenzweig will describe as the mode of vocativity.[42] As Dorothea Glöckner argues in relation to the framing of *Works of Love* by the prayer with which it opens, this can also be read as showing Kierkegaard's understanding of the God-relationship as a relationship of call and promise.[43] It is the word of one who knows that his ultimate flourishing lies in being able to call on the God who, he believes, is manifest in the gospel word of love that also calls out to him.

Following the prayer, the discourse proceeds to reject a religious approach that is content merely to lament the transience and inconstancy of life in time and find consolation in the thought that one day it will all come to an end. Such a lament is untrue to the text, which, in pure 'joy and gladness', speaks only of God's unchangeability and not of human changeability. Perhaps these are not so easily separable as Kierkegaard here implies and his own analogy of the mountain and the traveller, arguably the defining image of the discourse as a whole, might seem to endorse such a negative dialectic. 'Imagine a traveller', Kierkegaard writes,

> who has been brought to a halt at the foot of a great mountain he cannot scale. Yet…everything he wishes and longs for, everything he desires is on the other side. All that is needed is that he gets to the other side. 70 years pass, but the mountain remains unchanged, impassable. Let him live another 70 years and the mountain will remain unaltered, still blocking his path, unchanged, impassable. Perhaps he himself will be altered, and what he once wished and longed for, what he once desired, will no longer matter and perhaps he scarcely knows himself any more. A new generation

[40] Schleiermacher, *The Christian Faith*, 206.

[41] These notes also indicate that Kierkegaard was aware of the classic teaching that knowledge of God is always characterized by the interrelationship of the three ways of analogy, eminence, and negation, reflected in his own later insistence that all language about God and about spiritual existence is metaphorical or 'transferred' (SKS9, 212f./WL, 209f.).

[42] See the section 'The Eternal' in Chapter 9.   [43] Glöckner, *Das Versprechen*, 2ff.

finds him there, altered, sitting at the foot of the mountain that is itself unchanged, impassable. 1000 years pass, and all that is left of this so alterable man is his legend, but the mountain remains unchanged, impassable. And now imagine the one who is eternally unchangeable, for whom 1000 years are as a day…as a mere 'Now', really as if they didn't exist at all—if you want to make progress by any other way than the one he allows—woe betide you! (SKS13, 332/M, 273)

The contrast between human changeability and divine unchangeability and the implicit further contrast between human temporality and divine eternity could scarcely be more strongly stated. But is it that simple?

Reminding his listeners/readers that the discourse is commenting on the words of an apostle, Kierkegaard then, briefly, addresses God directly, stating that **'we shall therefore talk, both to alarm and to calm, if that is possible,** *about You, You the Unchangeable One,* or *about your Unchangeability'* (SKS13, 330/M, 271). The shift to second person address is not sustained, but it is force-fully highlighted in the bold font in which the discourse's subject-matter is set out, as if to add a further reminder that, despite its third-person form, what follows is to be understood in accordance with the principle of vocativity: that God can only be spoken of rightly as one who is called upon, in the manner of address.

In the first instance, Kierkegaard suggests that the thought of God's unchangeability is frightening, 'sheer fear and trembling' (SKS13, 331/M, 272)—another reprise of a familiar Kierkegaardian theme. And what is espe-cially frightening is precisely God's silence. This might be construed as a sign of God's non-existence, leading us to forget God and therefore also to forget that He forgets nothing—which, Kierkegaard suggests, is a truly frightening thought. Think of a situation, he says, in which a man speaks rather thought-lessly about his ambitions for life while the one he is talking to says nothing, quietly listening. Time passes. The speaker himself has entirely forgotten what he said and has developed his life in quite another direction from the one his words had suggested. But the listener has taken what was said to heart and even made it the basis of his own life. What a mockery of human constancy, muses Kierkegaard. Such analogies suggest a tragic view of the human heart as 'a graveyard of forgotten promises, resolutions, decisions, great plans and little bits of plans and "God knows what"—Yes, that's how we human beings talk, for we rarely think about what we are saying' (SKS13, 335/M, 277). But, of course, Kierkegaard thinks that God does indeed 'know what': 'What has become changed in your memory, He knows as unchanged—no, He knows it as if it was today' since He has 'an all-knowing memory, an eternally unchangeable memory from which you cannot escape, least of all in eternity' (SKS13, 336/M, 277). Everything is 'eternally equally present' and

neither the shifting shadows of forgetfulness or excuse-making change Him. No, nothing is in the shadows for Him. And if we, as it is said, are shadows, He is eternal clarity in its eternal unchangeability. If we are like shadows that flee, then, my soul, take care, for whether you wish it or no, you are fleeing towards eternity, to Him, and He is eternal clarity. (SKS13, 336/M, 278)

In these terms, God's very existence already *is* the last judgement: He is the light in which everything is seen for what and as it truly is and *His eternity is the luminosity that reveals how the world really is.*

And yet, Kierkegaard continues, if this is frightening it is also consoling, since (against the Augustinian view of time as an infinite vanishing), it means that time—the unchangeable time of God's abiding and insistent presence—is the gift that makes it possible for us to choose and to hold fast to God, that is, to live in the tension of call, prayer, and praise. As Kierkegaard has said earlier in the discourse: 'He gives time, and He can do that because he has eternity, and he is eternally unchangeable' (SKS13, 333/M, 274[44]). Reading this back into the claim made in *The Concept of Anxiety* that the moment of vision is an atom of eternity, this further suggests that this is not so much a matter of some fixed metaphysical structure but is to be understood in terms of time being given and being given precisely as the possibility of the (temporal) human being's entering into an abiding relation to the eternal God. The world (that is, the temporal world or life in time) comes to be interpretable as sheer possibility. As Kierkegaard puts it earlier in the discourse: 'In every moment [God] holds everything actual in His almighty hand as possibility; in every moment He has everything in readiness, in a flash he can change everything, human beings' opinions and judgments, human greatness and even lowliness, He changes everything—Himself unchanged' (SKS13, 330/M, 271). But this unchanging condition of possibility includes precisely and pre-eminently the possibility of human life being transformed from despair to hope. And this change itself is premised on the call to hope found in the divine promise of forgiveness.

It is easy to overlook the theological radicalism of the statement that everything actual is really possibility unless we remember that, following scholasticism's incorporation of the Aristotelian categories of possibility and actuality into fundamental theological reflection, the mainstream of Western Christian thought has consistently conceived of God as *actus purus*, pure actuality, and that this pure actuality is the necessary foundation of all that is: actuality determines potentiality or possibility, not vice versa. Furthermore, on this view, creatures draw closer to God by virtue of and to the degree that they actualize their given possibilities, shedding the penumbra of non-being that always clings corrosively to possibility as such. Here, however, Kierkegaard is saying that the world as a whole—'everything actual'—is, in the event, purely and

---

[44] The Hongs' translation has 'he takes his time', which seems an unnecessary departure from the plain meaning of the text.

solely possibility: because God exists as the eternally unchangeable one, reality is a field of open possibilities and that knowing it—knowing ourselves—as possibility is the condition of the God-relationship. And to know the world in this way is also to know it as temporal.[45]

All of this brings us to the point at which we can start to see some crucial connections between the eternal, time, and love. For if time is the gift of the eternal and is the mode in which the eternal opens a field of possibilities for human action and decision then it is not simply the purely formal condition of a freely-chosen God-relationship but also the condition of a God-relationship worthy of being called a relationship of love. Moreover, such a relationship can be a 'love'-relationship both from the divine and from the human side. From the divine side, a God who consents to relate to creatures in time is a God who, as Kierkegaard emphasized at the start of the discourse, moves and can be moved. From the human side, love, truly to be love, must involve an element of free choice or acceptance that may also be glossed as responsiveness to the word of forgiveness—but without time there can be no possibility of choice.

Yet, as Kierkegaard also notes in this discourse, the reality is that human beings often find it difficult to accept love, not least because we don't always know what is good for us and therefore interpret gestures or words of love as threatening our self-image. Socrates experienced his interlocutors' anger at being stripped of their illusions, and the help that is offered by an eternally and unchangeably loving that God is likely to be even more disturbing to human self-images. It is difficult for us to to submit meekly and without argument to such a love. And it is especially difficult if we implicitly experience time as mere infinite vanishing, since our aspirations to permanence are wedded to such structures as we ourselves have developed to help us to hold out against time—like the aesthetic personalities analysed in the second Part of *Either/Or*, who cling to success or beauty as a kind of prophylactic against the corrosive power of time but who, precisely by the way in which they attempt to resist change, make themselves vulnerable to time's corrosiveness (SKS3, 175-87/EO2, 180–92). But if time is the gift of the eternal and the gift of the eternal is to be found only in time, then we do not need to hold out against time in order to find something that abides in the midst of change. Time gives us the possibility of experiencing our lives otherwise and possibility is therefore the means by which, in, with, and

---

[45] This might seem to conflict with Climacus' argument in the *Postscript* that existence is directed towards actuality and that, existentially, actuality is to be privileged over possibility. However, Climacus' concept of 'actuality' is itself novel, since when actuality is identified with existence (in Climacus' sense of the word) it is identified with what is characterized by temporality and the possibility (indeed, the ineluctability) of change, whereas there is no possibility of temporal change or movement in pure actuality conceived according to Aristotelian principles. See also my *God and Being: An Enquiry* (Oxford: Oxford University Press, 2011), Chapter 7.

under the limitations of temporal life, the eternal becomes present to us. In these terms, possibility and time combine in the experience of hope. As we have heard Kierkegaard put it in *Works of Love*: 'By means of possibility, eternity is always sufficiently *near* to be at hand and yet sufficiently distant to keep a person moving forwards, towards the Eternal, in motion, progressing. Using possibility in this way is how eternity entices and attracts a person onwards, from the cradle to the grave—if only we choose to hope' (SKS9, 252/WL, 253).

The discourse on the unchangeability of God closes with another parable, which we might call the parable of the traveller and the desert spring. As with the parable of the traveller and the mountain this seems initially to invite being understood in terms of the contrast between human transience and divine eternity. A traveller in the desert is refreshed by a cooling spring. When he revisits the spot years later, he finds the spring dried up. In contrast to such a transient source of comfort and refreshment God is like a refreshing spring that accompanies us throughout our journey through life and, in describing this, Kierkegaard turns again to vocative mode: 'But You, O God, You unchangeable One, You are unchangeable and are always to be found, and always let Yourself be found unchanged; no one can travel so far, whether in life or in death, that You are no longer to be found and are not there; for You are everywhere' (SKS13, 338–9/M, 280–1). Even in relation to the dried-up spring, when there is no longer anything to be grateful for, the traveller still has the possibility of becoming grateful and therefore of returning to a relation of call-and-response to God and *precisely in this way* revealing 'that there is something unchangeable in a human heart' (SKS13, 338/M, 280).

Gratitude, freely given, opens a path towards the One to whom it is given and is a continuing condition of a movement of mutual love.[46] And if the possibility of a fundamental ontological gratitude opens a way towards the God-relationship from the human side, the final figuration of God given in the discourse is as a refreshing stream that not only accompanies us throughout our earthly wandering but actively seeks the one it is to refresh: 'You, You are like a spring that itself seeks those who thirst and who are lost, something that has never been said of any other spring' (SKS13, 339/M, 280–1). We always have the possibility of being grateful, even when, in economic terms, there is no longer anything to be grateful for: and God is always seeking to give us something for which to be grateful, even before we need it, always opening

---

[46] Kierkegaard's gratitude, I have suggested elsewhere, is not to be seen as a case of economic exchange but is rather to be interpreted in the spirit of Nietzsche's '*schenkende Tugend*', i.e. as a manifestation of power, and life and therefore as a mode of freedom. See my *Kierkegaard and the Theology of the Nineteenth Century*, 118. See also Jankélévitch's account, discussed in the section 'Eternally Irrevocable' in Chapter 5 of this volume.

up a new horizon of possibility. Exploding the logic of economic exchange, Kierkegaard thus portrays the God-relationship as a relationship of mutual and hopeful freedom enacted in the open space of temporal possibility that is itself the gift, the continually renewed and constant gift, of the eternal. Or, which is to say the same, the eternal is present to us and can only be present to us as the abiding and unchanging divine presence that awakens us, in time, to possibilities of hope and love—which, for Kierkegaard, is precisely the event of the gospel word of promise and challenge. And all this is possible because God Himself is no unmoved mover or Pure Act but a God who, moved by prayer, moves towards us in word and love.

FROM HEGEL TO HEIDEGGER

Although I have elsewhere described Kierkegaard as seeming like 'a lonely and forbidding mountain in the cultural landscape', it is clear that his thought is extensively and deeply intertwined with his reception of the heritage of classical German philosophy and that, for us, 'Kierkegaard' is also entangled in a complex effective history in which the recurrent comparison with Nietzsche and the transformation of his religious thought in the godless philosophy of *Being and Time* are major elements.

   With regard to Hegel, much will of course depend on whether we read Hegel (with the British Idealists) as teaching a kind of Spinozan view of temporal existence *sub specie aeternitatis* or (with Kojève) see him as anticipating Heidegger by thinking through the implications of a radically temporalized view of human being. Without engaging the overall question of 'Kierkegaard's relations to Hegel', we can see how each of these tendencies plays some role in Kierkegaard's treatment of time and eternity. In a work such as the *Postscript* the polemic against Hegel is directed precisely against what Kierkegaard sees as the claim of speculative philosophy to reach a view on existence *sub specie aeternitatis*. As we saw in Chapter 4 he launches a double-pronged attack on this claim, both contesting its logical foundations and subjecting it to a merciless barrage of satire, culminating in the declaration that such a philosophy is fit only for moon-dwellers.[47] But—and although he rejects Hegel's own reinterpretation of the Aristotelian categories of possibility and actuality—the 'temporal' Hegel who sought to integrate the passage of time into philosophical reflection seems surely to be part of the context for Kierkegaard's own attempts to think the existence of the human subject as a thoroughly temporal existence. As we have been

---

[47] See the section 'God is in Heaven, We are on Earth' in Chapter 4.

seeing, this involves Kierkegaard too in calling upon the eternal as a means of accounting for what is at stake in such temporal existence, leaving him open to the Heideggerian complaint that he is still thinking time by reference to eternity. We shall return to this complaint, but immediately note that if Kierkegaard, like Hegel, thinks time in the perspective of eternity there is nevertheless a great and, I suggest, decisive difference in the manner of how they set up this perspective. For Hegel, what is being sought is a form of theoretical knowledge, namely, cognition of the absolute appearing in time. But whatever else is going on in Kierkegaard's writings, it is not a cognitive relation to the eternal that is at issue (and here, he seems to be in basic agreement with Schleiermacher). 'Knowledge' is not the defining feature of the God-relationship, which is instead constituted in terms of such existential categories as faith, hope, and love.[48] Thus, if the attack on Hegel's claims to cognition of the absolute constitute the negative aspect of his thinking about the relations of time and eternity it is these existential categories that provide the positive element—only we must be careful to interpret them in the light of the whole development of his authorship and not just in relation to a few arbitrarily extracted killer quotes about irrational leaps of faith.

If the centrality of freedom in Kierkegaard's account of the self's relation to eternity may seem to bring him into proximity to Schelling and Nietzsche, Kierkegaardian choice does not aspire to the kind of sovereignty they proclaim, since it is always essentially responsive. This is not simply or immediately responsiveness to an abstractly conceived 'God' but to a God who can only be believed in and hoped on in the context of the whole richly structured weave of human life, aspirations, values, and relationships. This is also why, for *Works of Love*, hope—the medium of our relation to the eternal—is inseparable from the twofold relationship of freedom to self and responsibility towards and for others: our freedom to choose ourselves in relation to the eternal is the obverse of how we choose to be in our relations to others. In later religious works, this is also developed in terms of the drama of forgiveness exemplified in the encounter of Christ with the 'sinful woman' of Luke 7.[49] Against Heidegger's category of 'being-with' this therefore places Kierkegaardian freedom in the situation of a responsibility that is 'face-to-face' with the other and, in every case, qualified by an answerability that stretches beyond the deliverances of its own conscience.

But this still leaves open the question as to whether Kierkegaard's eternity is in any sense identifiable with Boethius' timeless eternity. In fact, despite the apparent emphasis on timelessness in *The Unchangeableness of God*,

Kierkegaard's God has been shown to be capable of reaching into time and becoming involved in temporal relations, in other words, God is a 'God in time'. This, however, would seem to have the further implication that such an 'eternal' being may also be described as having duration if it is not to simply dissolve into vanishing time. But if this would seem to be the force of the analogy of the mountain in *The Unchangeableness of God*, Kierkegaardian duration is not a primarily metaphysical attribute separable from God's freedom to be constant, faithful, and patient 'to usward'. As the title of another of the discourses in *Works of Love* has it, 'love abides'—but, note, *love*, that is, a work of freedom, is what abides. If it is possible to speak of God's timelessness in a Kierkegaardian perspective, this will therefore not be as an answer to questions about the divine attributes but as responding to the vision of time as more than time and therefore allowing for the possibility of a kind of love and faithfulness that does not pass with the passing of time, the kind of possibility that might therefore be believed on in terms of divine promise. When such a promise is taken as definitive of the God-relationship, our access to God and to God's time can only be revealed in the ever-renewed possibility of our having time to hope and time to love. The eternal comes to be in time as the other of time, as that which, giving time, shows itself as beyond all measure of time. Heidegger is therefore right in saying that Kierkegaard defines time by means of the eternal, but misses the decisive point that Kierkegaard's 'Eternal God' is not to be understood in a speculative or metaphysical horizon, but solely and exclusively in relation to the call that promises the possibility of love as ever-returning beyond all experiences of envy, betrayal, and misunderstanding.

WHO CALLS?

Kierkegaard, then, depicts human beings as called to know God in the temporally-inflected possibility of living lives of love—but who, precisely, calls us? What is the character of the word that could awaken and nurture such a possibility?

Such questions point us to the centrality of issues of communication and language in Kierkegaard's authorship. Fully to explore these issues would lead us away from the central concern of this study and, to the extent that I have not dealt with them extensively elsewhere, must largely be postponed for future research. Here, I shall limit myself to merely outlining what I see as a Kierkegaardian approach.[50]

---

[50] On the negative aspect of 'the poet' see my *Kierkegaard: The Aesthetic and the Religious* (2nd edn, London: SCM Press, 1999), especially Chapter 2 'The Genealogy of Art', 35–62.

One candidate for speaking the saving word might be 'the poet' and we have seen how Theunissen could portray Pindar as one such poet, bestowing a word that might empower his listeners to a more hopeful orientation towards time. On Kierkegaardian premises, however, the strictures on the aesthetic extend also to the poet who is, in effect, the epitome of both the power and the tragedy of the aesthetic. As he put it in a statement quoted earlier in this chapter, 'It is through the negation of the imperfect reality that poetry opens up a higher reality, extends and transfigures the imperfect into the perfect, and thereby reconciles the deep pain which seeks to darken all things. Thus far poetry is a kind of reconciliation but it is not the true reconciliation for it does not reconcile me to the reality in which I live'.[51] Like the aesthetes examined above, the poet is essentially a sufferer and his songs are likened by Kierkegaard to the screams of those being roasted alive in the bronze ox of Phalaris, a device that transformed the tyrant's victims' cries into a semblance of sweet music.[52] Yet in Kierkegaard's post-Romantic age of cultural Christianity this is what the majority crave and are satisfied with, since to want more would be to face the challenge of self-transformation that they experience as too demanding and therefore, in effect, typically avoid experiencing at all. In the late discourses on *The Lily of the Field and the Bird of the Air*, Kierkegaard returns to the poet as exemplifying the despair of his age and here the issue is said to be precisely that the poet speaks as if he *wants* to live as simply as the lily and as freely as the bird of the air—only he is not prepared to *will* it, as the Gospel requires.

Key to the poet's problem is that, cut off from the truth of the eternal in time and letting himself be trapped in an illusory idealistic eternity, he is unable to let himself be open to the gift of a true relation to the eternal that God is constantly offering in the life of creation. 'So, the fact that you came into the world, that you exist, that "today" you have got what you need in order to exist, that you came into the world, that you became a human being, that you can see—just reflect on the fact that you can see—that you can hear, that you can smell, that you can taste, that you can feel... is this nothing to be joyful about?' (SKS11, 43–4/WA, 39–40), Kierkegaard asks. In short, we have all the possibilities we need to be the creatures that we are—if only we paid attention to them. But separated from the truth of life lived in a temporally inflected relation to the eternal the poet is thrown back on his own 'genius', that is, on the possibilities associated with his solitary ego. These may, in some cases, be considerable (for Kierkegaard's age, Goethe provided an exemplary case of what genius could achieve)—and are likely to sufficiently overawe those less well endowed. But a word that has its sole root and ground in individual genius is by definition not a word that arises from a responsive and responsible relation to others. It is therefore consequently not a word capable of giving

---

[51] SKS1, 330–1/EO2, 297.   [52] SKS1, 27/EO1, 19.

any possibility of genuine relationship. In the end, the poet speaks with and to himself, his genius, his muse and does not address his audience as one responsible individual to another.

It is for such reasons that in the first of the discourses on the lily and the bird, Kierkegaard asks his readers to learn silence, in the specific sense of becoming silently attentive to God. He writes:

> If...you did indeed pray with real inwardness...you had less and less to say, and finally you became entirely silent. You became silent and, if it is possible that there is something even more opposed to speaking than silence, you became a listener. You had thought that praying was about speaking: you learned that praying is not merely keeping silent but is listening. That is how it is. Praying is not listening to oneself speak, but is about becoming silent and, in becoming silent, waiting, until the one who prays hears God. (SKS11, 17/WA, 11–12)

This passage suggests that the call to a true relation to the eternal can come to each individual from God, more or less immediately. In the larger context of these three discourses, it is clear that, as has been said, it is in fact a call mediated by reflection on the gifts of God in creation rather than a direct mystical experience of God. Is this enough? Elsewhere, Kierkegaard opposes the apostle to the poetic genius, arguing that only the authoritative word of an apostle can free us from the tyranny of time for a living relation to the eternal.[53] The further logic of this view is that the power of such an apostolic word is itself derived from the word spoken by God in time as the Word made flesh. There is a certain tension here that I have glossed elsewhere in terms of a theology of creation and a theology of redemption.[54] In between the world of the lilies and the birds and the pure Word of God, however, are the world and words of human intercourse. Here, in the light of what we have learned from Kierkegaard's teaching on love, we might say that any human word that is spoken with the intention of love and that calls us, for our part, to greater love will be a word that is also calling us to the possibility of something eternal in time.

Must we choose between these three kinds of call, respectively embodied in creaturely life, the revealed Word of God, and the words of human life together? The question is, in essence, the question previously posed in terms of the choice between a view of the eternal as opposed to and contradicting the world of time and a view that sees time itself as bearing possibilities for release from bondage to time into another time of creative living. If strong currents of twentieth-century theology tended to privilege the Word of God at the expense of the words of creation and human communication, the tendency of the present study is that, as regards the relations of time and eternity, only a

---

[53] See 'The Difference between a Genius and an Apostle' in SKS11, 95–11/WA, 91–108.
[54] See my *Kierkegaard and the Theology of the Nineteenth Century*. One could again think of this as the tension between a Barthian and a Tillichian reading of Kierkegaard.

word that is spoken in, with, and under the temporal conditions of human life in the world will also be a word capable of transforming our time-relationship and making possible a life in relation to the eternal. That this is not just a matter of 'language' but of mutually responsive and responsible discourse is, I suggest, a position to which Kierkegaard's linking of time and love points us. That such a position, at least as regards its overall tendencies, is also Kierkegaard's, goes beyond what can be argued within the parameters of a chapter such as this. I claim here only that it is a position that Kierkegaard's writings make possible. However, as such it immediately raises the further question as to whether, in this case, we might look to a certain rehabilitation of the poet. Why? Because if, as I am suggesting, human speech can serve the opening up of a relation to the eternal, then the possibilities of poetic discourse cannot be limited to those versions of poetic discourse that express only the tyranny of time. Albeit from an angle scarcely considered by Kierkegaard, the distance between the poet and the apostle begins to contract. Perhaps poets too can and might become witnesses to the power of eternal love?

9

Rosenzweig: The Eternal People

BETWEEN PHILOSOPHY AND THEOLOGY

Kierkegaard's account of how we might come to know the eternal God supposes that love for the neighbour is integral to any such knowledge. Nevertheless, he consistently focuses on the individual's role in building up a community of love, inviting the familiar charge of excessive individualism. If this charge is misplaced, it is still clear that he regarded contemporary society as fundamentally alienated from a saving relation to the eternal and as having unreservedly given itself over to the tasks and distractions of temporal life. Still worse, he saw this denial of eternity as becoming institutionalized in the emergent modern democratic state, even when—especially when—this incorporated a 'national' religion, since such a 'religion' will all the more occlude the possibility of a genuinely individual relation to the eternal that would be capable of radically transforming life in time. A very different view of the role of community is found in the religious philosophy of Franz Rosenzweig, which, as we shall see, has certain affinities with Kierkegaard but also envisages salvation from time in time—saving time—in terms that are more strongly and explicitly communal. In the case of Rosenzweig, this is primarily the communal life of the worshipping Jewish community but the model that he proposes has wider significance and, as we shall see, also allows for a critical reserve vis-à-vis political forms of community such as nation or state. Rosenzweig's doctoral thesis on *Hegel and the State* already broaches the question as to the relationship between individual, community, and the eternal and *The Star of Redemption* too shows how Jewish identity is intertwined both with the fate of the ancient pagan world, with the development of Christianity, and with the modern experience of a profound rupture in the relationship between time and eternity.

Like Kierkegaard, Rosenzweig rejects speculative philosophy's theoretical claims whilst also attempting to develop an alternative model of the relations between time and eternity that involves rethinking both. In doing so, he also anticipates Heidegger's suggestion that a new way of thinking time might yield a new way of thinking eternity—beyond the alternatives of timelessness and sempiternity. As we shall see, the manner in which he does so brings him into a perhaps surprising proximity to Heidegger's own 'poetic' politics.[1]

Rosenzweig, then, belongs firmly amongst the inheritors of German Idealism and its critics, and each of Hegel, Schelling, Kierkegaard, and Nietzsche are amongst those who play a significant role in the evolution of his thought as a whole and *The Star* in particular. Goethe too would have to be mentioned in any list of major influences from the German tradition. Yet—and the disjunction reminds us of the inescapable and immense intellectual effects of the catastrophe that he himself did not live to see—if Rosenzweig's thought is 'German' it is also 'Jewish' and has a special if debatable place in the narrative of modern Jewish religious thought. But are we speaking of a philosophy of religion, perhaps, or even a metaphysics, or of a theology? Or some kind of mixture of all of them? And, if it is the latter, is it merely some confused jumble or the development of a 'new thinking' that demands the revision of all existing views of 'philosophy' and 'religion' in relation to the central existential questions of what Rosenzweig would call God, Human Being, and World?

These questions can be focused by reference to Rosenzweig's place in a certain lineage of modern Jewish thinkers. We might sketch this lineage with the names Hermann Cohen, Martin Buber, and Emmanuel Lévinas and, in doing so, might be tempted to construct a narrative in which Rosenzweig is situated as the pupil of Cohen, the collaborator of Buber, and the inspirer of Lévinas.[2] This has some merit—but what kind of pupil, what kind of collaborator, and what kind of inspirer are we talking about? Do we read this as a straightforwardly philosophical lineage? Or do we read Rosenzweig as offering a theological interpretation of Cohen that is then reclaimed for a radicalized version of philosophy by Lévinas?[3] Or does the connection to Rosenzweig qua theologian actually reveal that Lévinas too is, in the end, more of a theologian than his philosophical style might suggest?[4] And perhaps Cohen—the later

---

[1] See the section 'But What Abides, That is What the Poets Found...' in this chapter.

[2] A lineage we could continue on to Derrida. See Dana Hollander, *Exemplarity and Chosenness: Rosenzweig and Derrida on the Nation of Philosophy* (Stanford, CA: Stanford University Press, 2008).

[3] Rosenzweig himself says that his 'new thinking' is not a theological thinking, although he acknowledges important theological impulses. See 'Das Neue Denken', in Franz Rosenzweig, *Zweistromland: Kleinere Schriften zu Glauben und Denken*, ed. Reinhold and Annemarie Mayer (Dordrecht: Martinus Nijhoff, 1984), 152.

[4] In the Preface to *Totality and Infinity*, Lévinas commented that Rosenzweig was 'too often present in this book to be cited' (Emmanuel Lévinas, *Totalité et Infini: Essai sur l'extériorité* (4th edn, Paris: Kluwer Academic, 2001), 14). Amongst the various ways of reading the Rosenzweig/ Lévinas relationship we find those that emphasize the similarities, such as Robert Gibbs, who writes that 'Levinas should be read as a Jewish thinker in the class of Rosenzweig and others, and that Rosenzweig should be read as a philosopher—specifically a postmodern philosopher' (Robert Gibbs, *Correlations in Rosenzweig and Levinas* (Princeton, NJ: Princeton University Press, 1992), 10). Leora Batnitzky, however, sees Lévinas as essentially thinking out of the Christian-inflected tradition of modern German philosophy, as opposed to Rosenzweig, who

Cohen whom Rosenzweig most specifically endorses—might also turn out to have been more of a theologian than his Neo-Kantian reputation would lead us to believe?[5] And where, in this narrative, do we fit Buber, the proponent of a seemingly less confessional and more universal version of Judaism? What is 'philosophical', what is 'theological' in each of these views? Is it a question of theology presupposing what Hegel might have called the 'positivity' of religious teaching as opposed to the universality of philosophy? Or might there be a kind of 'positive' religion—Rosenzweig's Judaism, perhaps—that was universal precisely on the basis of its positivity? Perhaps that is, after all, not so alien to Hegelianism itself, if we read that as having attempted to construct a universal hermeneutic of history on the basis of Christianity's 'positive' revelation.[6]

This chapter will move towards a reading of the theme of the Eternal in *The Star of Redemption*. In the course of this reading, we shall return to many of the questions that have just been posed and that serve to alert us to what readers schooled in Christian theology and German philosophy are likely to think of as crucial points of interpretation. This will lead us via a comparison with Heidegger's account of the poetic basis of community to reflect on different ways of construing a (or even *the*) human community as 'eternal'. However, by way of preparation, I shall briefly consider two aspects of Rosenzweig's thought that provide some general orientation to what *The Star* will say about the eternal. These are the significance of Hegel's view of the state, and the role of time and language in what Rosenzweig called 'the new thinking' that he

thinks from the particular historical context of Judaism. See Leora Batnitzky, 'Levinas between German Metaphysics and Christian Theology', in Kevin Hart and Michael A. Signer (eds), *The Exorbitant: Emmanuel Levinas Between Jews and Christians* (New York: Fordham University Press, 2010), 17–31. Stéphane Habib thinks that it is not a question of Lévinas being influenced by or radicalizing Rosenzweig, but simply that they are just different. See Stéphane Habib, *Lévinas et Rosenzweig : Philosophies de la Révélation* (Paris: Presses universitaires de France, 2005). (Disentangling Jewish and Christian threads in each of Buber, Rosenzweig, and Lévinas would, of course, further complicate the question as to the relationship between philosophy and theology in their respective bodies of work.)

   [5] Although Cohen's 'religion of reason' is argued with specific reference to the textual 'sources' of Judaism, these sources are read according to criteria that are derived from an analysis of reason itself, rather as Lévinas develops his ethics out of what starts as a seemingly Cartesian analysis of self-consciousness. Amongst the themes that might be seen as running from Cohen through Rosenzweig to Lévinas are the inseparability of knowledge of the 'I' from ethical responsibility for the suffering other as 'neighbour'; God—and, in relation to God, the human subject—as uniquely singular; a critique of views that see God's Being in terms of substance (although Cohen accepts the interpretation of the 'I am' of Exodus 3 as Being (*Sein*)); and, not least, the consummation of prophetic responsibility for the other in a horizon of universal Messianism. See Hermann Cohen, *Die Religion der Vernunft aus den Quellen des Judentums* (2nd edn, Frankfurt am Main: Kaufmann, 1929). For a good overall view of the Rosenzweig–Cohen relationship see Chapter 1 of Hollander, *Exemplarity and Chosenness*, 13–39. Hollander also draws links between Rosenzweig and Lévinas in her article ' "A Thought in Which Everything Has Been Thought": On the Messianic Idea in Levinas', *Symposium: Canadian Journal of Continental Philosophy* 14/2 (Fall 2010), 133–59, especially 141 n. 7 and the sections 'Anticipation' and 'Accomplishment', 143–9.

   [6] See also n. 2, with reference to Cohen.

believed was represented by his and Buber's work, as well as by the later Cohen and Heidegger. Finally—before coming to *The Star* itself—we shall examine a short text, 'The Eternal', in which Rosenzweig makes explicit some of his key assumptions in approaching this issue.

HEGEL, THE STATE, AND THE ETERNAL

In the *Encyclopaedia of the Philosophical Sciences*, in the third sub-section of the section on 'Ethical Life' that is itself a part of the Division devoted to 'Objective Spirit', Hegel defines the state as 'self-conscious ethical (*sittliche*) substance'. As such, it is 'the unification of the principle of the family and of bourgeois society' and its essence 'is the reasonableness of a will that is in and for itself universal, but which is also utterly subjective, knowing itself and acting as *one* individual'.[7] In this sense the state is also both 'living Spirit'[8] and 'justice as it exists as the actuality of freedom'.[9] Although connected with the national life of a people, it is not identical with this. The nation remains too entangled with what is merely natural and not yet historical—a point that Hegel drives home by commenting that it is entirely correct that Walter Scott's (and others') so-called 'historical' novels focus on the deeds and passions of individuals as opposed to genuine historical events, precisely because they are *novels* and not philosophical interpretations of history.[10] But, against the novelists, a people can only enter the stream of world-history by virtue of the form of the state, and this world-history is, as Hegel several times emphasizes, 'the last judgement'.[11] But this also implies that the state is the means by which it becomes possible for human beings to have a full relation to the eternal, although this relationship is only fully known by the philosopher who reads the exigency of the now in such a way as to reveal the presence of the eternal, the rose of reason in the cross of present suffering.[12]

---

[7] G. W. F. Hegel, *Werke*, x: *Enzyklopädie der philosophischen Wissenschaften III* (Frankfurt am Main: Suhrkamp, 1970), 330.

[8] Hegel, *Enzyklopädie der philosophischen Wissenschaften III*, 331.

[9] Hegel, *Enzyklopädie der philosophischen Wissenschaften III*, 332.

[10] Hegel, *Enzyklopädie der philosophischen Wissenschaften III*, 350.

[11] Hegel, *Enzyklopädie der philosophischen Wissenschaften III*, 347. The reference is to Schiller's poem 'Resignation'.

[12] These remarks refer to the Introduction to the *Philosophy of Right*. For comment see Franz Rosenzweig, *Hegel und der Staat* (Berlin: Suhrkamp, 2010), 356ff. As Rosenzweig emphasizes, Hegel's ambition here stretches only as far as cognition of the rational core of the present actuality, which the same Introduction refers to as the retrospective 'grey in grey' of the Owl of Minerva that only spreads its wings at the end of the day. This does not mean that history is devoid of future possibilities, merely that these are not knowable in the same way as the past is knowable.

But how does the path to the eternal that passes through the world-historical phenomenon of the state relate to religion and religion's faith in the eternal?

To ask this is to raise the enormous question as to the overall role of religion in Hegel's thought, in relation to art and philosophy as well as to social and political life. In relation to this last, and as Rosenzweig's own study makes clear, Hegel passes through a number of phases. Rosenzweig himself played a remarkable role 'in the reconstruction of this history when he discovered a short text subsequently published as 'The Oldest Programme for a System in German Idealism', now dated to 1796 or 1797 and also sometimes ascribed to Schelling or to Hölderlin. This text not only reveals Hegel's political hopes, but also his view of the connections between politics, religion, and art. Rosenzweig sees it as setting the agenda—perhaps never fully completed—for the whole subsequent development of German Idealism. But we can also see it as greatly significant for his own future work.

We recall that the fragment dates from the decade of the French Revolution and its call to 'liberty, equality, and fraternity', which is also the decade of the birth of Early Romanticism (in which, of course, Hegel's friend Schelling would play a significant role), as well as of Schiller's *Letters on the Aesthetic Education of Mankind* (1794). In other words, it is a time when radical youth looked to surpass what they saw as the narrow intellectualism of the Enlightenment and its division of heart and head by developing a new holism, a development in which the poets would emerge as the true legislators of mankind. All of this is precisely the atmosphere captured in Hegel's brief notes.

The notes begin with the words 'An ethics', which Hegel immediately glosses in terms of all future metaphysics being a development of Kant's discovery of the primacy of practical reason, so that even physics will need to be seen in the light of the question as to how the world must be formed if it is to be a world in which genuinely free and moral beings can emerge. The state is rather peremptorily dismissed as 'something mechanical', whereas 'What we call an idea can only be something [capable of being] the object of freedom'.[13] The state treats human beings as cogs in a machine and, since that ought not to be, it, the state, 'ought to cease'. Hegel's intention is to develop a history in which he will show how the state acquired such power over human life, 'stripping it to the skin', how the higher ideas of morality, divinity, and immortality emerge and how, versus superstition and priestcraft, these ideas are not to be sought in anything external but in the intellectual world itself. Finally, he comes to how 'the highest act of freedom' is an aesthetic act that also embraces truth and goodness: 'The philosophy of spirit is an aesthetic philosophy', he declares. Poetry thus becomes 'the instructress' of humanity. But this not only fulfils the aims of philosophy, it also speaks to the situation that the masses, incapable of abstract thought, will always need a mythological religion. What the poetry of the future will offer, however, is precisely 'a mythology of freedom' that provides a common symbolic system for the whole of society, the enlightened and the unenlightened. 'Then eternal unity... [u]niversal freedom and equality of spirit will reign. A higher spirit, sent from heaven, must establish this new religion among us, which will be the final and greatest work of humankind.'

Brief as it is (and even if it was actually conceived by Schelling or Hölderlin), the fragment articulates a set of commitments that remain important in Hegel's subsequent political thought. For even if the text seems to drift away from politics towards poetry, art, and religion, the task assigned to these is precisely the task of countering and healing the harm wrought by mechanistic conceptions and practices of statehood. It is not a matter of poetry withdrawing from the tough demands of common life but of poetry and religion (a new mythology) providing the link that allows individuals to feel the common life as genuinely expressive of their own. The future society that Hegel here anticipates is one in which freedom is manifest as the spontaneous common purpose of each and all.

But if this provides a thread we can follow through Hegel's own later development, it undergoes many permutations, not least with regard to the role of religion and, specifically, Christianity. In the early period of Hegel's thought reflected in his so-called 'Early Theological Writings' the ideal attained its nearest historical realization in ancient Greece, in comparison with which the development of Catholic Christianity marked a progressive alienation from authentic religious life. Religious positivity—the demand to believe particular and exclusive dogmas—supplants the spontaneous expression of common feeling and priestcraft and obscurantism gain increasing power over hearts and minds. By the time of *The Phenomenology of Spirit*, however, Hegel is assigning a more constructive role to Christianity as the absolute religion and the precondition for Spirit becoming conscious of its own infinity of powers. Yet even here, the earlier idealization of the ancient Greek city-state continues to echo in the account of 'the religion of art'. As Hegel writes,

> The dwellings and halls of the gods are for the use of human beings. The treasures stored in them are, in case of emergency, their own. The honour that [the god] enjoys in being thus ornamented is the honour of the richly artistic and high-minded people [*Volk*]. In the festivals, these likewise decorate their own dwellings and costumes as well as their daily tasks with graceful accoutrements. In this way, they receive a return from the gods for their gifts and the evidences of the god's favour. In this way both [gods and the people] are bound together through their work, not in the hope of some subsequent event but as having the immediate enjoyment of their own wealth and brilliance as they do [the god] honour in bringing [him] their gifts.[14]

---

[14] G. W. F. Hegel, *Werke*, III: *Phänomenologie des Geistes* (Frankfurt am Main: Suhrkamp, 1970), 524–5; Eng. *Phenomenology of Spirit*, tr. A. V. Miller (Oxford: Oxford University Press, 1977), 435.

And, in the immediately following section he continues 'The people who draw near to their god in the cultus of the religion of art are an ethical people, who know their state and its actions as what they themselves will and accomplish'.[15]

This idyllic state of affairs comes to an end with the triumph of Rome over Greece. And although this is largely portrayed in negative terms, as fragmenting the living unity of the socio-religious bonds of the city-state, it eventually provided a more global context for the future development of history. In this new context, Christianity took on the creative role of animating the otherwise empty and abstract universality of Roman law and Roman imperial religion. Thus the way is prepared for the state to become what it is in *The Philosophy of Right* and *The Encyclopaedia*, namely, living Spirit and the actuality of freedom. Hegel will continue to be suspicious of what he sees Catholic attempts to intervene between the citizen and the state. The Church has its place as complementing the public and universal life of the state by virtue of its role vis-à-vis the individual, and the state should tolerate a range of different ecclesiastical communities; but woe to the Church if it starts to claim a jurisdiction rivalling that of the state—which is what Hegel sees Catholicism as constitutionally prone to doing.

Clearly, these few comments raise more questions than they answer, but with regard to what we shall see in Rosenzweig's account of Judaism as 'the eternal people' they are sufficient to flag both the significance that Hegel—and I believe Rosenzweig—gives to the interplay of art and worship in constituting the identity of a people and also the question that follows from that, namely, the degree to which a people, thus constituted, may also take on a political identity and perform political tasks.

TIME, LANGUAGE, AND THE NEW THINKING

We come next to two salient features of what, in an article from 1925 aimed at elucidating what was said in *The Star*, Rosenzweig called 'the new thinking'. These are time and language. Whereas conventional philosophy abstracts from living experience and asks about the 'essence' of its objects and their attributes, the new thinking sees that each of God, the human being, and the world is irreducible; each, as Rosenzweig puts it, is 'substance' 'with the entire weight of this expression'.[16] What distinguishes them is not that God has the attribute of being transcendent in relation to the world or to human being (for example), but

---

[15] Hegel, *Phänomenologie des Geistes*, 525 (Eng. tr., 435–6).
[16] Rosenzweig, 'Das Neue Denken', 144.

that they are, simply, different entities. Consequently, their mutual relations can only be explicated by narrating how they come to be in relation to each other—an approach pioneered in Schelling's *Ages of the World*. But this means that time becomes constitutive of what each is. Time is not merely the medium in which things become what they are: it is how they are, as temporal. 'It is not what happens in [time] that happens, but it, it itself happens'.[17] Whether we are speaking of God, the human being, or the world, each 'is' its happening. There is no timeless being or knowing—and in this regard, Rosenzweig believes that what he is doing is simply returning philosophy to the standpoint of 'healthy human understanding' that naturally sees and knows whatever it sees and knows in this way.

> What God has done, what he does, what he will do, what has happened to the world, what will happen to it, what has happened to human beings, what they will do—none of this can be separated from their temporality, as if one could know the coming Kingdom of God in the same way that one knows the created world or as if one might regard the creation as one regards the future Kingdom.[18]

Times are not interchangeable. 'Reality as a whole…has its past and future, indeed an everlasting past and an eternal future'—which means that they can only be known temporally.[19] But this also means that only the past, what has happened, can ever be fully narrated. As for what is happening in the present, this cannot be narrated but only developed in the immediacy of dialogue—there is no third-person point of view that might give a definitive version of what is now happening in our actual present. In the present or 'now' we have only speech and attention. Consequently, the new thinking is thinking that occurs in and as language, that is, as speech. 'Speaking', Rosenzweig writes, 'is bound to time, nourished by time, and does not want to abandon its nourishing ground…It lives entirely from the life of the other, whether that other is the listener to the story or the respondent in a dialogue or the fellow speaker in a chorus.'[20] Rosenzweig therefore must and does reject any approach that might seem to offer an understanding of the eternal as outside of time or timeless. The eternal is meaningful only as it is manifest in past, present, and future tenses and only as it is articulated in the modes of speech and attention.

These presuppositions are reflected in how *The Star* rejects the 'mathematical' symbolism of German Idealism and its attempted systematic deduction of knowledge from the principle A=A (which Rosenzweig calls 'a science of dumb signs'), turning instead to grammar, as the 'science of living sounds'.[21]

---

[17] Rosenzweig, 'Das Neue Denken', 148.   [18] Rosenzweig, 'Das Neue Denken', 149–50.
[19] Rosenzweig, 'Das Neue Denken', 149–50.
[20] Rosenzweig, 'Das Neue Denken', 151. 'Dialogue' reminds us that Rosenzweig worked closely with Martin Buber, whose definitive work of dialogical philosophy, *I and Thou*, appeared in 1923.
[21] Franz Rosenzweig, *Der Stern der Erlösung* (Frankfurt am Main: Suhrkamp, 1988), 139; Eng., *The Star of Redemption*, tr. William W. Hallo (Notre Dame, IN: University of Notre Dame Press,

The Idealists' 'scientific' ambitions are also connected to the their quest for an aesthetic intuition beyond or before language, but Rosenzweig claims that this Romantic quest is merely 'An ersatz for the lost garden' (S, 162/146). It is language that is 'the higher mathematics' that truly reveals the miracle of the path we are travelling in life, a miracle that 'in us and around us and not different when it comes from "outside" than when it comes from "within" and responds to what is "outside." The word is the same as heard and as spoken' (S, 167/151). Indeed, even though 'God's ways and human ways are different... the Word of God and the human word are the same. What human beings perceive in their hearts or their own human language is the Word that comes from the mouth of God' (S, 167/151). Instead of 'logic' as it is conventionally understood, the new thinking is therefore from the outset a meditation on what Rosenzweig calls 'Ur'-words, primal words.[22]

This twofold emphasis on time and language seems to be the basis for Rosenzweig's belief that Heidegger too should be counted as a representative of the new thinking. His most explicit statement in this regard is in a 1929 review of the second edition of Cohen's *Religion of Reason*, entitled 'Exchanged Fronts'. The implied metaphor is a military one, suggesting a change in the basic configuration of the intellectual battleground with regard to the heritage of Cohen's work. Apart from *The Religion of Reason* and other later Jewish writings, Cohen was widely regarded as the quintessentially Neo-Kantian theorist, offering a quasi-idealist and aprioristic account of knowledge. As such he might also be thought to epitomize the kind of position against which Heidegger's own pursuit of a fundamental ontology was directed. Yet, Rosenzweig suggests, Cohen's turn in this late work to the 'individual as such', to the human subject in its finitude and unique singularity, anticipates Heidegger's own turn to existent *Dasein* as the starting-point of his own philosophical investigations. Here, Cohen can be seen stepping out of his own time and into 'our own'. The Neo-Kantian 'school' has turned Cohen into a schoolmaster. The school, as schools always do, will die, and the schoolmaster will die with it. But the master that Rosenzweig hails Cohen to have been, lives on.[23] And, paradoxically, it is perhaps precisely in the anti-Neo-Kantian,

---

1985), 125. (Further references are given in the text as 'S' followed by the page numbers of the German and English editions.)

[22] Again, compare with Buber's 'basic words' (*Grundworte*) that, in *I and Thou* are seen as constitutive of human being. See M. Buber, *Ich und Du* (Stuttgart: Reclam, 1995), 3; Eng, *I and Thou*, tr. Walter Kaufmann (Edinburgh: T. & T. Clark, 1970), 53. In the first two decades of the twentieth century both Buber and Heidegger reveal a penchant for the prefix *Ur-*, especially when writing about religious experience.

[23] Franz Rosenzweig 'Vertauschte Fronten', in *Zweistromland*, 235–7. For a full discussion of the Davos dispute and Rosenzweig's interpretation of it see Peter Eli Gordon, Chapter 6 '"An Irony in the History of Spirit": Rosenzweig, Heidegger, and the Davos Disputation', in *Rosenzweig and Heidegger: Between Judaism and German Philosophy* (Berkeley. CA: University of California Press, 2003), 275–304.

Heidegger, that 'the leap into Dasein' is made the starting-point of philosophical investigation.

THE ETERNAL

A useful approach to Rosenzweig's discussion of the eternal is the short essay entitled 'The Eternal: Mendelssohn and the Name of God' (1929). The essay largely concerns the translation of the divine name with which God names himself to Moses at Exodus 3:14, familiar from the King James translation I AM THAT I AM, which in turn reflects the Vulgate EGO SUM QUI SUM. Rosenzweig, of course, is oriented towards Jewish translation traditions and, as the essay title suggests, it is the translation 'the eternal' favoured by Mendelssohn, with its precedents in the Septuagint *ho Aiōnios* and the 1588 Geneva Bible *l'Éternel*. Mendelssohn himself translates the text (in a literal translation from Mendelssohn's own German) as 'God said to Moses: I am the being (*Wesen*, or essence), which is eternal. For as he said, You should say thus to the children of Israel "The eternal being (*Wesen*), who calls himself: I am eternal, has sent me to you".'[24] Mendelssohn explains this translation in terms of the simultaneity of past and future as present to God so that he is at all times to be called upon with the same name. But, as he says, 'There is no word in German that combines the meaning of being at all times, with the meaning of necessary existence and the meaning of providential foresight, as this holy name does... That is why we have translated it as "the Eternal" or "the eternal Being".'[25]

In Rosenzweig's view, however (as the reference to necessary existence reveals), Mendelssohn is unduly influenced by the rational theology of the eighteenth century with its roots in scholastic Aristotelianism, whereas the biblical context shows the entire irrelevance of 'a lecture on God's necessary existence'. More urgent is the Israelites' need for assurance that God will be with them. Thus a more adequate translation would reflect not God's being eternal in the sense of possessing the metaphysical attribute of eternity but his 'being contemporary, his being for you and with you and his continuing to be with you'.[26] The name must be able to be spoken in the mode that Rosenzweig calls 'vocativity'. It is not the naming of an object but a name to be called upon in prayer. In this connection Rosenzweig commends an older Jewish commentator's remark that 'He calls himself I AM THERE and we call him HE IS THERE'.[27]

---

[24] Rosenzweig, 'Der Ewige: Mendelssohn und der Gottesname', in *Zweistromland*, 804. The translation 'the Eternal' is also familiar from James Moffatt's English translation.

[25] Rosenzweig, 'Der Ewige', 804. [26] Rosenzweig, 'Der Ewige', 806.

[27] Rosenzweig, 'Der Ewige', 807.

In this respect the Christian 'Lord' (*Dominus*) might seem an improvement, although Rosenzweig sees this as having connotations of domination rather than helping. These connotations are perhaps softened in the translation of Adonai as '*my* Lord', which 'maintains this inward resonance, with an undertone of vocativity, of address and call...for a moment ['the flash of an eye'] it somehow flashes forth from the midst of the sentence towards heaven'.[28] The name and the event of revelation are bound together, and Mendelssohn must be credited with discovering this, even if his own translation choices ultimately obscured the point.

As Rosenzweig explains, 'Biblical "monotheism" does not consist in knowledge of the unity of the divine Being...But what is distinctive in biblical faith is that it presupposes this "pagan" unity—as Kasari says, the God of Aristotle—but knows this God in its unity with what is most personally and immediately experienceable—again as Kasari says, the God of Abraham'.[29] What matters is not being able to define the essence of God but to relate to God in the tension of distance and proximity that constantly characterizes the biblical account of Israel's God-relationship. That is to say, it is 'the correlation of the God of creation with the God who is present to me, to you'.[30] The Tetragrammaton is thus both a name and an epithet and equally both—an idea that is perhaps akin to Nelson Pike's notion of the divine attributes as 'title-names'.[31] Since the name is thus inseparable from its meaning, the worst possible translation choice is simple to reproduce it as Yahweh, which, for most readers, is simply a meaningless name, a name that says nothing and therefore, if applied to God, idolatrous.[32] The rendering of the name must not represent God as 'congealed in his essence' but as 'inclining towards us by being-there (*Dasein*) and as present (*An-Wesenheit*)'. The grammatical awkwardness of Exodus 3:14 is not a cause for embarrassment, but underlines the point that the text 'bears witness to a moment of revelation that is now repeated and renewed by the reader in a thousand moments of knowing'.[33]

Is 'the Eternal', then, simply to be dropped as too Aristotelian and insufficiently vocative? This might seem to be the direction of Rosenzweig's argument, and yet he ends by conceding that, nevertheless, 'For we transient beings "eternal" is the word of the longing that is the final word of our "Song of the Earth." Our heart can wish nothing more than this'.[34] Yet, even in relation to this 'final word' the experience and hope of God's becoming present in this world weakens the desire for any eternity other than the eternity of the divine

---

[28] Rosenzweig, 'Der Ewige', 808. [29] Rosenzweig, 'Der Ewige', 809–10.

[30] Rosenzweig, 'Der Ewige', 809–10. [31] See Chapter 1.

[32] Rosenzweig, 'Der Ewige', 813. This is the choice made by the Catholic Jerusalem Bible.

[33] Rosenzweig, 'Der Ewige', 814.

[34] Rosenzweig, 'Der Ewige', 814. The reference to the 'Song of the Earth' is presumably to Mahler's song-cycle of Chinese poems that ends with the song 'Farewell', a meditation on approaching death.

presence itself: 'In face of time that has come alive our human longing for eternity falls silent'. And, in a final twist, the naming of God as eternal is redeemed when it is interpreted in terms of providence and redemption, as it is in Cohen, Mendelssohn, and even Calvin.[35]

Despite his own later cautions, 'the eternal' did, nevertheless, have a special place in *The Star*, as in the closing lines of Part III: 'He is truly the first and the last. Before ever the mountains were born and the earth was brought forth in travail, you were God. And were from eternity what in eternity you will be: truth' (S, 464/417)—although here too we should note the 'vocative' turn from 'He' to 'You'. But the essay did not deny divine eternity, it merely warned against reading 'eternity' in the prism of Aristotelian, scholastic, or other metaphysical forms of thought, admonishing us to read it as a call only and ever to know God as the one on whom in all our exigent need we must repeatedly call in the temporally inflected language of human existence. If Rosenzweig's own translation avoids the word 'eternal', this is not to deny that God is the one who eternally comes to the aid and comfort of his people but to affirm that its way of being is precisely to be eternally, that is, ever and again, to be there for them 'to all eternity':

> But God said to Moses,
> I will be there as the one I will be.
> And said:
> You should speak thus to the sons of Israel,
> I am there sends me to you.

THE STAR OF REDEMPTION: 'FROM DEATH...'

It is first in Part III that *The Star of Redemption* reveals itself to be a work essentially concerned with the eternal, as can be seen from the sub-titles of its three sections: the eternal life, the eternal way, and the eternal truth, which relate, respectively, to Judaism, Christianity, and, finally, to God himself. However, the organic and narrative quality of *The Star* means that ascribing 'eternal life' to the blood-community of Judaism is only really intelligible in the light of what has gone before. Consequently, whilst the following account of the teaching of *The Star* will focus most intensely on Part III, I shall begin with a brief survey of the preceding argument. Especially, I shall look somewhat closely at the account of revelation in Part II, where Rosenzweig presents a view of the moment of revelation that is integral to how he understands the Eternal God whom the moment reveals.

---

[35] Rosenzweig, 'Der Ewige', 815.

The Star famously opens with the words 'From Death', words that establish a narrative arc that is only completed nearly 500 pages later with the book's closing words 'into life'. They also provide a dramatic introduction to the evocation of the First World War battlegrounds in which Rosenzweig began to write *The Star*. 'Everything mortal lives in this anxiety of death…Unceasingly, the tireless womb of the earth brings forth new beings that are, each of them, given over to death, each of them awaiting the day of its departure into darkness with fear and trembling' (S, 3/3). The reference becomes yet more explicit as Rosenzweig proceeds to underline the difference between how philosophy responds to this predicament and the natural response of the embodied mortal human being:

> Man may creep into the folds of the earth in face of the hissing and whizzing volleys of relentless death and may there be forced without any means of avoidance to sense what he will never otherwise sense: that his I would only be an It were it to die; and with every shriek that is still in his throat he may shriek out his I against the threat of such a relentless and unthinkable annihilation—philosophy smiles its empty smile at all this need and with outstretched finger points the creature whose limbs are quivering with anxiety in face of its [fate in] this world towards a 'beyond' that it is not at all interested in knowing about. (S, 3/3)

This staging of the confrontation between the isolated existing individual and his death may prompt us to think of Heidegger, but although Rosenzweig would claim kinship with Heidegger, it is nearly ten years till *Being and Time* will be published. And, in fact, it is at least arguable that these lines may be critically applied as well to Heidegger as to the idealist philosophy against which they are explicitly directed.[36]

Rosenzweig next sketches his own immediate philosophical lineage, giving decisive roles to Hegel, Kierkegaard, Schopenhauer, and Nietzsche. Hegel aimed to demonstrate the unity of the laws of thinking and of being and of reconciling 'heaven and earth', revelation and history in the unity of a single 'All'. This was immediately countered by Kierkegaard's insistence on the impossibility of incorporating revelation and the individual's first-person experience of existence into any philosophical system. Yet Kierkegaard is not yet representative of the new thinking itself, only of the challenge that it has to face. And whereas Kierkegaard challenged philosophy from outside, Schopenhauer and Nietzsche challenged it from within, thus pointing the way to the new thinking proper. Schopenhauer is said to be distinctive because of how his thought does not attempt to offer an account of the world as the totality of what is knowable but concerns the meaning or 'value' it

---

[36] For discussion see my *Heidegger on Death: A Critical Theological Essay* (Farnham: Ashgate, 2013).

has for the 'unique individual and his own',[37] that is, the living human being whom we encountered in the opening paragraphs creeping into the folds of the earth so as to escape the onrushing volleys of death. Yet, Rosenzweig's argument suggests, Schopenhauer still maintained a certain distance between philosophy and the philosopher himself,[38] and it is only in Nietzsche that the philosopher's own life truly becomes the matter of his philosophy. Thus thinking becomes a matter of establishing and asserting a 'life-view' rather than a mere 'world-view' (S, 6–11/5–9).

But if such formulations invite us to see Rosenzweig's project as essentially subjectivist and maybe even somewhat poetic, this would be entirely to miss the point that he is not seeking to do away with the world, and still less with God, but to see these in their real independence of each other and of human cognitive constructions. In the short work *Understanding the Sick and the Healthy*, Rosenzweig conceived of an imaginative project for curing those suffering from a sickness induced by too much philosophy of the wrong kind. There are, he says, three basic realities amongst which the path of life winds its way: God, human beings, and the world. He presents this cure in a fictional exchange of letters with the director of a sanatorium that stands at the exact geometrical centre of three mountains, taken to represent the three primordial facts of God, the world, and the human being. These mountains are reached by a series of criss-crossing paths that, although different, share a similar pattern. In the early days of his stay the patient cannot expect a simultaneous view of all three peaks. That will come not through theorizing but only through practice and experience, as he becomes familiar with the paths and is able to find his way to the correct viewpoint.[39]

THE ELEMENTS

The image of the three mountains provides us with a useful frame in which to set the argument of Book 1 of *The Star*, which deals with God, the world, and the human being under the three rubrics of metaphysics, metalogic, and metaethics. These terms might seem to suggest that Rosenzweig himself is

---

[37] An allusion to the title of Max Stirner's *Das Ich und sein Eigentum* that Rosenzweig (perhaps oddly) sees as as epitomizing Schopenhauer's intellectual orientation.

[38] Although Rosenzweig would seem not to have endorsed Kierkegaard's view that Schopenhauer was, in the end 'nevertheless, a German thinker, eager for recognition' and not prepared to reduplicate his teaching in his life (SKS25, 356 (NB29:95)/JP, 3877).

[39] See F. Rosenzweig, *Das Buchlein vom gesunden und kranken Menschenverstand*, ed. Nahum Norbert Glatzer (Düsseldorf: Joseph Melzer, 1964); Eng., *Understanding the Sick and the Healthy: A View of World, Man, and God*, tr. Nahum Norbert Glatzer (Cambridge, MA: Harvard University Press, 1953).

now engaging in some kind of a priori construction, but this is not the case. He himself claims that his method is indebted to Schelling's *The Ages of the World*. There, we will recall, Schelling attempted—versus Hegel—to develop what Tillich would call an 'a priori' or 'metaphysical' empiricism that does not deduce the systematic development of the idea of God, the world, and human life from a purely mathematical principle of identity (A=A), but, in Tillich's words, regards 'the world-process' 'as the continuous self-revelation of the *Unvordenkliche* (that which all thinking must presuppose)'.[40] As we saw, Schelling presented this as a descriptive account of the emergence of the world from a primordial nature devoid of form or light. This, in effect, offers a kind of phenomenology of representation, reaching back into the abyssal 'Ur' dimension that preceded the advent of any form, divine, worldly, or human. As material for this account, Schelling took the history of mythology in which he saw the progressive self-revelation of human self- and world-consciousness. But this means that religion is no longer conceived of as a particular set of distinctive objects (God or the gods) but, instead, as a mode in which being (*Seyn*) as such is revealed. Yet the history of mythology only gives us knowledge (*Wissen*) of how consciousness came, in the past, to be. If we are to cognize it (*Erkennung*) in its present actuality, in its 'being...there' (*Da-sein*), we must turn instead to the analysis of human freedom, whilst the future will only ever yield itself as the matter of premonition (*Ahnung*). Yet the relation of past, present, and future is not merely linear, since the divine pre-history continues as the unthinkable ground of freedom's self-revelation, whilst freedom is eternally realized in itself prior to and beyond all manifestation.[41]

Although *The Star* is doing more than simply applying the Schellingian blueprint, there is clearly some analogy to Rosenzweig's own method. Thus, when Rosenzweig begins each strand of his narrative with the claim that, in the beginning, we know 'nothing' of God, world, or human being, this is not the 'nothing' of Hegel's logic that is reached by an act of total abstraction from all concrete content. Instead, it is the 'nothing' that appears in the moment of human consciousness's primordial emergence, when, vis-à-vis itself, its world, or its gods, it knows: nothing. Starting from this nothing, Part I, Book 1 (metaphysics) offers a history of mythology that is presented as a key to the emergence of the human phenomenon and it tracks this history through to the polytheism of the Greek pantheon as revealing the living world of nature in its infinite but incoherent vitality and multiplicity, a vibrant monism that is, nevertheless, just that: a monism. Part I, Book 2 (metalogic) presents a similar story, but now focusing on the history of ancient cosmology, climaxing in the

---

[40] Paul Tillich, 'Existential Philosophy: Its Historical Meaning', in *Theology of Culture*, ed. Robert Kimball (New York: Oxford University Press, 1959), 86–7.

[41] See the section 'Ur-time and Eternity' in Chapter 2 for further discussion of these clearly problematic claims.

philosophy of Plato and Aristotle. Here the world comes to be represented as a whole, but not, as Rosenzweig puts it, as an 'All'. It is self-contained and self-sufficient, devoid of what a later philosophical generation would call alterity. Finally, Part I, Book 3 (metaethics) looks to the emergence of the human subject, culminating in the hero of Attic drama, a 'personality' who, according to Rosenzweig, is essentially silent, incapable of development through dialogue and fated to be what and as he is. He is a self, but without soul, fixed in a posture of defiance (*Trotz*), a maximum of self-assertion that demands immortality and even a certain 'eternity' but that falls short of being free, personal existence.

In each of these spheres (metaphysics, metalogic, and metaethics), the ancient world arrives at a threefold structuring of life that, as in *Understanding the Sick and the Healthy*, shows three distinct and separate developments that do not, as yet, stand in any essential relation to each other: the living world of myth is not the world known by philosophy, nor is the solitude of the tragic hero redeemable by either God or world. Each is what and as it is in its isolation from the others. This is why Rosenzweig refers to them precisely as 'the elements'. They do not constitute the world in an emphatic sense as an 'All' inclusive of divinity, cosmos, and humanity. They are only the first and, as yet, mutually unknown processes of theogony, cosmogony, and anthropogony between which there is no order of hierarchy or even lateral positioning. As Rosenzweig will say in a transitional section leading from Part I to Part II, what brings them together is and can only be the movement of 'the single stream of world-time', uniting them as 'elements' of a unifying historical movement leading, in Rosenzweig's words, from the morning of the world, through its noon-day and on into evening. But, precisely with regard to its historical character, the resulting unification is not an 'All' but—in an allusion to the prophetic 'day of the Lord'—'the one world-day of the Lord' (S, 96/87). Only in the perspective of this day of the Lord does the threefold of self-enclosed and silent elements become revealed as creation, a revelation that Rosenzweig describes as 'miraculous'.

As Rosenzweig's vocabulary of revelation, creation, miracle indicates, we are now passing historically from the Greek world to the world of biblical revelation. And, as we progress through Part II, it is no longer Platonic mathematics or the logical symbols of German Idealism but language and grammar that carry the argument forward. Revelation can occur only in and as language, as a *word*. Consequently, whilst Rosenzweig insists on this word being miraculous, he equally insists that its miraculous character is not to do with some supposed breach of the laws of nature but that even what are called nature miracles are interpreted in the Bible as signs that foretell what is to come. Nor does such foretelling itself involve abrogating or suspending the laws of temporality: the future remains the future, but re-envisaged as being under the sway of God's foresight or providence. Every hair of our heads is counted!—and thus, contrary to the terms of popular debate, the miracle of revelation reveals the conformity of the world qua creation to a unitary law.

Counter-intuitively perhaps, Rosenzweig also claims that in the particular historical situation at which we have now arrived, a situation marked by the failure of nineteenth-century attempts to ground faith in the historical reconstruction of, for example, the life of Jesus, it is philosophy that is best placed to recall theology to its own presuppositions, that is, to creation, miracle, and the possibility of revelation.[42] But, we recall, Rosenzweig does not see philosophy in terms of the construction and development of a priori categories of thought. It is not a system of concepts, but deals with 'reality as present to hand' (S, 119/108).[43] In this regard, philosophy becomes the 'Old Testament' [*sic*] or 'Sybil' to theology's New Testament by showing how the miracle is to be understood as signifying divine providence. And, in the spirit of the 'new thinking', it does so by reflecting on language itself. If the elements constituted a set of *Urworte* or primordial words, these were still inaudible, secret, and hidden: now the focus must be on language as spoken and thinking as spoken in living language.[44] And this, in turn, is possible precisely because the original divine word that is simultaneously a word of creation and revelation is the word 'Let there be light': 'and the light of God, what is it?' Rosenzweig asks, before answering: 'The human soul' (S, 123/111). Again, as both Heidegger and his theological collaborator Rudolf Bultmann will maintain, the human capacity for understanding what is said is itself the original *lumen naturalis*, the inborn light, and, for Bultmann at least, therefore also the presupposition of meaningful revelation.[45] As he will later say, language is the 'higher mathematics' that reveals the miracle of the path we are to travel and we can trust it

---

[42] In this regard, Rosenzweig shows himself to be responding to the same historical context as early dialectical theology but doing so in an exactly opposite way.

[43] Rosenzweig uses the term '*vorhanden*' that Heidegger will later use to characterize a worldly or reified form of knowing. However, the context suggest that Rosenzweig means something more like Heidegger's '*existentiell*'.

[44] Of course, the theogonies, cosmogonies, and anthropogonies of the ancient world were transmitted in and as language, as oral and written poems and treatises. Surely the world of Hesiod, Homer, Pindar, Sophocles, Plato, and even Aristotle was not a world devoid of language? But the point is that biblical thought offers an entirely different understanding of language from what we find amongst the Greeks. Without Rosenzweig ever spelling this out, it seems that this difference is something like this: that the Greeks understood language as voicing a mimetic response to aesthetic experience of the world whereas the Hebrew thinks more of non-mimetic word-acts, 'he spoke and it was done'. This seems to be the thrust of Rosenzweig's subsequent characterization of the root-words of philosophical reflection as words are immediately expressive of the original '*Ur-ja*' (Primordial Yes) of creation, pure properties that are beyond comparison, like the blue of the sky as 'drunk' by the eye of the artist, and that point beyond mere things to the relational forms of listening, bestowing, and thanking (S, 140–3/126–7).

[45] See Martin Heidegger, *Gesamtausgabe*, II: *Sein und Zeit* (Frankfurt am Main: Klostermann, 1977), 171 (Eng. *Being and Time*, tr. John Mcquarrie and Edward Robinson (Oxford: Blackwell, 1962), 133) and Rudolf Bultmann, *Das Evangelium des Johannes* (Göttingen: Vandenhoeck and Ruprecht, 1950), 22.

because 'it is in us and around us, and no different when it comes from "outside" than when it comes from "within" and responds to "outside." The word is the same as heard and spoken. God's ways and human ways are different, but the Word of God and the human word are the same. What human beings perceive in their hearts or their own human language is the word that comes from the mouth of God' (S, 167/151). But this revelation will also, and for the first time, reveal God, world, and human beings in their interconnection, revealing the world to human beings as God's creation, revealing them to themselves as creatures and revealing God as their creator.

CREATION, REVELATION, AND REDEMPTION

Part 2, then, sets out from the interrelated themes of creation and revelation, but in doing so it must and will also embrace redemption. Again following the thread of language, Rosenzweig asserts that against the static 'pure thought' of Idealism with its A=A and its 'Ersatz' creation (that is really only an emanation of thought), the word in which God creates is, in the beginning, only a beginning and, as such a *promise*. If the creation prior to the creation of human beings is at each particular act marked by the comment in the past tense that 'It was so', indicating that the act is complete in its enactment, the creation of human beings is recounted in a different mode. God's 'Let us…' opens way to a view comprising more than an I and an It, a view towards a 'You', and, in naming Adam, God creates the first proper name. Furthermore, in asserting that what is now created is not only good but 'very good', God opens up the prospect of something 'supra-worldly in the world'. But, in what is a surprising and unargued—though important—move, Rosenzweig immediately connects this to the thought of death. Implying that humans are distinctive amongst creatures in their capacity for knowing their own death, he suggests that death indicates the advent of a creature whose life requires a fulfilment that is not given with the simple fact of its biological existence. Of course, the opening page of *The Star* already warned us against any philosophy that would smilingly dismiss the animal terror of death and talk of some other world 'beyond' this. The clue provided by death is not to be followed in the direction of some kind of afterlife or narrative of immortality. Yet what death does is quietly to reduce the previous strata of creation to the past and point the human being towards the question of its future. Thus, says Rosenzweig, it becomes 'the quiet, constant foretelling of the miracle of [creation's] renewal' (S, 173/155).

Moving into Part II, Book 2, 'Revelation', Rosenzweig continues via the biblical saying 'Love is strong as death' (Song of Songs 8:6) to show that, in human understanding (guided by the words of biblical revelation), death is also inseparable from love. Death may be 'the seal of creatureliness' but in the experience of love 'The capstone of the dark arch of creation becomes the foundation stone of the illuminated house of revelation' (S, 174/157). And if the narration of creation is always the narration of what was done in the past, the power of love 'is a fateful power over the heart in which it awakens and is yet so newborn, so...without a past...so self-originating' (S, 178/160). Whereas the fate that hung over the pagan world (and even over the pagan gods themselves) manifested the unchanging power of Moira (Fate), the fatefulness of love strikes suddenly, in the moment, as an event (*Ereignis*) that is unrepeatable, distinct and individuated.[46] As such, 'It is a glance (*Blick*) falling upon the creation created in God's image and likeness' (S, 178/160).[47] If creation is already, in a sense, 'revelation', the revelation of all that is as God's good creation, the illumination cast by the divine look of love is revelation in a more specific sense. The God revealed in this look emerges, as it were, from his concealment behind creation and shows himself as lover. But the love of God is not an abstract 'metaphysical' property: love is love only as event, only as moment (*Augenblick*), the moment in which the look of love flashes forth, and so, as Rosenzweig expresses it, love is not the form of God's countenance but the light that lightens its features; it is both 'always today' and yet also 'the eternal triumph over death' (S, 182–3/164). Rosenzweig distinguishes this kind of love, that he regards as authentically biblical, from the divine compassion taught by Islam: as he sees it Allah's mercifulness is a universal, unchanging, timeless attribute that is entirely indifferent to the variations in its objects (S, 184/165).

But—in a move paralleling the kind of emphasis on 'the hearer of the Word' found in such Christian theologians as Bultmann and Rahner—Rosenzweig next asks what it could mean for a creature to become the recipient of such a revelation. Whereas the silent self-assertion of the tragic hero culminated in a kind of defiant self-affirmation in despite of the fate that broke him (*Trotz*), the faithful response to love is a kind of pride (*Stolz*) that is at the same time humble and reverent, a state in which the beloved 'rests in the feeling of being sustained; he knows that nothing can happen to him' (S, 187/168). The relation between God and the soul is non-reversible: God loves, the soul is beloved. Yet whilst God, the divine lover, loves by constantly renewing his love in each moment, it is precisely the assurance of being loved that enables the beloved to interpret this love 'as eternal, for ever and eternal'. Its only true response is, in a probable allusion to the story of Jesus and the sinful woman in Luke 7, 'to sit silently at the feet of the love of the lover', not even thanking but

---

[46] Cf. Theunissen's references to archaic poetic images of fate 'hanging over' human beings and to the idea of the god's suddenness in breaking through this form of time's tyranny (in the section 'Chronos' in Chapter 7).

[47] The reference to the glance (*Blick*) sets up an implicit link to the moment of vision (*Augenblick*) that is to follow.

simply loving: the only possible response to love is love (S, 188/169).[48] The tragic hero's defiant self-assertion is transformed into the determined hold on the vision revealed in the look of love.

If the look of love is an event of the moment, its being received as the assurance of love allows us to see it also as enduring. But more: in the moment of becoming aware that I am loved I also become aware that my preceding state was not one of love. In encountering this new power I learn that our human possibilities are such that we could have done and been differently and that because we have allowed ourselves to fall short of this exceptional and eminent possibility of existence we are and have been 'sinners'. Yet precisely this awareness of sin leads the beloved both to a more intense knowledge of what it means for God to be the God of love (the God who forgives sins and accepts even the undeserving sinner) and also to understanding him- or herself in the light of a history or process through which he or she passes from one kind of life to another.

This is not just a history in which a certain past flows into a certain present. The present revelation of God's love as transformative of the history of the recipient is by the same token the revelation of love as a possibility for the beloved's future. Consequently, the act in which the beloved takes hold of this possibility and sees their own fate as bound up with it, is defined by Rosenzweig as prayer. Why? Because all prayer is in essence prayer for the coming of the Kingdom in which God's love will finally reign over and in all creation. There is therefore a sense in which prayer is already the fulfilment of revelation, since it brings the future into the trajectory of past and present assurances of love. But comprised within the prayer for the coming of the Kingdom is the fact of the soul's addressing itself to God as to a 'You' and knowing both God and itself in the uniqueness of their individual relationship. As a word of address—recall what was said in the essay on 'The Eternal' about vocativity—the response of prayer shows that revelation is also the revelation of the divine name, that is, a revelation of God in which the primary content is the giving of a name by which he can be called upon. And it is precisely this character of knowing by name that is theology's extra, so to speak, in relation to Idealist philosophy. The latter can know God as having certain attributes, perhaps inclusive of eternity, but it does not know God by name. 'In that [love] speaks', says Rosenzweig, 'it has already become superhuman' (S, 224/201).

The Second Book of Part 2, 'Revelation', finishes with a difficult but suggestive interpretation of the Song of Songs. Rejecting modern interpretations

---

[48] As mentioned in the previous chapter, the 'love scene' between Jesus and the sinful woman is a defining feature of Kierkegaard's theology. See my *Kierkegaard and the Theology of the Nineteenth Century: The Paradox and the 'Point of Contact'* (Cambridge: Cambridge University Press, 2012), 150–61 and *Kierkegaard and the Quest for Unambiguous Life: Between Romanticism and Modernism: Selected Essays* (Oxford: Oxford University Press, 2013), 206–18.

that strip the Song of reference to God by making it merely a collection of love songs, Rosenzweig twists the closing words of Goethe's *Faust* to show a different light on the text: ' "Everything transitory is but a parable," but love is not "but" but utterly and essentially parable, for it is only apparently transitory, but in reality eternal' (S, 224/201). The 'I' spoken by the lover is a temporal and transitory 'I', but the persistence of the love that it expresses manifests also a power of endurance. This endurance beyond the momentary desire of love is indicated by two lines of the Song. The first is 8.6 'Love is strong as death', the second 8.1 'O that you were like a brother to me'. Of the first, Rosenzweig comments that these are the only words spoken about love in the third person and thus the only 'objective' statement about what love is in the book. As we have already seen, he gives a special place to death as showing the need for creation itself to be transcended in its aspect as mere givenness, but, as he now adds, these words show that 'Death is the final consummation of creation—and love is as strong as it is' (S, 225/202). Does this imply that the movement of transcendence that death sets in motion is itself potentially transcended in and through love? And how might this happen? The clue, for Rosenzweig, is provided by the speaker's wish that her lover was also her brother. This, he says, makes concrete love's own longing for something 'more' than the mere present moment of love, that is, for love to reveal its power as more than transient rapture. As he puts it, 'it is an eternalizing of love than can never spring from the ever-temporal present of feeling; that is, it is an eternalizing that does not just grow [in the space] between I and You but that longs to be grounded in face of all the world' (S, 228/204). And this longing is fulfilled in marriage: 'Marriage', he writes, 'is not love. It is infinitely more than love'. In marriage love makes the passage from Eros to life in the world (S, 228/204).[49] And in marriage, love becomes the matter of obligation: 'As He loves you, so you must love', that is, the divine love becomes not only the ground but also the model for human, worldly love (S, 228/204).

These reflections mark the transition to Book 3, 'Redemption or the eternal Future of the Kingdom', that, on the basis of married love, now turns to the commandment to 'Love your neighbour'. This commandment presupposes the freedom of the human being—that we can indeed choose to love—but also that the possibility of loving is not an invention on the part of human freedom since it presupposes the experience of being gifted with ontological security by the revelation of divine love. 'God must first turn to the human being before the human being can convert to God's will' (S, 239–40/215).

---

[49] Cf. Schleiermacher's discussion of how it is love—exemplified in the story of Adam's first encounter with the newly created Eve—that first reveals not only the beloved but also the world; Friedrich Schleiermacher, *On Religion: Speeches to its Cultured Despisers*, tr. and ed. Richard Crouter (Cambridge: Cambridge University Press, 1988), 119.

That it is the neighbour I am to love, the neighbour who is the representative of all and any, takes me decisively beyond the circle of erotic self-gratification into the world, where my love must deal with opposition and failure. Here love is not just the irruption of a power: it is a work, and in and through this work the world becomes the creation that it was created to be but which it has not yet become. The world, creation in the full sense, is not a given ensemble of things and relations—it is still to come: 'It is what ought to come. It is the Kingdom' (S, 245/219). And it is not here yet—the tear has not yet been washed from every eye (S, 244/219).

Rosenzweig sees two aspects to the coming of the Kingdom. The first is that it is growing: as the fulfilment of creation it is an organic growth in continuity with the life of the world. It is the *life* of the world that is to be redeemed—and we think back to the opening pages and the sheer terrible desire of the one faced with death to *go on living*, to remain, to abide. What life seeks, in other words, is eternity, the eternity of the Kingdom. That life cannot give itself this eternity is indicated by the fact that anything living is, as such, also and always exposed to death: it can cease to be. But 'Once it has entered the Kingdom, a being (Dasein) can never again fall away from it; it has been marked by a once-for-all and has become eternal' (S, 250/224). Yet, at the same time as the Kingdom appears as fulfilling the living growth of the world, what the world waits for in waiting for its fulfilment in eternity is precisely love or, more precisely still, for the works of love that embrace the world and bring it to completion as what it should be. But love is not something that just 'grows'. Love has to be enacted. It is therefore waiting and working, the life of the world and human action that, together, constitute redemption (S, 254/228).

And here we may note a significant development in Rosenzweig's vocabulary. In Part I the 'world' of the three isolated elements of God, world, and the human was described as 'ever-enduring'. In Part II creation, revelation, and redemption are collectively called 'the world being at all times renewed'. It is only when redemption is accomplished that he allows himself to introduce 'eternal' (*ewig*) in the strong sense. However, Rosenzweig is also clear that it is not a matter of the timeless eternity of idealist philosophy. The Kingdom '...is always future, but it is future always...It comes eternally. Eternity is not a very long time but a morning that could just as well be today. Eternity is a future that without ceasing to be future is nevertheless present. Eternity is a today, but a today that is aware of being more than a today' (S, 250/224).[50]

---

[50] We may see here a structure of future-in-present-and-present-in-future that is already found in Kierkegaard's account of the moment of vision and that Theunissen, citing Buber, saw as characteristic of the redemptive effect on time, enriching one-dimensional linearity with the co-presence of all three time dimensions. Nor is it far from the relationship between eternity and futurity set out in Hegelian terms by Bradley and McTaggart.

Again Rosenzweig's distinctive focus on language is reflected in the move from creation through revelation to redemption. Where creation is manifested in the existence of substantives—earth, sea, plants, animals, etc., and revelation in the dialogue of I–You, redemption is expressed in the unity of a sentence or, more precisely, in the 'and' (that is *not* the idealist '=' of identity) that links substantives and persons in a relationship in which each continues in its independence whilst interacting with the others. If creation is expressed by listing the sequence of God's creative acts, and if revelation is marked by dialogical exchange, redemption can only be voiced by words that are communal and, as such, bind God and humans together in community. These are eminently the words of communal hymns of praise. The language of the hymn is neither indicative nor imperative but a call: to sing, to thank, and to confess that 'He is good'; it is the call to 'Give thanks' and to 'Give praise', not as instruction but as in the optative imperative 'Let us give thanks'. In such hymnody, we give thanks for what has been and is, but our thanksgiving reaches out also to embrace all that shall be. The real theme of the common hymn is the future that, in redemptive action, is even now effective: 'This anticipation, this today, this eternity of thanks for God's love—that it "endures eternally"—... that is the real melodic content of... common song' (S, 261/234).

The communal hymn is by definition the expression of a 'we' composed of distinct individuals, each of whom is also the neighbour to whom I owe the obligation of love. But because the neighbour can be each and any human being, the maximal extension of the 'we' of hymnody is 'we all', the entire historical human community in its quest for a kingdom of all-embracing love and justice. The Scottish metrical rendering of Psalm 100 perhaps renders Rosenzweig's thought well at this point: 'All people that on earth do dwell | Sing to the Lord with cheerful voice'. Or we might think again of Hermann Broch's appeal to 'the voice of comfort and hope and immediate love: "Do thyself no harm! for we are all here!" '.[51] And yet the hymn is not just the self-affirmation of the 'We'; qua word it is a word that must be spoken and being spoken it is a word that is spoken 'to', that is, to God, and it is in the dative of its singular address that 'all we' find our final unity—remembering that, as Rosenzweig also warns us, the You to whom our hymns are addressed is also the one who comes in judgement; we are each of us objects of divine justice and we cannot be released into our eternal life through love without also coming under judgement.

However, although the hymnodic 'we' is, ultimately, 'we all' (i.e. all of we human beings), Rosenzweig now argues that there is, nevertheless, a distinctive 'we' in whom the possibility of such Kingdom-anticipating-praise is

---

[51] Hermann Broch, *The Sleepwalkers*, tr. Willa and Edwin Muir (London: Quartet, 1986), 648.

already fulfilled, namely the Jewish community qua 'the eternal people'. That this is a community defined by blood is justified, he says, on the grounds that all human relationships are, qua created, blood-relationships; therefore the community of the Kingdom must be a community conscious of its blood-relationship as the basis of its growth towards the Kingdom, whilst (he implies) remembering the love commandment that binds it to all. Only in such a community, conscious of its blood-tie, can we speak of a 'we' that is also 'eternal'. For the Jew does not have to become a Jew through time and choice. Unlike the Christian, who must receive the second birth of baptism in order to enter the community of faith, the Jew is born a Jew, or simply *is* a Jew. Jewish identity is not an identity bestowed by time and therefore Rosenzweig can say that 'The We is eternal; in the face of this triumphal shout on the part of eternity death plunges into nothingness. Life becomes immortal in redemption's eternal song of praise' (S, 281/253).

That these assertions may prove problematic for non-Jews and, for that matter, for versions of Judaism that do not in this way stress the blood-tie is clear. Why? Because they seem to restrict the achievement of life's ultimate goal to a people defined by biological descent. We shall return to this issue in more detail later, and for the present note just two points. The first is merely the comment that before immediately leaping to ideas of 'race' we might interpret what Rosenzweig is saying as a way of emphasizing the role of the family in the moral and religious formation of the child. This, for example, is the most natural reading of the closing words of his introduction to an edition of Cohen's works, referring to Cohen's dedication of *The Religion of Reason* to his father. Cohen, Rosenzweig remarks, finally revealed his work to be other than the product of a school (such as neo-Kantianism) but as dedicated 'to the memory of the man to whom he owed thanks for his embodied and ensouled connection with this home that can never be lost, the home of blood and Spirit: the father'.[52] The second point is simply to remark that, positively, any philosophy or theology that is concerned to take fully seriously the incarnate, bodily reality of human life must, in some way, deal with the role played in human community-formation by kinship networks in which sexual reproduction, understood as the basis of the 'blood-tie', is a necessary ingredient.[53]

---

[52] Rosenzweig, 'Einleitung in die Akademieausgabe der Jüdischen Schriften Hermann Cohens' in *Zweistromland*, 223. Of course, in a post-feminist context, this immediately raises a whole new set of questions, but these are not our immediate concern.

[53] Of course, whether technological developments commencing with the contraceptive pill, moving through artificial insemination, and on to genetic manipulation will, with other social changes, finally break the bonds linking sexual reproduction and community may be becoming an open question and one of far greater dimensions than the focus on the 'ethics' of reproduction can reasonably hope to contain.

THE ETERNAL LITURGY

We return to the thread of Rosenzweig's argument, as he makes the transition to Part III in an extended Introduction entitled 'On the Possibility of Entreating the Kingdom'. The communal worship that gives thanks for what God has done and that also looks towards the future fulfilment of His Kingdom is not itself a realization of that Kingdom. But what role does humanity and, especially, human freedom then have in relation to the coming of the Kingdom? Or, as Rosenzweig provocatively puts it, can human beings 'tempt' God to hasten its coming? At its simplest, this is also, again, a matter of prayer, since, if prayer is anything more than a matter of assent to the divine will, it must imply a role for human freedom in the counsels of God. But this in turn suggests that, in some sense, God does tempt human beings, so as to find out whether they are willing to use the freedom with which he has endowed them: 'He must make it difficult for human beings, no, impossible... truly to believe and trust in Him, that is to do so in freedom' (S, 296/266). Even if human beings are limited in other respects, they are to be free in relation to God. Despite conventional assumptions about Jewish legalism, the law is 'freedom on tablets' (S, 297/266). And, above all, it is freedom to love; redemption is essentially nothing other than God 'in his love, freeing souls for the freedom to love' (S, 297/267).

But if the human freedom manifest in our freely loving God is itself willed by God, how might we talk of it as 'tempting' God? Rosenzweig's answer is that although prayer is always prayed by the individual, it reaches out beyond the individual to the one for whom he or she prays and therefore becomes a prayer for the coming of a world-order in which those others might flourish and be fulfilled. Thus it both tempts God to act precipitately, before the right time has come, whilst also running the risk of neglecting to thank God for what he has already done and for the fulfilment of his will that is manifest in the present qua present.

In all of this we see that human freedom must learn to negotiate the difference between divine time and human time. 'For God the future can in no way be anticipated. He is eternal and He alone is eternal, he is the Eternal in an absolute sense; in his mouth "I am" is the same as "I will be" and only becomes manifest therein' (S, 303/272). God—and only God—is both in and beyond time. Human beings, however, being temporal, seek assurance regarding a future that they cannot know other than by anticipation. The justification of prayer, then, is not just to be decided with regard to its content but also its timing. Prayer, truly to be prayer, must be prayer at the right time, a view that reintroduces the idea of fate.

Rosenzweig himself introduces the idea of fate via Goethe's prayer: '"Create" the daily labour of my hands, O high fortune, that I might accomplish it'— which, as he also comments, is essentially addressed to Goethe's own 'fortune' (S, 306/275). He further links this to Goethe's reflection that '"Perhaps I am the only Christian today"' (S, 308/277). In the light of Goethe's reputation as a 'pagan' and of the way in which he directs his prayer to his own 'fortune', this may seem a surprising claim, but Rosenzweig nevertheless offers a defence both of Goethe's prayer and of his claim to be 'the only Christian'. Let us briefly see how he does so.[54]

Prayer must be at the right time and, as such, must be conformed to or grow out of the embodied life of the soul: it cannot be a purely intellectual thing imposed onto life. Historically, however, Christian theology early on took the path of refuting the pagan philosophers without adequately addressing pagan philosophy itself. But theory cannot be refuted by love alone. German Idealism attempted for the first time to incorporate philosophy and theology into a unified system based on Christian principles, but it did so solely at the intellectual level and was not open to the fullness of individual and common life.[55] And this is where Goethe corrects or supplements Hegel and by rooting his thought in the progress of his own life, is able to pray to his own fortune. This is 'pagan'—but in the specific sense that the conflict between paganism and faith is now, since Goethe, fully internalized. Christianity is no longer in pursuit of another world but turns towards the question as to how each individual and even nation engages with its own historical moment. Therefore the Christian mission is no longer directed outwards, but inwards: it is himself the Christian must now convert. This new situation is the dawn of the Johannine Church, following on the Petrine (Roman) and Pauline (Protestant) Churches of the past—and Goethe is the first 'Father' of this Church.[56] The prayer for the fulfilment of one's own fortune marks the ripening of the self to the point at which the eternal divine life can break in to human existence (S, 318/286).

The demand of the moment, the right time, is now acknowledged. Whereas the Petrine Church measured time in terms of the history of its own spatial expansion and the Pauline Church ignored time, the Johannine Church grasps time as living. At the same time, and precisely as such, it embraces the twofold truth that, on the one hand, time needs the eternal if it is not to flow away into nothingness, whilst, on the other, 'Life and all of life must become entirely living before it can become eternal life' (S, 320/288). But, because eternity itself is both present now and still to come, it is always possible that it might come today—otherwise it would not be eternity but some ontologically indeterminate future state. If the power of the eternal were in no way and never *now*, it would not be the power of *the eternal*. Prayer, Rosenzweig suggests, is always

---

[54] This also brings the discussion back into the orbit of Nietzsche and what Rosenzweig sees as his singular service in making the personal destiny of the philosopher the matter of his philosophy.

[55] Rosenzweig seems to imply that medieval Christian philosophy was marked by an implicit acceptance of a theory of two truths that were never finally unified. This is, of course, contestable.

[56] Rosenzweig considers but rejects the view that the Russian Church—specifically as represented by Alyosha Karamazov—might be this Johannine Church. See S, 317/285.

intrinsically a way of eternalizing what is being prayed for, offering it (or him or her) towards its future in the will of God; but if it is always future-directed, the fact that the one to whom it is offered is the Eternal means that it is also directed towards a non-transient moment, rising to existence in the same moment that it vanishes away. Consequently, this prayed moment is both future and present in intention. It is also a moment that is both constant and constantly new, constant because the eternal to which it is directed is always the same, but also constantly new because (and here Rosenzweig once more plays on the etymology of 'the moment' or *Augen-blick* qua 'moment of vision' or 'glance of the eye') 'each moment that our eyes open and look out, they see something new. [But] the new thing that we are seeking must be a *Nunc stans*, that is, no mere fleeting entity but an "abiding" [moment of] vision. By way of contrast to the moment of vision we call such a "standing now" "the hour" ' (S, 322/290).[57] 'The hour' is consequently a moment of vision 'transformed into a *Nunc stans*, into eternity' (S, 322/290).[58]

In this context the 'hour' is more than a numerical measure of time pass-ing—sixty minutes or one twenty-fourth of a day. Rather, he uses the term emphatically, as in the notion of prescribed 'hours' of prayer. But, in this sense, an 'hour' can also extend to days, weeks, or months, as when we talk of 'Britain's Finest Hour', which was actually a period stretching over several months. However (and remembering that this whole discussion builds on the centrality of common worship to religious life), Rosenzweig's focus is not on exceptional historical events but on exactly the opposite, on the cycles of solar and lunar time and of the agricultural year with its allotted festivals that make the repetition of the eternal in time visible. As in the promise of a never-failing seedtime and harvest in the Noachic covenant, 'in the ever-repeated daily and yearly service of the earth, human beings, in fellowship with one another, come to feel their earthly eternity' (S, 323/291).

However, to the extent that this cycle is determined by the rhythms of cos-mic time, it is not yet an adequate sign of the relation of human freedom to the eternal. This is provided by the week and by the purely human cycle of work and rest as constituted in the biblical understanding of creation. Thus understood, the week becomes an image of eternity in miniature, in which the beginning coincides with the end and the 'today' of the Sabbath is endowed with abiding presence. Of course, the week is itself temporal and can therefore only be a 'parable' of the eternal—but insofar as the weekly ritual is construed as divinely ordained it is also more than a parable. Here, in the keeping of the

---

[57] The first part of this is rather freely translated to try to express the point of Rosenzweig's word-play.

[58] Rosenzweig refers neither to Kierkegaard nor to Nietzsche in this discussion. However, as suggested above (see n. 54), the emphasis on fate or destiny might point to his idea of the moment as being less Kierkegaardian and more Nietzschean. There are also strong affinities to his contemporary Tillich, likewise influenced by each of Schelling, Kierkegaard, and Nietzsche.

Sabbath ritual, 'the Eternal not only enters "Today" in a parabolic sense, but in reality too' (S, 324/292).

To sum up: 'In the daily-weekly-yearly repetition of the cycle of cultic prayer, faith makes the moment into a 'hour' and time is made ready for eternity, which, as it is received into time itself becomes—as time becomes' (S, 324/292). Such cultic or liturgical prayer is not the more or less arbitrary and isolated prayer of the individual but the prayer of a 'people'. As such it relates to the whole course of that people's history as well as embracing the neighbour, constantly recalling them to the view of the eternal that becomes manifest in the obligation of love. Its moment is the moment 'between' revelation and redemption.

From this analysis of the week, Rosenzweig now expands his account to take in the whole annual cycle of festivals. We shall not follow all the details of this exposition, but will focus only on the points that are most relevant to the question of the eternal. Thereafter, I shall compare Rosenzweig's version of the mediation of time and eternity in the liturgical cycle with Heidegger's vision of the poetic foundations of community, a comparison that will enable us to identify some of the issues around Rosenzweig's use of the term 'eternal' and, especially, the question as to whether there could really be an 'eternal people'.

It is in this third Book that the defining image of the work as a whole, the star, comes to play a central role, which Rosenzweig analyses in terms of (a) its fiery life, (b) its outspreading rays, and (c) its abiding and entire eternal truth. All of these are, however, intimately connected and, as we shall see, Rosenzweig's account is also an account of the different ways in which Judaism and Christianity relate to the truth of the divine star. In fact, Rosenzweig returns immediately, at the start of Book 1, to the question of the community of blood.

Fundamental to the character of the star, to its being a star, is that 'its flames must be fed eternally from within…It must produce its own time; it must eternally thrust forward of itself. It must eternalize its life in the sequence of generations' (S, 331/298). Only so can past and future be bound into a living whole and therefore 'Only the community of blood senses the guarantee of its eternity already warmly coursing through its veins' (S, 332/299). If this may, in principle, be a universally human truth, it is, nevertheless true of the Jewish community 'in a quite particular way', since whereas the identity of other nations is based on their occupancy of a certain land, Jewish identity is based solely and exclusively on the continuity of Jewish life through the generations. In Judaism, even the idea of a 'holy land' exists only as an object of historical longing, not as a category of geopolitical identity.[59] Likewise, Judaism forgoes

---

[59] We recall that Rosenzweig is writing before the establishment of the State of Israel, although, of course, his views were already controversial in his own time, when the Zionist option was gathering momentum.

its own language for use in daily life.[60] Rather than the historical and temporal imprint with which other languages are marked, this language binds the Jew to eternity by virtue of its being exclusively scriptural and liturgical; the custom and law, the past and the future that it transmits are bound together in 'one unchangeable present' (S, 337/303). Other peoples know from experience that the occupancy of their land is only ever temporary, that sooner or later a new conqueror will come or the people itself move on and, similarly, that their language is heavy with the burden of its own transience. 'Eternity' can therefore never find defining form in their lives—only in Judaism is this possible, since only here is it true that 'the people is a people only through the people' and not through the transient possession of land or language (S, 333/300).

The unchanging 'eternal' language in which the faith is liturgically articulated bears or expresses the relation between people and eternity, but precisely because language is necessarily and intrinsically temporal, 'in eternity, the word passes into silence' and therefore liturgy too, 'as the lens that gathers the solar rays of eternity into the "little circle" of the year must introduce human beings to this silence' (S, 342/308).[61] However, true to his constant emphasis on the embodied character of thinking, Rosenzweig's conception of silence is itself approached through the actual practices of religious life. Liturgy teaches silence through collective listening to the reading of scripture, which, as opposed to preaching, is 'A hearing without answering back' (S, 342/309). Preaching too, unlike political oratory, is not primarily about persuading the listeners but, as exposition of the text, is also a way of keeping silent before Scripture (S, 342-3/309). But *living* silence—with all the emphasis we have heard Rosenzweig give to the idea of life—is not just a matter of listening to words. Attention to what reaches beyond language is therefore also taught by the ritual of the common meal, supremely the Passover Seder (S, 349-52/317-19). And prior even to the mutual acquaintance that is developed through the conversation that accompanies the meal, each participant welcomes the others and knows him- or herself to be welcomed in the silent welcoming greeting that each gives to each as they gather for the meal. In this possibility of silent mutual greeting lies the highest mark of true community.

But what kind of silent greeting could really unite all to all? How might such a thing be thinkable, let alone practicable?

Rosenzweig's answer is in terms of an analogy with military life. Although soldiers salute one another in the daily life of the unit, these occasional salutes are grounded in the ritual of the parade and saluting the flag, the symbol of what will outlast the individual. It is as expressive of a common allegiance to

---

[60] Again, this remark is controversial vis-à-vis the revival of Hebrew as a living language.

[61] This can be related to the 'apophatic' aspect of the final Book, where he speaks of a silence that reigns 'beyond' the word and of an abyss of Godhead in which 'God himself is set free from his own name' in ultimate silence (S, 426-7/383-4).

the flag that soldiers also salute each other. Similarly, it is in silently kneeling before the Lord of time, 'where all know each other and greet him without words, face to face' (S, 358/323), that the community 'greets' the one in whom the possibility of each member greeting the other as brother or sister is grounded. And if the weekly and annual cycle of liturgy establishes the year as the 'plenipotentiary representative of eternity', this is eminently condensed into the New Year festival and what is nothing less than a yearly 'return', as it were, of the Last Judgement (S, 360/325). In the New Year ritual, as Rosenzweig comments, the community says 'we' but each stands 'immediately, in his naked singularity, before God' and in the sheer simplicity not of his Jewishness but his humanity. That the worshipper is also ritually clothed in his shroud is further indicative that this involves a confrontation with the basic issue of human existence, namely, with the death that, from the moment of creation, marked humanity out as destined for a more than merely biological existence. In this regard the symbolism of the shroud is even more emphatic than when it is worn on the wedding day. 'Love is strong as death'—but in the eternity towards which we are called by virtue of the God-relationship, even human love must yield to the only love that can overcome the ritualized death of the solitary individual and re-start the cycle of time beyond that death, as symbolized in the New Year festival. This love is the love of God (S, 360–3/325–7). The cycle of time thus points beyond itself to messianic time, a time of redemption not only for Jews but for all humanity, since it is time freed from death.

In Book 2 of Part 3, Rosenzweig will turn to Christianity as represented by the rays of the eternal fiery star, spreading out into the world and, through its manifestation in art, music, and movement, gathering the peoples into the life of the star. But, for the Jewish people, the movement is always already complete. In the liturgical cycle this people has self-consciously reached its goal and knows that it has no need of further development. In this way, 'the eternal people ensures its eternity' (S, 369/332). Other nations, however, have no such inner principle of eternity. Both land and language, the standard criteria of national identity, are palpably transient. If, then, the state seeks—as, on the Hegelian understanding, it must—to create a kind of eternity for its citizens, then it too, in its own way, must also seek to impose some kind of constancy on what is otherwise the sheer centripetal flow of time. But it can only do this by the exercise of force, and even if force is only ever exerted in the name of right, either for the preservation of existing rights (e.g. the rights of the sovereign) or the restoration or renewal of rights (e.g. the rights of the people), the state must therefore exist in a condition of continuous potential conflict. As Rosenzweig puts it: 'War and revolution are thus the only reality that the state knows and in a moment in which there is neither the one nor the other—even if only it is only a matter of thinking about either war or revolution—it would no longer be a state' (S, 370/333–4).

Here we are returned once again to the break with Hegelianism that was so dramatically marked in the opening pages of *The Star*. There, the issue was Idealist philosophy's failure to take seriously the individual's terror of death. Now, it is more in the manner of a political objection. Because Idealism is unable to give an adequate account of reality as temporal it must therefore impose its own constructions on the reality of lived time and, in the case of common life, the supreme example of such a construction is the state. The state will therefore by its very nature always seek to impose permanent order on what is intrinsically not susceptible to such ordering. Consequently, no matter how lofty its ideals (and perhaps *especially* when they are most lofty), the state will only ever be able to ensure its stability by doing violence to the real ebb and flow of living temporal relations.

Another aspect of the complex of aims that Rosenzweig has woven into *The Star* now becomes clearer. Where the focus on the blood-tie and the unique-ness of the Jewish liturgical cycle might seem to mark it out as a work of Jewish apologetics (albeit one that is grafted onto a Schellingian cosmogony), we can now see that it is also an attempt to conceive of how the human future might free itself from the possibilities that were realized in the catastrophic events of 1914–18. Rosenzweig is not a pacifist and scorns the pacifist bid to impose the peace that is the end of history on a process of historical development that is not yet ready for it. Yet *The Star* does offer grounds for hope, namely, a way of seeing the world that allows us to affirm that God's will for human beings is good and that it is possible to hope for a transformation of human rela-tions that would inaugurate a Kingdom of peace: we are not only free to pray, we may pray, we may entreat the coming of the Kingdom, and we may do so with a confidence that is better grounded than utopian optimism because it is grounded in the lived covenant of the eternal people's liturgical life with God.

Nevertheless, we cannot but ask what is there to stop this vision from sliding into what Rosenzweig himself had before the war described as 'atheistic theol-ogy'. He then considered Christian theology's quest for the historical Jesus to have been a primary example of such theology, seeing its focus on the 'per-sonality' of Jesus as an attempt to provide a purely human basis from which to think about God. Rather than theo-ology in the strict sense it was a question of ideal humanity. And he also saw something similar in in nineteenth-century Jewish thought. In the face of Hegelianism's assertion of the transience of each historical community, Judaism had attempted to justify its claims to be an 'eternally abiding' people by invoking the new 'scientific' notion of race or blood.[62] But this is precisely 'atheistic theology', Jewish-style. Jewish life and practice is separated from the continuing urgency of the divine call and 'instead [of God's] descent to the mountain where the Law was given [we have]

---

[62] Rosenzweig, 'Atheistische Theologie' in *Zweistromland*, 691.

the autonomy of the ethical law'.[63] But whilst Rosenzweig will insist that God and human life can and must be thought within a single frame of reference and that this 'stands at the entrance to any knowledge whatsoever of our faith',[64] the event of revelation also calls on us to think the *difference* between God and human beings.[65] But the question now is whether his own constructive effort isn't itself exposed to the charge he levelled against both Christian and Jewish liberalisms, namely, that it places a human community or a human ideal in a position that should and can only be occupied by God—creator, revealer, and redeemer.

We have seen that Rosenzweig glosses the matter of the blood-tie less in terms of nineteenth-century racial theory and more in terms of (patriarchal) family relations. We have also noted that while he sees the Jewish community of blood as distinctive, it is also portrayed as a particular example of a universal human phenomenon. His account of the liturgical identity of Judaism further qualifies this account (1) by a messianic dimension that points towards a fulfilment of God's purposes in time for all peoples, (2) by an insistence on the inherent peaceableness of a people whose life accepts its self-containment within its familial and liturgical cycles, and (3) by a constant and variously articulated reminder of the divine difference. All of these mean also that this is not a community that can ever be sublated into a Hegelian state, but will always remain in history as a counter-cultural voice of opposition to the intrinsic belligerence of the state. Nor is it necessarily defined in terms of opposition to any other religious community—as Part III, Book 2 will go on to argue, the exclusion of Christianity from the self-enclosed cycle of eternal self-affirmation is by no means to exclude Christianity from a living bond to the life of the Star itself since Christianity is precisely the means by which the Star's rays are to spread out into the world and draw the world into its light. The difference between Jew and non-Jew is decisive, but not absolute, not final. It is a difference of historical existence, not of ontological constitution.

'BUT WHAT ABIDES, THAT IS WHAT THE POETS FOUND . . .'

We have heard that Rosenzweig regarded Heidegger as also representing the new thinking to which he himself was committed, and I shall therefore briefly compare Rosenzweig's account of the liturgically enacted unification of time and eternity with a strikingly similar element in Heidegger's Hölderlin lectures

---

[63] Rosenzweig, 'Atheistische Theologie', 693
[64] Rosenzweig, 'Atheistische Theologie', 696.
[65] Again, this is specially emphasized in Part III, Book 3.

of the 1930s and 1940s. Heidegger, of course, does not aspire to speak of eternity, although we have heard him acknowledge that the problem with familiar concepts of eternity is essentially a problem rooted in the experiences and concepts of time to which they relate. Yet he also speaks—with Hölderlin—of the poet as founding 'what abides' and doing so with particular regard to a social and historical order. Both, therefore see community as grounded in a word spoken to it from a 'beyond'.

The first of Heidegger's major lecture series on Hölderlin, on the poems 'Germania' and 'The Rhine', were given in the winter semester 1934–5. This seems to place them in the period subsequent to his resignation of the rectorship of Freiburg University but, on most accounts, prior to his final disillusionment with Nazism. Regrettably, there are elements in Heidegger's Hölderlin lectures that only too painfully permit being read in the horizons of National Socialist thought, but while that connection may be real, I do not think that the points I shall especially emphasize here amount to a distinctively or exclusively National Socialist vision. Rather, they show a vision of community that is poetically grounded and that, as such, enters into both a certain proximity and a certain rivalry with visions, such as Rosenzweig's, that, instead, ground community in revelation and religious life. It is precisely the proximity and the rivalry of the poetic and the religious as articulated by these two contemporaries that the turn to Heidegger at this point is intended to bring into closer view, and to do so with the aim of stating more precisely what, precisely, is specific to the religious.

The Introduction to the lectures starts with the citation of a line from 'Remembrance' to which allusion has already been made: 'But what abides, that is what the poets found'.[66] Whether 'what abides' will prove to have any relation to 'eternity' is not immediately and perhaps never explicitly addressed. But we are rapidly given notice that the matter of the poem is not some private poetic feeling but a question as to the real meaning of 'The Fatherland, our Fatherland Germany—mostly a forbidden topic, withdrawn from the rush of daily life and the noise of busyness. The highest, and thus the hardest, the last, because basically the first—the unspoken origin' (HH, 4). This suggests (perhaps) that the meaning of 'Fatherland' is not in the first instance a social or political question but one that concerns a more fundamental level of identity to which, in Heidegger's view, Hölderlin is the one best fitted to guide us. Heidegger quotes from Hölderlin's _Hyperion_ a vision of the life of ancient Athens as expressing a communal experience that is at once an experience of art, of religion, and of mutual love, a vision reflecting the ideals that, for a time, the poet shared with his friend Hegel (HH, 21). The poet, as Heidegger will go on to say, is one whose calling consists in 'enduring without flinching' 'the

---

[66] Martin Heidegger, _Gesamtausgabe_, xxxix: _Hölderlins Hymnen 'Germanien' und 'Der Rhein'_ (Frankfurt am Main: Klostermann, 1989), 3. (Further references are given in the text as 'HH'.)

thunder and lightning' that is 'the language of the gods' and, at the same time, making this language effective in the life of his people (HH, 31). But the poet does not teach in any direct or didactic sense. His own language is a language of hints, gestures, and waves (as in waving a welcoming or departing greeting to a friend or beloved).

But who is the 'we' invoked by the poet? And how does that relate to the 'we' who read his words today (i.e., we Germans, now in 1934)? Do we really know what it means to be this 'we'—or is our present historical moment only truly known in a temporal perspective very different from that of day-to-day political debate? That seems to be Heidegger's conclusion. 'We do not know our own proper historical time. The world-hour of our people is hidden from us. We do not know who we are when we ask about our own being, the being that is temporally proper to us' (HH, 50). Such considerations drive Heidegger back towards what he calls the 'original' or 'primordial' time of a people, and this is precisely the time in which the poet (together with the philosopher and the statesman) is active in founding the life of the people. Such originally creative spirits inhabit a time that Heidegger compares to the relations between mountain peaks rising above the clouds of everyday time to a certain proximity with the gods (HH, 52). Higher than the one-thing-after-another of everyday time, theirs is nevertheless a kind of time, time that is so little to be measured in terms of linear continuity as—to continue the mountain-peak analogy—to be marked by the abysses and precipices that fall away from the pinnacle at which each truly creative spirit lives.

It is at this point that Heidegger insists that what Hölderlin is speaking about is not a timeless or supra-temporal eternity, since the poet says explicitly that even God is 'nothing but time' (HH, 54). This is distinctive vis-à-vis the way in which 'One is otherwise accustomed to place the gods and the divine outside of time and to speak of them as the Eternal' (HH, 54). However, in a statement that can be read as epitomizing the impulse directing the entire structure of the present study, Heidegger continues: 'Only, the attribution of eternity is nothing in itself; rather, the representation of what we call eternity and its concept are always defined in accordance with the guiding representation of time' (HH, 54). The metaphysical representation of time generated two characteristic concepts of eternity, *sempiternitas* and eternity in the sense of the *Nunc stans*, which Heidegger glosses as 'the standing Now' and the 'ever-enduring present' (HH, 55).[67] However, both of these reflect the experience of time 'as the pure transience of the now in succession' (HH, 55). Hölderlin's view of divine time, by way of contrast, is of a time that is 'long' but in a sense quite different from the experience of time as long in the sense of tedious or boring. It is time that is essentially long because it is the time of preparing for what

---

[67] As in Boethius—see Chapter 1 of this volume.

is historically decisive, namely, the event of the revelation of being—and this time is not yet come (HH, 56).

I shall not now follow the development of Heidegger's exposition in further detail, but want to focus in more detail on the role of the poet in founding the history of his people by summoning them to the festival time of liturgical celebration. An important discussion of this is found in a later lecture series, from the winter semester 1941–2, on the poem 'Remembrance'. Before doing so, however, I mention two brief points. The first is that, in the further development of the 1934–5 lectures, Heidegger sees the poet as pointing towards a conception of the historic calling of the German people that is contrasted with the specific historic destiny of the Greeks. Each, German and Greek, has a decisive place in the history of being. The Greek lived 'in arousing proximity to the heavenly fire, touched [in his being] by the force of being [*Seyn*]'. His task, consequently, was how to bring this raging fire to form. The Germans, however, as inheritors of a certain ordering of being from the Greeks, have the task of rediscovering what it means to be 'touched by Being [*Seyn*]', which, in the first instance, means learning that there is a question to be addressed (HH, 292–4).

A deflationary version of what Heidegger is doing here is to see him as interpreting the teaching of the poet as pointing us back to the need for a more thoughtful reflection on the philosophical question of what it is for being—a being, any being—to be at all, the question he had already, on the first page of *Being and Time*, charged his readers with forgetting.[68] Even on such a 'deflationary' reading, however, we can still see a number of points of contact with what we have been reading in Rosenzweig. If we read the poetic word as an analogy of the word of biblical revelation, it too is concerned to unite the different times of the everyday and historical human worlds with a divine time that, in Rosenzweig, is eternal (though never timelessly so), and, in Heidegger, is 'long' and perhaps 'eternal' in a new sense that will correspond to the new kind of time-experience that the poet reveals. Both speak of this revelation as translating the burning flame of a heavenly fire into human language, and if the quotation from *Hyperion* is to be taken seriously, then this points to a new configuring of the interrelationship between art, religion, and society, infusing mere Being [*Sein*] with the life of living, fiery being [*Seyn*]. But where Rosenzweig sees this process finding its appropriate form in the living liturgical life of the blood-community, Heidegger seems to end by recalling us back to sober

---

[68] By 'deflationary' I mean taking a minimalist interpretation of the elevated and hyperbolic language in which Heidegger expounds the poetic word, and reading him also at a certain distance from a presumed 'history of being' that some commentators see (entirely wrongly in my view) as offering a kind of neo-mythological cosmology.

philosophical reflection. In 'Remembrance', however, we will see a different emphasis that brings us closer again to Rosenzweig.[69]

The second point relates to the 'Remembrance' lectures' extensive treatment of the motif of 'greeting' prompted by the lines 'Go now and greet | The beautiful Garonne'. Heidegger acknowledges that daily life provides us with a wide range of experiences of greeting, from the emptiest to the most significant—so much so that we mostly don't even think about what it really means to greet someone. Heidegger does not emphasize the silence of the greeting in the same way that Rosenzweig does, but he clearly sees it as 'saying' more than what is said in the words in which it is articulated. 'What is said in the discourse does not have need of our words. But, on the other hand, we have need of some hint.'[70] A greeting affirms the worth and identity of the one being greeted so as to let him or her be what and as he or she is: it is a kind of contact or touch that does not involve an attempt to grasp the other. It reaches across the distance between individuals whilst recognizing their distance and distinctness, as the creative poets, each atop their individual mountain-tops, might greet one another in recognition of a common spirit (A, 50–1). Through being in this way greeted for what and as it is, the one being greeted attains its appropriate splendour (A, 52).[71]

The 'simplest and yet most inward' greeting is one in which something that we have once known is greeted as if revealing itself for the first time (A, 51). But this shows also that the phenomenon of greeting is laden with significance for time-experience. The poet 'greets' the region he has once loved, from a distance that is temporal as well as spatial. In doing so, he not only makes it present in the mode of recollection but also marks it as significant for his future—it is a memory with which he is not yet finished (A, 54).[72] In this way the greeting condenses what is at issue in a poem concerning 'remembrance', which, in the

---

[69] It is also striking that the quest inaugurated in the 'Germania' lectures will lead Heidegger to give crucial roles to poems that—contra Rosenzweig's claims regarding the privilege of the blood-community—define the historical place of the German community in terms of the land marked out by its great rivers, the Rhine and the Danube. However, Heidegger interprets 'land' in essentially temporal terms that are rather different from Nazi evocations of 'blood and soil'. The land is 'land' precisely by virtue of the manner in which it is inhabited by the mortals who make it their home. Nevertheless, this also indicates that, despite the possibilities of dialogue between Heidegger and Rosenzweig, there is unlikely to be a final fusion of horizons.

[70] Martin Heidegger, *Gesamtausgabe*, LII: *Hölderlins Hymne 'Andenken'* (Frankfurt am Main: Klostermann, 1992), 51. (Further references are given in the text as 'A'.)

[71] Of course, those eager to uncover a National Socialist message at the heart of everything Heidegger wrote will see here an allusion to the '*deutsche Gruß*' (i.e. the 'Heil Hitler!' salute) that became expressive of the new community into which the German people were being summoned at this time. Yet even if this allusion is intentional, it does not follow that the idea of such a greeting is solely Nazi in meaning—we have already seen how Rosenzweig himself emphasizes the role of silent greeting in the founding of community and Christians might think of, for example, the ritual kiss of peace.

[72] Cf. Jankélévitch's comments on the essentially temporal nature of nostalgia for place.

poem, is named as *Andenken*, with its connotations also of 'thinking' (as in 'thinking of you'). These connotations allow Heidegger to muse that perhaps thinking and poetic creation are 'almost' the same (A, 55).

Heidegger sees the temporal significance of the greeting as especially pertinent to times of festival, when 'The brown women | walk on silky ground | in march-time | when day and night are equal'. How so? The kind of festival time meant here, he reminds us, is far more than the modern experience of 'time off' from work. For the true spirit of festival is not a matter of time off but of time for giving ourselves to what truly merits celebrating (A, 64–6). The radiance associated with festival-time is not mere glitz but originates in 'the luminosity and shining-forth of what is essential' (A, 66). In this connection the poem's allusion to the equality of day and night is not just an observation about the facts of the spring equinox but suggests how this light, the light of what is essential, is not limited by the alternation of day and night. It is another light, and it is by this light that the festival truly becomes time for dancing and music A, 67).[73] True festival time is not simply time set aside within a pre-existent calendar since, at the deepest level, the calendar is itself determined by the cycle of feast-days. It is the feast that, in this sense, orders our time-experience. As Heidegger puts it ' "The feast" is itself the ground and essence of history' (A, 68). As such it is 'the marriage feast' of 'humans and gods'. What is truly festal, what calls most for celebration, is just this: that human beings are called into their proper humanity on the basis of the mutual greeting of humanity and divinity. 'As the wedding-feast the festival is the event of the originating greeting. This originating greeting is the hidden essence of history. This originating greeting is *the* event, *the* beginning' (A, 70). In these terms it is the advent of the holy and of the possibility that gods and humans might live in a fellowship of mutual help (A, 70). And, since this fellowship is grounded in an event of nuptial encounter, it cannot be assigned to either gods or humans as some kind of attribute but exists for each and for both together as fate or destiny (*Schicksal*): 'Festival and destiny belong together in one' (A, 91). But, as Heidegger elsewhere says, the 'testing' of our relation to the gods 'must have passed through the knee. Self-centredness must bend down and vanish away' and if human beings are to enter into this relationship they must kneel before their destiny.[74] But, according to a logic we have already encountered in both Nietzsche and Rosenzweig, 'destiny' is itself the occasion for what is unique and individual freely to affirm itself for what and as it is and to establish itself in its own distinctive manner of being. But the kind of abiding in being that is grounded in such festal/fateful freedom is very different, Heidegger says,

[73] Cf. Theunissen's comments on Pindar's Olympian Ode 10 in the section '*Chronos*' in Chapter 7.

[74] Martin Heidegger, *Gesamtausgabe*, IV: *Erläuterungen zu Hölderlins Dichtungen* (Frankfurt am Main: Klostermann, 1996), 196.

from duration, metaphysically conceived. It is 'the abiding of the unique individual'—not the enduring of an entity.[75]

Leaving Heidegger's subsequent discussion of how this relates to the specific destiny of the German people in a historic comparison with the Greeks, let us return to how it might help us in interpreting Rosenzweig's claims regarding the distinctiveness of 'the eternal people'.

We have already noted a number of seemingly analogous themes. For each, the common life of a people is grounded in a cycle of religious feast-days. The communal nature of this founding event is not in the first instance established in some kind of verbal social contract but in the mutual non- or pre-verbal greeting between humans and between humans and their God or gods (or 'destiny'). But this greeting itself is performed as a free response to a unique and fateful call that we can only receive when we are ready to kneel in silent attention to the one who calls. And, in both cases—crucially—it is a matter of re-ordering time, redeeming the empty time of transient existence and daily life by the advent of a time capable of giving endurance and of binding past and future into the demand of the present. Rosenzweig calls this the time of the eternal. Heidegger refrains from using this term, clearly anxious to avoid what he sees as its inappropriate connotations. But even where Rosenzweig embraces the *Nunc stans* that Heidegger specifically eschews, it is clear that Rosenzweig is using this in a distinctive way and precisely not as intended to suggest a timeless eternity, since it is this *Nunc stans* that is to give the abiding temporal constancy of the liturgical cycle and this in turn has the concrete, specific, and unique character that Heidegger ascribes to the (non-eternal) 'other' time of the festival.

In what way, then (if any), does Rosenzweig offer *more*—logically, metaphysically, *and religiously*—than Heidegger? When he says 'eternal', are we required to think of more than 'historically enduring in an exceptional degree'? 'Degree' is perhaps not insignificant, and one might argue that understanding the call as the call of a God who founded a three-thousand year history (the history of the Jewish people) is more impressive than the call of a 'poetic' Fatherland that failed to achieve any abiding historical form. But this only takes us so far—perhaps only as far as a humanistic 'atheistic theology'. For what is at issue is not the relative vigour of actual historical communities, but the possibility of one or other or any community—the actual historical community of the Jewish people or the poetic Fatherland of a Hölderlin—becoming able rightfully to claim the epithet 'eternal'. What might make us see Moses on the mountain of revelation (or, for that matter, Jesus on the mountain of Transfiguration) as essentially 'more' than the creative mountain-top poets of whom Heidegger spoke—or more, even, than Pindar, glimpsing the radiance of Zeus falling

---

[75] Heidegger, *Hölderlins Dichtungen*, 93.

on mortal achievements? Can we really distinguish the maximum point of a heightened time-experience at which we start to speak of an 'other time' from an eternity that would be truly and properly more than or other than time? And if we posit both an 'other' time and a divine eternity, how might these be both unified and distinguished, whether in relation to each other or to our lived historical time? And how, most importantly, do such questions relate to Rosenzweig's principle of vocativity, according to which it is not the possession of metaphysical attributes that makes the Eternal 'eternal' but the constancy of his coming to help, nurture, and guide those who call upon his name? And if this principle holds, does it not mean that whatever community of blood a human being may belong to, Buber was essentially correct in his view that whoever 'addresses with the entire devotion of his being the You of his life, that is, that which cannot be limited by any other, he addresses God'?[76] The moment of invocation, the praying of 'Thy Kingdom come', reaching out in obedience to the commandment of love—is it not enough that these continue to be real possibilities for human beings? Isn't the point of the new thinking to make us attentive to a certain call rather than to justify any particular religious, historical, or metaphysical claim regarding the Eternal? And isn't this also the force of Heidegger's concept of the poet as one called to call his people to a decisive choice in which they and their community might find enduring historical continuity? But is there or could there be any human community that would be 'eternal' in the sense not just of outliving its individual members but of being genuinely eternal and undying? What community is possible in the face of the traumas of historical life, including not only the deaths of individuals, the collapse of empires and civilizations, but the wounds of deep, unconsoled, and unreconciled wrong?

In considering such questions, we might think further about the nature of the respective calls by which the poet and prophet are called to their task of calling the people to the encounter with their God. Both can speak of the voice that calls as thunder, whether, like Heidegger's Hölderlin (or Theunissen's Pindar) this is the thunder of Zeus or whether, as for Rosenzweig, it is the thunder of Sinai. But are these the same? I have argued elsewhere that Heidegger's thunder seems, in the end, to reduce to the bare uncanniness of existence, thrown towards death. There is no one who calls. Behind the thunder of Zeus is only the thunder of Olympus, stripped of the pathetic fallacy that seems to humanize it.[77] But for Rosenzweig the thunder of Sinai is essentially more than a mighty cosmic event engendering fear and trembling. The thunder of Sinai reveals the voice of God.[78] But this is, in the first instance, merely an

---

[76] M. Buber, *Ich und Du*, 72; *I and Thou*, 124.

[77] See my *Heidegger on Death*, 135–7.

[78] As, according to Theunissen, the Pindaric poet listens to Zeus and not to Zeus's thunder (see the section 'The Poet' in Chapter 7).

unsubstantiated claim—you call it death, I call it God. In what ways might we move the issue between Heidegger and Rosenzweig beyond assertion and counter-assertion?

In the first instance, it might prove helpful to turn aside from the directly theological or metaphysical question as to who it is that is calling, and think more about those who are called and the manner of their hearing. For Rosenzweig it is axiomatic that the dynamic of calling and being called is inseparable from the life of a community that is bound together by something deeper and broader than its willingness to respond to the call of a particular historical moment. Israel is not called to be or to become a people, but *as* a people. Furthermore, its calling is not just to national self-assertion but, in biblical language, to be a light to the nations. That is, its calling identifies it as the bearer of a promise that is essentially and necessarily universalizable such that it can serve as instigator of a messianic hope for the world.

This brings us back once more to Rosenzweig's problematic emphasis on the blood-tie. Yet even if this invites being read in terms of ethnicity, the comparison with Heidegger allows us to see that it also builds a number of elements in to Rosenzweig's account that counter the decisionistic tendency of Heidegger's 'call'. These include: (a) the attempt to do justice to the embodied nature of sexually reproducing human life; (b) an affirmation of the trans-generational nature of community; and (c) the acknowledgment that community is not primarily founded on moral intentions—the greeting that welcomes each individual into the community is not conditional on that individual's moral virtue or ability to respond in an appropriate way.[79] Moral accounting may be integral to any 'last judgement' but—as Dostoevsky's parable of Mary's visit to hell suggested—we can envisage a quality of mercy beyond the finality of judgement; (d) it opens an alternative approach to the meaning of death.

In relation to this last, we may think again of Berdyaev's account of an 'eternal memory' 'in which each of us in the depths of his spiritual experience achieves a victory over the corruption and disintegration of history'.[80] And, as we have heard, the supreme expression of this victory is nothing less than resurrection.

> Memory of the past...carries forward into eternal life not that which is dead in the past but what is alive...Society is always a society not only of the living but also of the dead; and this memory of the dead...is a creative dynamic memory.

---

[79] Rites of passage relating to the new-born indicate that bestowing membership of a people and recognition of the child as fully human are, in an important sense, pre-moral actions, even if they also call for fulfilment in and through moral life.

[80] Berdyaev, Nicholas, *The Meaning of History*, tr. George Reavey (London: Geoffrey Bles, 1936), 71–2.

The last word belongs not to death but to resurrection. But resurrection is not a restoration of the past in its evil and untruth, but transfiguration'.[81]

This, I suggest, complements Rosenzweig's own comments regarding death as the mark of genuinely spiritual life. Our relation to the dead is integral to the recognition of ourselves as mortals, marked out for death. But it also highlights especially the importance of the blood-tie, since grief is nowhere more powerful than when it is grief at the death of those who are closest to us in the bonds of family life—parents, brothers, sisters, children. We too 'die a little' in the deaths of those whose bodily substance we share. Other deaths too afflict us, and do so in manifold ways, sometimes more sharply, but the point is not to measure different categories of grief in a merely quantitative sense. Rather, it is a question of the impact of grief in revealing the mortality consequent upon our existing as bodily beings and it is therefore natural that those with whom we are most physically intimate will also be those whose deaths most immediately and ineluctably recall us to our own deaths and make us realize what mortality might mean for human life as such.

It is a cliché, perhaps the most banal of all clichés concerning death—but no less eloquent for that—to say that 'to live in the hearts of those we love is not to die', and it is similarly banal to inscribe funeral notices or memorials with variations on the phrase 'Eternal memory'. Yet such sayings point us towards what Berdyaev too, in his more elevated manner, is expressing. At the very least, we may say, a truly eternal community would be a community in which the dead had a continuing place alongside the living. In remembering them, we not only remember them in terms of the individual occasions we shared with them in life, but we remember that what we are as a community now is inseparable from what flowed through them to us. The 'existential link' that underwrites the anticipated triumph of resurrection over oblivion is less a matter of remembering detailed achievements and more like Heidegger's sense of a 'greeting' that recognizes, accepts, and welcomes the other in his or her irreducible otherness. He or she never became 'mine' in any exclusive or possessive sense (a lover, or part of my family, tribe, or nation), but what I am and forever will be is inseparable from the fact that he or she lived—even if we were never known to each other otherwise than in the prayer that Zosima taught Alyosha to pray for all the dying or that Rosenzweig spoke of as the prayer that beseeches the hastening of the Kingdom.

In an increasingly post-religious society, public memory of the dead is most intense in relation to those who die in war. Such acts of remembrance are the most symbolically powerful means of expressing and underwriting the perpetuity of the state or nation for which 'the supreme sacrifice' is the ultimate

---

[81] Berdyaev, Nicholas, *Slavery and Freedom*, tr. R. M. French (London: Geoffrey Bles, 1943), 111.

guarantee. As such they ritually exemplify the Hegelian sublation of the individual into the eternity of the state. But they may also and sometimes do look beyond the remembrance of 'our' dead to embrace the fallen enemy, acknowledging sometimes as well that perhaps the cause for which each died was no cause at all, or no good cause.[82] When this happens, we are lifted above both the merely personal and the interests of the single state or nation to a universal solidarity in the face of a death towards which all, heroes and enemies, ultimately go as victims—victims of our own embodiment as well as of the malice of foes or negligence of friends.[83] It is this solidarity that in turn provides the basis for the future hope that it need not be ever thus; that for all the seeming ease with which war continues to seduce the imaginations and wills of human communities and individuals, the true community is a community beyond friendship and enmity—even if, in Rosenzweig's terms, the world of time and history has not yet grown to the point of being ready for the full and final incarnation of that community. And if the universal reach of eternal memory is most obviously reflected in acts of communal remembrance, it is implicit too even in the private and personal memories of our own 'loved ones'. In each case it is the affirmation we have heard condensed in Hermann Broch's invocation of 'the voice that binds our loneliness to all other lonelinesses', calling on us to do ourselves no harm, ' "for we are all here!" '.[84]

Christian theology might at this point want to invoke the idea of the mystical body of Christ, the community of those who will be found in Christ on the last day, inclusive of the departed, those now living, and those to come. Yet (and perhaps this is one lesson we owe to the traumatic memories of modern history) a Christian universalism based on faith in a singular historical manifestation of the God in time is, in its own way, as particular and as exclusive as a supposed Jewish exclusivity. The bodily reality of all human relationships that Rosenzweig's privileging of the blood-tie highlights is not only a sign of exclusivity but is also integral to the solidarity to which we are called. As in Broch's confession of faith, it is not the mystical body of one religious community any more than it is the blood-community of one historical people. The resurrection to which creative memory guides us is one in which 'we are *all* here'. And in order to be a hope that is truly hope for all, messianic hope, born within Judaism, propagated by Christianity, must point beyond the histories of particular and contingent historical communities. In Abraham, many nations are to be blessed.

---

[82] For example, the War Memorial in King's College Chapel, Cambridge, records the name of a former Fellow of the College who was killed fighting for the Central Powers in World War I.

[83] This is the thrust of Simone Weil's extraordinary essay on *The Iliad*, 'The Iliad, Poem of Might', in Simone Weil, *Intimations of Christianity Among the Ancient Greeks*, ed. and tr. Elisabeth Chase Geissbuhler (London: Routledge and Kegan Paul, 1957), 24–55.

[84] Broch, *The Sleepwalkers*, 648.

But does such memory and such hope—*can* such memory and such hope—really embrace and incorporate all that has been lost in the historical passage of the nations? Can each forgotten victim find recompense in such memory and such hope? Will each dark corner of individual and collective trauma ever give up its metaphorical and literal dead? Just how far back and how deep can memory go? And what future possibilities can hope really encompass? Can 'eternal memory' ever be more than a rhetorically heightened expression for acts of memory and hope that are always, ever, and only situated in fragile and transient interludes of luminous time edged on all sides by oblivion? And can remembrance be reconciled with the demands of justice in such a way as to allow us to envisage not only resurrection but also reconciliation?

Questions remain. What, outside the moment of vocation and invocation, can we, need we, must we say? Do we still require the hypothesis of an eternal being existing independently of and apart from the temporal universe, or may the experiences we can have in time, in the temporal time of our own historical lives, be enough to justify us in invoking an eternal memory and beseeching the coming of the eternal Kingdom? Such questions (which have special urgency with regard to the twentieth-century Jewish experience) invite metaphysical answers—but perhaps it is just these that, for the sake of vocativity, must be resisted. To be sure they point us towards a hope that, within the horizons of worldly historical time, must be a virtual impossibility, a hope that can scarcely be hoped for and that is perhaps unhoped-for. Yet if, as I suggested in Chapter 1, the hallmark of 'the eternal' is that eternal time is time that cannot be measured against the knowledge we have of historical time, can we ever finally say 'memory goes this far and no further' or 'hope reaches this far and no further'? Should we not rather say that when it is projected onto immeasurable time, memory reaches out—in hope—towards eternal memory, the eternal memory of the people of whom it can justly be said 'We are all here'.

Conclusion

One of the threads running through this study has been the tension between a speculative or metaphysical approach to the question of divine eternity and a more exclusively soteriological approach. Since the characteristics of the metaphysical God of early and medieval Christianity were precisely those that fitted him for delivering human beings from the changes and chances of temporal and finite life, these two approaches were almost inevitably intertwined for much of the first two millennia of Christian theology. However, subsequent to Hegel, the two approaches have separated out in such a way that metaphysics can no longer be easily seen either to answer or to ground soteriological questions. Conversely, the contemporary human quest for happiness—unlike Augustine's *beata vita* or blessed life—seems no longer to require or to endorse metaphysical buttressing. But whilst a soteriological approach is probably not able to presume upon *shared and uncontroversial metaphysical commitments* it may nevertheless push forward into the domain of *open metaphysical questions* and develop concepts, themes, and images that help articulate and explain its distinctive methods and concerns to those for whom they are not immediately intelligible.

In the case of divine eternity I suggested that the notion of immeasurability may serve as a kind of bridge between metaphysical theories of divine timelessness and approaches directed by an essentially soteriological interest. Responsive to elements in the metaphysical interest in establishing a timeless divine eternity, immeasurability also reflects both the ancient and the modern religious interest in 'the Eternal'. I have also claimed that the religious or soteriological interest in the question of eternity is well served by what I have loosely called a phenomenological approach. Such an approach especially leads to the reframing of the religious question in the context of phenomenological philosophy's distinctive preoccupation with the character of human temporality. Here too the notion of immeasurability helps relate the idea of eternity to the possibility that the time of human existence is not simply the vanishing linear time of conventional Augustinian theology, radicalized in Heidegger's account of Dasein's thrownness towards death, but may also be experienced as the coinherence of multiple temporalities. As such it signals towards more, not less, of time. On this view, the present is no longer just a fugitive and empty moment but, as the fullness of time, may also become the site of a presence or *Parousia* that is at once both anticipation and remembrance. Perhaps this may be experienced in terms of a 'timeless moment' or 'eternal now', though it need not take any such directly mystical form. In any case and as such, it opens up a relation to the immemorial and unhoped for that cannot be measured by any standard measure of time and that grounds a counter-movement to the otherwise relentless temporal annihilation of beings that Dostoevsky saw as climaxing in the nihilistic cult of oblivion. This countermovement to oblivion is therefore also a movement of hope, whether this takes shape as the utopian excess of time over fulfilment (Bloch) or as the existential possibility of faith in divine promise (Moltmann). In either case, hope points forward to messianic time, whether that is a time that is no time (Derrida) or the uni-dual *kairos* of the Messiah's constant arrival (Agamben). This is always a *kairos* for the individual, but—as we have seen in Kierkegaard (yes, even in Kierkegaard!), Rosenzweig, and Theunissen— the phenomenon of hope is such that it can never be limited to the individual. Hope is, will, and must always be hope for the maximum number, 'for all', and beyond all measure—even when its immediate focus is the life of a particular community.

But isn't this just rhetoric? How far can hope really extend? Even if there is much to commend hopefulness as both a private and a civic or political virtue, and even if we are persuaded that there is more to time than 'mere' time, doesn't a visit to any graveyard immediately demonstrate that time remains 'what it ever was'—the destroyer of individuals, communities, and nations? Even as it brings forth new futures, doesn't this same temporal productivity also bring about the irreversible wreckage of all human possibilities? Where is the ambition, the love, or the achievement that can really be called 'immortal'?

This study has not even begun—it was never its aim—to argue for the 'survival' of the individual centre of self-consciousness beyond death. In that sense it has nothing to say about 'eternal life'. However, the phenomenon of remembrance, as figured in Berdyaev's invocation of an eternal memory, reveals the hope that, whatever else we may say of it, death is not the exclusive measure of human life and that both the root and the flowering of our common life extend beyond death. Eternal memory does not bring back or reanimate the dead, but it does reconfigure our relation to time, to our own individual time and to 'our time', the time or age that now is and that, as such, is always also a common time. Neither with regard to the past nor with regard to the future can time, our time, be its own measure. It is always and only what and as it is by virtue of its relation to other times and, ultimately, to times that forever escape our capacity for self-conscious reflection. Beyond all measure available to us, human solidarity is itself the measure of our humanity and we are none of us really here until 'we are all here'— and every one of us singly.[1]

But what is it that moves us or might move us to embrace such solidarity? Several of the key figures of this study have spoken of a 'call'. But can we ever really distinguish between the call that is simply the resonance of our own subjectivity and the call that places us under a real obligation to the other, or between the god whose thunder is no more than the thunder of Olympus and the thunder of Sinai that bespeaks the presence and promise of the one, eternal God? And if we cannot or cannot easily make such a distinction how certain is our solidarity or our hope?

'We are all here' may be taken as a non-negotiable element in the conclusion of this study. But if the voice that calls us to solidarity, remembrance, and vigilance is, as Broch stated, 'the voice of man and of the tribes of men', is it really a voice capable of speaking a word of promise, that is, a word that is in some way more than the demand for truth, justice, and peace or even a 'certain Messianism'? Is it not just here that a dividing-line has to be drawn between a hopeful and even a messianic humanism (perhaps the messianic humanism of a 'warm' and non-'scientific' Marxism) and a theology that begins with faith in a promise both coming from and reaching beyond history? For only such a promise, surely, could also give assurance that even our best efforts at remembrance will not, in the end, be drowned in oblivion? And could anything less call us to what religion calls 'eternal life'? But, as we have seen, it is not easy to articulate such a 'more', faith's excess over historical materialism, without succumbing to an ontology that, in its own way, would once more constrain the openness, freedom, and possibility of faith.

Here, perhaps, Dostoevsky can again help us, if we recall that central to the hope beyond hope that was suggested by the Virgin's visit to Hell was the possibility that our ultimate, immemorial and unhoped-for destiny may not be decided solely by the record of our temporal identity and historical acts but *by the intercession of another*. Whatever we have been and done, there is—there may be—also the hope that others have for us. For Dostoevsky, the Virgin is perhaps that voice of prayer and hope for all who are without or beyond hope, those who have been even forgotten by God. But such prayer and such hope are also a task that he sees as something to which we are all, now, called. Recalling that 'every hour and every moment thousands of people leave their life on this earth and their souls come before the Lord—and so many of them

---

[1] To which I would also add—although this would be the beginning of a whole other discussion—that nurturing such solidarity is and ought to be a greater human priority than pursuing currently fashionable fantasies of redeeming ourselves through the construction of future post- or trans-humanity.

part with the earth in isolation, unknown to anyone, in sadness and sorrow that no one will mourn for them, or even know whether they had lived or not', the elder Zosima calls upon his disciple, Alyosha, to pray without expecting ever to know what deliverance his prayer might effect.[2]

Promise—knowledge—prayer? Perhaps, however we define the space indicated by the interplay of these three words, what is truly decisive is the miracle that such a possibility of hope is open to us. In the end, perhaps, the question is less about the origin of the promise or how it is categorized in terms of the ontological categories of immanence and transcendence and more about how this—miraculous—possibility of being called to hope reveals something essential to human life. Every life cries out to be remembered. Every departing soul is to be prayed for. And every life should, through loving remembrance and patient attention, be helped to become worthy of remembrance. In this all-encompassing solidarity, remembrance becomes also a ground for hope, assisting us to look towards a measure other than time, as well as a spur to ethical action directed towards ensuring that fulfilment really will mean hope for all. All shall be well, for all.

Even as we remember our important dead, however, we must acknowledge that many of our relationships have been left broken or unreconciled at the point of separation. Perhaps there are those fortunate enough never to have such experiences, but even if we abstract from the horrors of the twentieth century, they are, I suspect, a minority. Most families have some experience of broken relationships left unreconciled at death. Even so, to believe in an eternal reconciliation *nevertheless* and *in spite of* what has become impossible in time may be an existential need if we are to live well in time. Such faith may well be accused of sentimentality and wish-fulfilment, but this (real) risk is, I think, countered if faith is also qualified by the tension of remembrance and hope, including remembrance of all the ill that has been witnessed and hope that is focused on concrete stretegies of redress.

What is unreconciled at the level of time is unreconciled. But, even though a direct view onto eternity is, in this life, denied us, we are not denied the possibility of another time in time, a 'turn' in time that moves us towards what life in time itself will never give: the Eternal. Such time, I suggest, is saving time—and it is to this that the remembrance of what is to be hoped for directs us. Even in the face of broken and unreconciled relationships and in a world that continues to be disfigured by the ongoing history of war and the multiple uncalled-for deaths and breakdowns of human relationships that war will always bring about, such remembrance and such hope enable us to go on living and go on turning and turning again to one another with love and goodwill. If the time that grants such hope is experienced as eternal time, then, in

    [2] Dostoevsky, Fyodor, *The Brothers Karamazov*, tr. Constance Garnett (London: Heinemann, 1912), 318.

that eternal time, we may also find faith—once more (and faith must be always 'once more')—in a saving God.

We may seem to have come a long way from questions concerning the kind of eternity enjoyed by God and whether we human beings are capable of having knowledge of the divine eternity. Yet affirmation of the solidarity of all those living and suffering the vicissitudes of historical time does at one point illuminate something central to the Christian vision of God. If we look at any of the great depictions of heaven in the art of the Middle Ages and early Renaissance, we may, of course, find ourselves regretting the anthropomorphism involved and the careless abandon of the apophatic markers that should accompany all good theology. But we might also note that the artists—Fra Angelico, for example—had a clear sense that the vision of God in glory was always also a vision of God encompassed by an angelic and human multitude beyond number. If, ultimately, divine eternity is beyond human conception or imagination, any conception or image we do have of the eternal God will nevertheless be more rather than less adequate if it is able to incorporate some notion of the community of heaven. And the solidarity to which eternal and creative memory calls us is a minimum condition of such heavenly community.

And yet we must remind ourselves once more that the meaning of looking towards an eternal God and the hope of an eternal life can only become actual in and through what happens to us in the context of life in time. For Hegel this could mean the exhaustive recovery of all that has been significantly said and done in history in the light of Spirit's own eternal logic. Few subsequent thinkers have dared to stake as much on time's fragile materials. If there is a post-Hegelian consensus it is that time will not yield knowledge of the Eternal. Yet the aspiration towards the Eternal, the need of, the hope for, the dream of the Eternal and of the coming of a future world in which time would be 'forever' free from all the terrors of historical existence—none of this is quelled by the mere surrender of knowledge claims. In particular, it has been the argument of this study that the perennial turning towards a better—a *different*—future is the eminent way in which the possibility of the Eternal, of eternal life, takes shape in time—that is, as a self-realizing event of hope. There is nothing here that prevents such claims from being classified as wishful thinking, and reductionists will always be ready with their mantra of 'nothing but'! But if the essential interest of religion is the possibility of salvation, that is, the possibility of human beings coming to experience their lives as being borne away from sin and death and towards abundance of life, the occurrence of such hope, no matter how fragile, obscure, or even 'everyday' its concrete forms may be, will always be lived as the miraculous possibility of a new life, a new beginning beyond the end of the old. And perhaps the first intimation of such a possibility may be a momentary glimpse, a 'timeless moment', a passing sense of 'the eternal now' that is gone as soon as we try to reflect on it—but that, in eschatological hope, finds the form that can make it existentially and historically effective.

Bibliography

Adorno, Theodor W., *Kierkegaard: Konstruktion des Ästhetischen* (Frankfurt am Main: Suhrkamp, 1974).

Agamben, Giorgio, *Il sacramento del linguaggio: Archeologia del giuramento* [The sacrament of language: The archaeology of oath-swearing] (Bari: Laterza, 2008).

Agamben, Giorgio, *The Time that Remains: A Commentary on the Letter to the Romans*, tr. Patricia Dailey (Stanford, CA: Stanford University Press, 2005).

Andler, C., *La dernière philosophie de Nietzsche: Le renouvellement de toutes les valeurs* (Paris: Bossard, 1931).

Arendt, Hannah, *Eichmann in Jerusalem: A Report on the Banality of Evil* (New York: Penguin, 1964).

Augustine, *Saint Augustine's Confessions*, tr. William Watts (London: Heinemann, 1968 [1912]).

Augustine, 'The Soliloquies', in *Augustine: Earlier Writings*, ed. and tr. J. H. S. Burleigh (London: SCM Press, 1963).

Barr, James, *Biblical Words for Time* (London: SCM Press, 1962).

Barth, Karl, *Church Dogmatics* ii.1 *The Doctrine of God*, ed. G. W. Bromiley and T. F. Torrance (London: T. & T. Clark, 2004).

Barth, Karl, *Church Dogmatics*, iii.1 *The Doctrine of Creation* (London: T. & T. Clark, 2004).

Barth, Karl, *The Epistle to the Romans*, tr. E. C. Hoskyns (Oxford: Oxford University Press, 1933).

Barth, Karl, 'Rudolf Bultmann—An Attempt to Understand Him' in Hans-Werner Bartsch (ed.), *Kerygma and Myth: A Theological Debate*, tr. R. H. Fuller (London: SPCK, 1972), ii. 83–132.

Benjamin, Walter, 'Über den Begriff der Geschichte', in *Gesammelte Werke*, I.2 (Frankfurt am Main: Suhrkamp, 1974), 691–704; Eng. 'Theses on the Philosophy of History', in *Illuminations*, tr. H. Zohn (London: Pimlico, 1999).

Berdyaev, Nicholas, *The Beginning and the End*, tr. R. M. French (London: Geoffrey Bles, 1952).

Berdyaev, Nicholas, *The Divine and the Human*, tr. R. M. French (London: Geoffrey Bles, 1939).

Berdyaev, Nicholas, *The Meaning of History*, tr. G. Reavey (London: Geoffrey Bles, 1936).

Berdyaev, Nicholas, *The Russian Idea*, tr. R. M. French (London: Geoffrey Bles, 1947).

Berdyaev, Nicholas, *Slavery and Freedom*, tr. R. M. French (London: Geoffrey Bles, 1943).

Berdyaev, Nicholas, *Solitude and Society*, tr. George Reavey (London: Geoffrey Bles, 1938).

Berdyaev, Nicholas, *Sub Specie Aeternitatis: Ópyty Filosofskie, Sotsialnye i Literaturnye* 1900–1906 (St Petersburg: M. P. Pirozhkova, 1907).

Bergson, Henri, *L'évolution créatrice* (Paris: Presses universitaires de France, 1941).

Blake, William, 'Auguries of Innocence', in *Poetry and Prose of William Blake*, ed. Geoffrey Keynes (London: Nonesuch, 1967).

Batnitzky, Leora, 'Levinas between German Metaphysics and Christian Theology', in Kevin Hart and Michael A. Signer (eds), *The Exorbitant: Emmanuel Levinas Between Jews and Christians* (New York: Fordham University Press, 2010), 17–31.

Bloch, Ernst, *Atheismus in Christentum: Zu Religion des Exodus und des Reichs* (Frankfurt am Main: Suhrkamp, 1973).

Bloch, Ernst, *Avicenna und die Aristotelische Linke* (Berlin: Rütten and Loening, 1952).

Bloch, Ernst, *Man on his Own: Essays in the Philosophy of Religion*, tr. E. B. Ashton (New York: Herder and Herder, 1970).

Bloch, Ernst, *The Principle of Hope*, tr. Neville Plaice, Stephen Plaice, and Paul Knight, 3 vols (Oxford: Blackwell, 1986).

Böhme, Jakob, *The Aurora*, tr. John Sparrow, ed. C. J. Barker and D. S. Hehner (London: James Clarke, 1960).

Boethius, *The Consolation of Philosophy*, tr. Victor E. Watts (Harmondsworth: Penguin, 1969).

Bradley, F. H., *Appearance and Reality: A Metaphysical Essay* (Oxford: Clarendon Press, 1930 [1893]).

Broch, Hermann, *The Sleepwalkers*, tr. Willa and Edwin Muir (London: Quartet, 1986).

Buber, Martin, *The Prophetic Faith*, tr. Carlyle Witton-Davies (London: Macmillan, 1949).

Buber, Martin, *Ich und Du* (Stuttgart: Reclam, 1995); Eng. *I and Thou*, tr. Walter Kaufmann (Edinburgh: T. & T. Clark, 1970).

Bultmann, Rudolf, *Das Evangelium des Johannes* (Göttingen: Vandenhoeck and Ruprecht, 1950).

Bultmann, Rudolf, *History and Eschatology: The Gifford Lectures, 1955* (Edinburgh: University of Edinburgh Press, 1957).

Cappelørn, Niels Jørgen, Lore Hühn, Søren R. Fauth, and Philipp Schwab (eds), *Schopenhauer—Kierkegaard: Von der Metaphysik des Willens zur Philosophie der Existenz* (Berlin: De Gruyter, 2012).

Caputo, John D., *The Mystical Element in Heidegger's Thought* (New York: Fordham University Press, 1986).

Caputo, John D., *The Tears and Prayers of Jacques Derrida: Religion without Religion* (Bloomington, IN: Indiana University Press, 1997).

Carlisle, Clare, *Kierkegaard's Fear and Trembling: A Reader's Guide* (London: Continuum, 2010).

Carlisle, Clare, 'Kierkegaard and Heidegger', in John Lippitt and George Pattison (eds), *The Oxford Handbook of Kierkegaard* (Oxford: Oxford University Press, 2013), 421–39.

Carlisle, Clare, 'Spinoza on Eternal Life', in *American Catholic Philosophical Quarterly* (forthcoming).

Chrétien, Jean-Louis, *L'inoubliable et l'inéspéré* (2nd edn, Paris: Desclée de Brouwer, 2000).

Cohen, Hermann, *Die Religion der Vernunft aus den Quellen des Judentums* (2nd edn, Frankfurt am Main: Kaufmann, 1929).

Cohn, Norman, *The Pursuit of the Millennium: Revolutionary Millenarians and Mystical Anarchists of the Middle Ages* (London: Secker and Warburg, 1957).

Crisp, Oliver D. and Michael C. Rea (eds), *Analytic Theology: New Essays in the Philosophy of Theology* (Oxford: Oxford University Press, 2009).

Danto, Arthur C., *Nietzsche as Philosopher* (New York: Macmillan, 1965).

Derrida, Jacques, 'D'un ton apocalyptique adopté naguère en philosophie', in Philippe Lacoue-Labarthe and Jean-Luc Nancy (eds), *Les fins de l'homme: À partir du travail de Jacques Derrida* (Paris: Hermann, 2013 [1981]).

Derrida, Jacques, *The Gift of Death*, tr. David Wills (Chicago: University of Chicago Press, 1995).

Derrida, Jacques, 'How to Avoid Speaking: Denials', tr. Ken Frieden, in Harold Coward and Toby Foshay (eds), *Derrida and Negative Theology* (Albany, NY: State University of New York Press, 1992).

Derrida, Jacques, *Politics of Friendship*, tr. G. Collins (London: Verso, 1997).

Derrida, Jacques, *Specters of Marx: The State of the Debt, the Work of Mourning, and the New International*, tr. Peggy Kamuf (London: Routledge, 1994).

Dodds, E. R., *The Ancient Concept of Progress and Other Essays on Greek Literature and Belief* (Oxford: Clarendon Press, 1973).

Donagan, Alan, 'Spinoza's Proof of Immortality', in Marjorie Grene (ed.), *Spinoza: A Collection of Critical Essays* (Garden City NY: Anchor Books, 1973).

Dostoyevsky, Fyodor, *The Brothers Karamazov*, tr. Constance Garnett (London: Heinemann, 1912).

Dostoevsky, Fyodor, *The Idiot*, tr. Richard Pevear and Larissa Volokhonsky (London: Granta, 2003).

Dostoyevsky, Fyodor, *The Possessed*, tr. Constance Garnett (London: Heinemann, 1946).

Dreyfus, Hubert L. and Sean D. Kelly, *All Things Shining: Reading the Western Classics to Find Meaning in a Secular Age* (New York: The Free Press, 2011).

Dupré, Louis, 'Of Time and Eternity in Kierkegaard's Concept of Anxiety', *Faith and Philosophy* 1/2 (April 1984), 160–76.

Eliade, Mircea, *The Myth of the Eternal Return: Or, Cosmos and History*, tr. Willard R. Trask (Princeton, NJ: Princeton University Press, 1971).

Eriksen, Niels Nymann, *Kierkegaard's Category of Repetition: A Reconstruction* (Berlin: de Gruyter, 2000).

Fedorov, Nikolai Fedorovich, *What Was Man Created For? The Philosophy of the Common Task: Selected Works*, tr. and ed. Elisabeth Koutiassov and Marilyn Minto (Lausanne: Honeyglen/L'Age d'Homme, 1990).

Frank, Manfred, *Der kommende Gott: Vorlesungen über die Neue Mythologie* (Frankfurt am Main: Suhrkamp, 1982).

Gage, John, *J. M. W. Turner: 'A Wonderful Range of Mind'* (New Haven, CT: Yale University Press, 1987).

Garrett, Don, 'Spinoza on the Essence of the Human Body and the Part of the Mind that is Eternal', in Olli Koistenen (ed.), *The Cambridge Companion to Spinoza's Ethics* (Cambridge: Cambridge University Press, 2009), 284–302.

Gibbs, Robert, *Correlations in Rosenzweig and Levinas* (Princeton, NJ: Princeton University Press, 1992).

Gilson, Etienne, *Being and Some Philosophers* (Toronto: Pontifical Institute of Medieval Studies, 1952).

Glöckner, Dorothea, *Das Versprechen: Studien zur Verbindlichkeit menschlichen Sagens in Søren Kierkegaards Werk Die Taten der Liebe* (Tübingen: Mohr Siebeck, 2009).

Gordon, Peter Eli, *Rosenzweig and Heidegger: Between Judaism and German Philosophy* (Berkeley. CA: University of California Press, 2003).

Gould, Warwick, and Marjorie Reeves, *Joachim of Fiore and the Myth of the Eternal Evangel in the Nineteenth and Twentieth Centuries* (rev. edn, Oxford: Clarendon Press 2001).

Grene, Marjorie, *Spinoza: A Collection of Critical Essays* (Garden City NY: Anchor Books, 1973).

Grøn, Arne, 'Time and History', in John Lippitt and George Pattison (eds), *The Oxford Handbook of Kierkegaard* (Oxford: Oxford University Press, 2013),

Grønkjaer, Niels, *Den Nye Gud* (Copenhagen: Anis, 2010).

Habib, Stéphane, *Lévinas et Rosenzweig: Philosophies de la Révélation* (Paris: Presses universitaires de France, 2005).

Hasker, William, *God, Time, and Knowledge* (Ithaca, NY: Cornell University Press, 1989).

Hauptmann, G., *Der Narr in Christo, Emanuel Quint* (Frankfurt am Main: Fischer Verlag, 1910).

Hegel, G. W. F., *Phänomenologie des Geistes*, ed. Gerhard Göhler (2nd edn, Frankfurt am Main: Ullstein, 1973).

Hegel, G. W. F., *Werke*, i: *Frühe Schriften* (Frankfurt am Main: Suhrkamp, 1970).

Hegel, G. W. F., *Werke*, ii: *Jenaer Schriften 1801–1807* (Frankfurt am Main: Suhrkamp, 1970).

Hegel, G. W. F., *Werke*, iii: *Phänomenologie des Geistes* (Frankfurt am Main: Suhrkamp, 1970); Eng. *Phenomenology of Spirit*, tr. A. V. Miller (Oxford: Oxford University Press, 1977).

Hegel, G. W. F., *Werke*, v: *Wissenschaft der Logik I* (Frankfurt am Main: Suhrkamp, 1970).

Hegel, G. W. F., *Werke*, vii: *Grundlinien der Philosophie des Rechts* (Frankfurt am Main: Suhrkamp, 1970); Eng. *The Philosophy of Right*, tr. T. M. Knox (Oxford: Oxford University Press, 1967).

Hegel, G. W. F., *Werke*, ix: *Enzyklopädie der Philosophischen Wissenschaften II* (Frankfurt am Main: Suhrkamp, 1970).

Hegel, G. W. F., *Werke*, x: *Enzyklopädie der philosophischen Wissenschaften III* (Frankfurt am Main: Suhrkamp, 1970).

Hegel, G. W. F., *Werke*, xvi: *Vorlesungen über die Philosophie der Religion I* (Frankfurt am Main: Suhrkamp, 1969).

Hegel, G. W. F., *Werke*, xx: *Vorlesungen über die Geschichte der Philosophie III* (Frankfurt am Main: Suhrkamp, 1970).

Heidegger, Martin, *Gesamtausgabe*, ii: *Sein und Zeit* (Frankfurt am Main: Klostermann, 1977), 171; Eng. *Being and Time*, tr. John Mcquarrie and Edward Robinson (Oxford: Blackwell, 1962).

Heidegger, Martin, *Gesamtausgabe*, iv: *Erläuterungen zu Hölderlins Dichtungen* (Frankfurt am Main: Klostermann, 1996).

Heidegger, Martin, *Gesamtausgabe*, v: *Holzwege* (Frankfurt am Main: Klostermann, 1977), 321–73; Eng. *Off the Beaten Track*, ed. and tr. Julian Young and Kenneth Haynes (Cambridge: Cambridge University Press, 2002).

Heidegger, Martin, *Gesamtausgabe*, vi.2: *Nietzsche II (1939–1946)* (Frankfurt am Main: Klostermann, 1997); Eng. *Nietzsche*, iii: *The Will to Power as Knowledge and as Metaphysics*, ed. David Farrell Krell (San Francisco: Harper, 1987).

Heidegger, Martin, *Gesamtausgabe*, xvi: *Reden und andere Zeugnisse eines Lebensweges* 1910–1976 (Frankfurt am Main: Klostermann, 2000).

Heidegger, Martin, *Gesamtausgabe*, xxiv: *Die Grundprobleme der Phänomenologie* (Frankfurt am Main: Klostermann, 1975); Eng. *The Basic Problems of Phenomenology*, tr. A. Hofstadter (Bloomington, IN: Indiana University Press, 1982).

Heidegger, Martin, *Gesamtausgabe*, xxxii: *Hegels Phänomenologie des Geistes* (Frankfurt am Main: Klostermann, 1980); Eng. *Hegel's Phenomenology of Spirit*, tr. Parvis Emad and Kenneth Maly (Bloomington, IN: Indiana University Press, 1988).

Heidegger, Martin, *Gesamtausgabe*, xxxix: *Hölderlins Hymnen 'Germanien' und 'Der Rhein'* (Frankfurt am Main: Klostermann, 1989).

Heidegger, Martin, *Gesamtausgabe*, xlix: *Die Metaphysik des deutschen Idealismus. Zur erneuten Auslegung von Schelling: Philosophische Untersuchungen über das Wesen der menschlichen Freiheit und die damit zusammenhängenden Gegenstände (1809)* (Frankfurt am Main: Klostermann, 2006).

Heidegger, Martin, *Gesamtausgabe*, lii: *Hölderlins Hymne 'Andenken'* (Frankfurt am Main: Klostermann, 1992).

Heidegger, Martin, *Gesamtausgabe*, lx: *Phänomenologie des religiösen Lebens* (Frankfurt am Main: Klostermann, 1995).

Heidegger, Martin, *Gesamtausgabe*, lxiv: *Der Begriff der Zeit* (Frankfurt am Main: Klostermann, 2004); Eng. *The Concept of Time*, tr. William McNeill (Oxford: Blackwell, 1992).

Heidegger, Martin, *Gesamtausgabe*, lxv: *Beiträge zur Philosophie (Vom Ereignis)* (Frankfurt am Main: Klostermann, 1994).

Helm, Paul, *Eternal God: A Study of God Without Time* (Oxford: Clarendon Press, 1988).

Hessayon, Ariel, and Sarah Apetrei (eds), *An Introduction to Jacob Boehme: Four Centuries of Thought and Reception* (New York: Routledge, 2014).

Hollander, Dana, *Exemplarity and Chosenness: Rosenzweig and Derrida on the Nation of Philosophy* (Stanford, CA: Stanford University Press, 2008).

Hollander, Dana, 'A Thought in Which Everything Has Been Thought": On the Messianic Idea in Levinas', *Symposium: Canadian Journal of Continental Philosophy* 14/2 (Fall 2010), 133–59.

Hugo, Victor, 'Le 5 juin 1832' in *Les Misérables*, ii: Book X (Paris: Gallimard, 1973/95, 395–421).

Huxley, Aldous, *The Doors of Perception* and *Heaven and Hell* (London: Granada, 1977).

James, William, *The Varieties of Religious Experience* (London: Collins, 1960).

Jankélévitch, Vladimir, *Forgiveness*, tr. Andrew Kelley (Chicago: University of Chicago Press, 2005).

Jankélévitch, Vladimir, *L'imprescriptible: Pardonner? Dans l'honneur et la dignité* (Paris: Seuil, 1986).

Jankélévitch, Vladimir, *L'irréversible et la nostalgie* (Paris: Flammarion, 1974).

Jarvis, Simon, *Wordsworth's Philosophic Song* (Cambridge: Cambridge University Press, 2007).

Jaspers, Karl, *Nietzsche: Einführung in das Verständnis seines Philosophierens* (Berlin: de Gruyter, 1936).

Jenni, Ernst, 'Das Wort 'ōlām im Alten Testament', *Zeitschrift für die alttestamentliche Wissenschaft* 65/1 (1953), 1–35.

Jüngel, Eberhard, *God's Being Is in Becoming: The Trinitarian Being of God in the Theology of Karl Barth*, tr. John Webster (Edinburgh: T. & T. Clark, 2001).

Kaufmann, Walter, *Nietzsche: Philosopher, Psychologist, Antichrist* (Princeton, NJ: Princeton University Press, 1968).

Kee, Alistair, *Nietzsche against the Crucified* (London: SCM Press, 1999).

Kierkegaard, Søren, *Søren Kierkegaards Skrifter (SKS)*, ed. Niels Jørgen Cappelørn, Joakim Garff, Johnny Kondrup et al. (Copenhagen: Gad, 1997–2013).

Individual works, in order of SKS volume number:

Enten-Eller (SKS2–3, 1997); Eng. *Either/Or*, tr. Howard V. and Edna H. Hong, 2 vols (Princeton, NJ: Princeton University Press, 1987).

Begrebet Angst (SKS4); Eng. *The Concept of Anxiety*, tr. Reidar Thomte (Princeton, NJ: Princeton University Press, 1980).

Frygt og Baeven (SKS4) and *Gjentagelsen* (SKS4); Eng. *Fear and Trembling* and *Repetition*, tr. Howard V. and Edna H. Hong (Princeton, NJ: Princeton University Press, 1983).

Opbyggelige Taler 1843, 1844 (SKS5); Eng. *Eighteen Upbuilding Discourses*, tr. Howard V. and Edna H. Hong (Princeton, NJ: Princeton University Press, 1990).

Tre Taler ved taenkte Leiligheder (SKS5); Eng. *Three Discourses on Imagined Occasions*, tr. Howard V. and Edna H. Hong (Princeton, NJ: Princeton University Press, 1993).

Afsluttende uvidenskabelig Efterskrift (SKS7); Eng. *Concluding Unscientific Postscript to Philosophical Fragments*, tr. Howard V. and Edna H. Hong (Princeton, NJ: Princeton University Press, 1992).

Opbyggelige Taler i forskjellig Aand (SKS8); Eng. *Upbuilding Discourses in Various Spirits*, tr. Howard V. and Edna H. Hong (Princeton, NJ: Princeton University Press, 1993).

Kjerlighedens Gjerninger (SKS9); Eng. *Works of Love*, tr. Howard V. and Edna H. Hong (Princeton, NJ: Princeton University Press, 1998).

Christelige Taler (SKS10); Eng. *Christian Discourses: The Crisis and a Crisis in the Life of an Actress*, tr. Howard V. and Edna H. Hong (Princeton, NJ: Princeton University Press, 2009).

Lilien paa Marken og Fuglen under Himlen (SKS11); Eng. *Without Authority*, tr. Howard V. and Edna H. Hong (Princeton, NJ: Princeton University Press, 1997).

Sygdommen til Døden (SKS11); Eng. *The Sickness unto Death*, tr. Howard V. and Edna H. Hong (Princeton, NJ: Princeton University Press, 1980).

Øjeblikket (SKS13); Eng. *The Moment*, tr. Howard V. and Edna H. Hong (Princeton, NJ: Princeton University Press, 2009).

Breve. Dedikationer (SKS28); Eng. *Letters and Documents*, tr. Henrik Rosenmeier, (Princeton, NJ: Princeton University Press, 1978).

Kierkegaard, Søren, *Journals and Papers*, tr. Howard V. and Edna H. Hong, 6 vols (Bloomington. IN: Indiana University Press, 1967–78).

Kirk, G. S. and J. E. Raven, *The Presocratic Philosophers: A Critical History with a Selection of Texts* (Cambridge: Cambridge University Press, 1971).

Kneale, Martha, 'Eternity and Sempiternity', in Marjorie Grene (ed.), *Spinoza: A Collection of Critical Essays* (Garden City NY: Anchor Books, 1973).

Kosch, Michelle, *Freedom and Reason in Kant, Schelling and Kierkegaard* (Oxford: Oxford University Press, 2006).

Koyré, Alexandre, *La Philosophie de Jacob Boehme* (New York: Burt Franklin, 1968 [1929]).

Kroeker, P. Travis and Bruce K. Ward, *Remembering the End: Dostoevsky as Prophet to Modernity* (Boulder, CO: Westview, 2001).

Lampert, Laurence, *Nietzsche's Teaching: An Interpretation of Thus Spoke Zarathustra* (New Haven, CT: Yale University Press, 1986).

Leftow, Brian, 'God's Impassibility, Immutability, and Eternality', in Brian Davies and Eleonore Stump (eds), *The Oxford Handbook of Aquinas* (Oxford: Oxford University Press, 2012).

Leftow, Brian, *Time and Eternity* (Ithaca, NY: Cornell University Press, 1991).

Lévinas, Emmanuel, *De Dieu qui vient à l'idée* (Paris: Vrin, 2004).

Lévinas, Emmanuel, *Le temps et l'autre* (Paris: Presses universitaires de France, 1983).

Lévinas, Emmanuel, *Totalité et l'Infini: Essai sur l'exteriorité* (4th edn, Paris: Kluwer Academic, 1971).

Lippitt, J., *Humour and Irony in Kierkegaard's Thought* (Basingstoke: Macmillan, 2000).

Lloyd, Genevieve, *Routledge Philosophy Guidebook to Spinoza and the Ethics* (London: Routledge, 1996).

Lossky, Vladimir, *In the Image and Likeness of God* (New York: St Vladimir's Seminary Press, 1974).

Löwith, Karl, *Nietzsches Philosophie des Ewigen Wiederkehr des Gleichen* (Stuttgart: Kohlhammer, 1956).

McTaggart, John McTaggart Ellis, *Philosophical Studies*, ed. S. V. Keeling (London: Edward Arnold, 1934).

McTaggart, John McTaggart Ellis, *Studies in the Hegelian Dialectic* (Cambridge: Cambridge University Press, 1896).

Mander, W. J., *British Idealism: A History* (Oxford: Oxford University Press, 2011).

May, Reinhard, *Heidegger's Hidden Sources: East Asian Influences on his Work*, tr. Graham Parkes (London: Routledge, 1996).

Mesch, Walter, *Reflektierte Gegenwart: Eine Studie über Zeit und Ewigkeit bei Platon, Aristoteles, Plotin und Augustinus* (Frankfurt am Main: Klostermann, 2003).

Michalski, Krzysztof, *The Flame of Eternity: An Interpretation of Nietzsche's Thought*, tr. Benjamin Paloff (Princeton, NJ: Princeton University Press, 2007).

Moltmann, Jürgen, *Theology of Hope: On the Ground and Implications of a Christian Eschatology*, tr. James W. Leitch (London: SCM Press, 1967).

Moltmann, Jürgen, *God in Creation: An Ecological Doctrine of Creation* (*The Gifford Lectures* 1984–1985), tr. Margaret Kohl (London: SCM Press, 1985).

Morgan, Michael L., *Discovering Levinas* (Cambridge: Cambridge University Press, 2007).

Mulhall, Stephen, *Routledge Philosophy Guidebook to Heidegger and Being and Time* (London: Routledge, 1996).

Nadler, Steven, *Spinoza's Heresy: Immortality and the Jewish Mind* (Oxford: Oxford University Press, 2001).

Nehamas, Alexander, *Nietzsche: Life as Literature* (Cambridge, MA: Harvard University Press, 1985).

Nietzsche, Friedrich, *Ecce Homo,* in *Werke in drei Bänden III,* ed. K. Schechta (Frankfurt am Main: Ullstein, 1969).

Nietzsche, Friedrich, *Also Sprach Zarathustra,* in *Werke in drei Bänden II,* ed. K. Schechta (Frankfurt am Main: Ullstein, 1972).

Nietzsche, Friedrich, *Götzen-Dämmerung,* in *Werke in drei Bänden III,* ed. K. Schechta (Frankfurt am Main: Ullstein, 1972).

Padgett, Alan G., 'The Difference Creation Makes: Relative Timelessness Reconsidered', in Christian Tapp and Edmund Runggaldier (eds), *God, Eternity, and Time* (Farnham: Ashgate, 2011), 117–25.

Pattison, George, 'Death', in Nicholas Adams, George Pattison, and Graham Ward (eds), *The Oxford Handbook of Theology and Modern European Thought* (Oxford: Oxford University Press, 2013), 193–212.

Pattison, George, 'Existence, Anxiety and the Moment of Vision: Fundamental Ontology and Existentiell Faith Revisited', in Anthony Paul Smith and Daniel Whistler (eds), *After the Postsecular and the Postmodern: New Essays in Continental Philosophy of Religion* (Newcastle upon Tyne: Cambridge Scholars Publishing, 2010).

Pattison, George, *God and Being: An Enquiry* (Oxford: Oxford University Press, 2011).

Pattison, George, *'The Heart could Never Speak': Existentialism and Faith in a Poem of Edwin Muir* (Eugene, OR: Cascade Books, 2013).

Pattison, George, *Heidegger on Death: A Critical Theological Essay* (Farnham: Ashgate, 2013).

Pattison, George, *Kierkegaard: The Aesthetic and the Religious* (2nd edn, London: SCM Press, 1999).

Pattison, George, 'Kierkegaard, Metaphysics and Love', in Heiko Schulz, Jon Stewart, and Karl Verstrynge (eds), *Kierkegaard Studies Yearbook* 2013 (Berlin: de Gruyter, 2013).

Pattison, George, *Kierkegaard and the Quest for Unambiguous Life: Between Romanticism and Modernism: Selected Essays* (Oxford: Oxford University Press, 2013).

Pattison, George, *Kierkegaard and the Theology of the Nineteenth Century: The Paradox and the 'Point of Contact'* (Cambridge: Cambridge University Press, 2012).

Pattison, George, *Kierkegaard's Upbuilding Discourses* (London: Routledge, 2002).

Pattison, George, ' "Water the Earth": Dostoevsky on Tears', *Litteraria Pragensia: Studies in Literature and Culture* 22/43 (July 2012): *Towards a Lachrymology: Tears in Literature and Cultural History,* 95–111.

Peters, Julia, 'Proust's *Recherche* and Hegelian Teleology', *Inquiry* 53/2 (April 2010), 141–61.

Pike, Nelson, *God and Timelessness* (London: Routledge and Kegan Paul, 1970).

Pindar, *Pindar,* i: *Olympian Odes; Pythian Odes* and ii: *Nemean Odes; Isthmian Odes; Fragments,* ed. and tr. William H. Race (Cambridge, MA: Harvard University Press, 1997).

Plotinus, *Ennead II,* tr. A. H. Armstrong (Cambridge, MA: Harvard University Press, 1966).

Plotinus, *Ennead III*, tr. A. H. Armstrong (Cambridge, MA: Harvard University Press, 1966).

Proust, Marcel, *Le temps retrouvé* (Paris: Gallimard, 1990).

Richardson, Alan, *History Sacred and Profane* (London: SCM Press, 1964).

DeRoo, Neal, and John Panteleimon Manoussakis (eds), *Phenomenology and Eschatology: Not Yet in the Now* (Farnham: Ashgate, 2009).

Rosenzweig, Franz, *Das Buchlein vom gesunden und kranken Menschenverstand*, ed. Nahum Norbert Glatzer (Düsseldorf: Joseph Melzer, 1964); Eng. *Understanding the Sick and the Healthy: A View of World, Man, and God*, tr. Nahum Norbert Glatzer (Cambridge, MA: Harvard University Press, 1953).

Rosenzweig, Franz, *Hegel und der Staat* (Berlin: Suhrkamp, 2010).

Rosenweig, Franz, *Der Stern der Erlösung* (Frankfurt am Main: Suhrkamp, 1988); Eng. *The Star of Redemption*, tr. William W. Hallo (Notre Dame, IN: University of Notre Dame Press, 1985).

Rosenzweig, Franz, *Zweistromland: Kleinere Schriften zu Glauben und Denken*, ed. Reinhold and Annemarie Mayer (Dordrecht: Martinus Nijhoff, 1984).

Roth, Leon, *Spinoza, Descartes and Maimonides* (Oxford: Clarendon Press, 1924).

Ryan, Christopher, *Schopenhauer's Philosophy of Religion: The Death of God and the Oriental Renaissance* (Leuven: Peeters, 2010).

Sambursky, S. and S. Pines, *The Concept of Time in Late Neoplatonism* (Jerusalem: Israel Academy of Sciences and Humanities, 1971).

Sartre, Jean-Paul, *Words*, tr. Irene Clephane (Harmondsworth: Penguin, 1967).

Schelling, F. W. J., *Philosophische Untersuchungen über das Wesen der menschlichen Freiheit* (Frankfurt am Main: Suhrkamp, 1975).

Schelling, F. W. J., 'Die Weltalter', in *Ausgewählte Schriften*, iv: 1807–1834 (Frankfurt am Main: Suhrkamp, 1985).

Schleiermacher, Friedrich, *The Christian Faith*, ed. and tr. H. R. Mackintosh and J. S. Stewart (Edinburgh, T. & T. Clark, 1928).

Schleiermacher, Friedrich, *On Religion: Speeches to its Cultured Despisers*, tr. and ed. Richard Crouter (Cambridge: Cambridge University Press, 1988).

Scholem, Gershom, *Sabbatai Sevi: The Mystical Messiah*, 1626–1676, tr. R. J. Zwi Werblowsky (London: Routledge and Kegan Paul, 1973).

Schopenhauer, Arthur, *The World as Will and Representation*, tr. E. F. J. Payne, 2 vols (New York: Dover, 1969).

Schürmann, Reiner, *Wandering Joy: Meister Eckhart's Mystical Philosophy* (Great Barrington, MA: Lindisfarne Books, 2001).

Schweitzer, Albert, *The Quest of the Historical Jesus*, tr. William Montgomery (London: A. & C. Black, 1954).

Shestov, Lev, *Dostoevsky, Tolstoy and Nietzsche* (Athens, OH: Ohio University Press, 1969).

Soloviev, Vladimir, 'Tri rechi v pamyat' Dostoevskogo' ['Three talks in memory of Dostoevsky'], in *Spor o Spravedliivostii* (Moscow: Eksmo-Press, 1999).

Sorabji, Richard, *Time, Creation and the Continuum: Theories in Antiquity and the Early Middle Ages* (Ithaca, NY: Cornell University Press, 1983).

Spinoza, Benedictus de, *Ethics; and On the Correction of the Understanding*, tr. Andrew Boyle (London: J. M. Dent [Everyman's Library], 1977).

Spinoza, Benedictus de, *Theological-Political Treatise*, ed. Jonathan Israel (Cambridge: Cambridge University Press, 2007).

Stambaugh, Joan, *Nietzsche's Thought of Eternal Return* (Baltimore, MD: Johns Hopkins University Press, 1972).

Steinberg, Diane, 'Spinoza's Theory of the Eternity of Mind', *Canadian Journal of Philosophy* 11/1 (March 1981), 35–68.

Stoudt, John Joseph, *Sunrise to Eternity: A Study in Jacob Boehme's Life and Thought* (Philadelphia, PA: University of Pennsylvania Press, 1957).

Stump, Eleonore, and Norman Kretzmann, 'Eternity', *Journal of Philosophy* 78/8 (August 1981), 429–58.

Sussman, Henry, *The Hegelian Aftermath; Readings in Hegel, Kierkegaard, Freud, Proust and James* (Baltimore, MD: Johns Hopkins University Press, 1982).

Suzuki, D. T., *Mysticism: Christian and Buddhist* (London: George Allen and Unwin, 1957).

Swinburne, Richard, *The Christian God* (Oxford: Clarendon Press, 1994).

Swinburne, Richard, *The Coherence of Theism* (Oxford: Clarendon Press, 1993).

Tanner, Michael, *Nietzsche*, Past Masters (Oxford: Oxford University Press, 1994).

Tapp, Christian, 'Eternity and Infinity', in Christian Tapp and Edmund Runggaldier (eds), *God, Eternity, and Time* (Farnham: Ashgate, 2011), 99–115.

Tapp, Christian and Edmund Runggaldier (eds), *God, Eternity, and Time* (Farnham: Ashgate, 2011).

Taubes, Jacob, *Occidental Eschatology*, tr. David Ratmoko (Stanford, CA: Stanford University Press, 2009).

Taylor, A. E., 'The Concept of Immortality in Spinoza's *Ethics*', *Mind* ns 5/18 (April 1896), 145–66.

Theunissen, Michael, *Der Begriff Verzweiflung: Korrekturen an Kierkegaard* (Frankfurt am Main: Suhrkamp, 1993).

Theunissen, Michael, *Negative Theologie der Zeit* (Frankfurt am Main: Suhrkamp,1991).

Theunissen, Michael, *The Other: Studies in the Social Ontology of Husserl, Heidegger, Sartre, and Buber*, tr. Christopher Macann (Cambridge, MA: MIT Press, 1984).

Theunissen, Michael, *Pindar: Menschenlos und Wende der Zeit* (Münich: C. H. Beck, 2000).

Theunissen, Michael, 'The Upbuilding in the Thought of Death: Traditional Elements, Innovative Ideas, and Unexhausted Possibilities', tr. George Pattison, in Robert L. Perkins (ed.) *International Kierkegaard Commentary*, ix: *Prefaces and Writing Sampler* and x: *Three Discourses on Imagined Occasions* (combined volumes) (Macon, GA: Mercer University Press, 2006), 321–58.

Thompson, Diane, *The Brothers Karamazov and the Poetics of Memory* (Cambridge: Cambridge University Press, 1991),

Tillich, Paul, 'Ideologie und Utopie: Zum gleichnamigen Buch von Karl Mannheim' (1929), in *Gesammelte Werke*, xii: *Begegnungen: Paul Tillich über sich selbst und andere* (2nd edn, Stuttgart: Evangelisches Verlagswerk, 1980),

Tillich, Paul, *The Interpretation of History*, tr. N. A. Rasetzki and Elsa L. Talmey (New York: Charles Scribner's Sons, 1936).

Tillich, Paul, 'Mensch und Staat', in *Gesammelte Werke*, xiii: *Impressionen und Reflexionen: Ein Lebensbild in Aufsätzen, Reden und Stellungnahmen* (Stuttgart: Evangelisches Verlagswerk, 1972).

Tillich, Paul, *The Protestant Era*, tr. James Luther Adams (London: James Nisbet, 1951).

Tillich, Paul, *Die sozialistische Entscheidung*, in *Gesammelte Werke*, ii: *Christentum und soziale Gestaltung. Frühe Schriften zum Religiösen Sozialismus* (Stuttgart: Evangelisches Verlagswerk, 1962),

Tillich, Paul, *Systematic Theology* (One Volume Edition, Welwyn Garden City: James Nisbet, 1968).

Tillich, Paul, *Theology of Culture*, ed. Robert Kimball (New York: Oxford University Press, 1959).

Vattimo, Gianni, *Il Soggetto e la Maschera: Nietzsche e il problema della liberazione* (Milan: Bompiani, 1974).

Walsh, Sylvia, *Living Christianly: Kierkegaard's Dialectic of Christian Existence* (University Park, PA: Pennsylvania State University Press, 2005).

Walter, Gregory, *Being Promised: Theology, Gift, and Practice* (Grand Rapids, MI: Eerdmans, 2013).

Ward, Keith, *Rational Theology and the Creativity of God* (Oxford: Blackwell, 1982).

Watts, Alan, *Behold The Spirit: A Study in the Necessity of Mystical Religion* (New York, Vintage, 1971).

Watts, Alan, *This is It and Other Essays on Zen and Spiritual Experience* (New York: Vintage, 1960).

Weil, Simone, *Intimations of Christianity Among the Ancient Greeks*, ed. and tr. Elisabeth Chase Geissbuhler (London: Routledge and Kegan Paul, 1957).

Wolfe, Judith, *Heidegger's Eschatology: Theological Horizons in Martin Heidegger's Early Work* (Oxford: Oxford University Press, 2013).

Wolterstorff, Nicholas, *Inquiring about God: Selected Essays, Volume 1* (Cambridge: Cambridge University Press, 2010).

Yates, John C., *The Timelessness of God* (Lanham, MD: University Press of America, 1990).

Index

The terms 'eternal'/'eternity'/ time, and God are too extensively discussed to be fully indexed and only the most salient themes are highlighted below.